Advances in Ecohydrology in Arid Inland River Basins

Advances in Ecohydrology in Arid Inland River Basins

Guest Editors

Lianqing Xue
Guang Yang

MDPI

Basel • Beijing • Wuhan • Barcelona • Belgrade • Novi Sad • Cluj • Manchester

Guest Editors

Lianqing Xue
College of Hydrology and
Water Resources
Hohai University
Nanjing
China

Guang Yang
College of Water and
Architectural Engineering
Shihezi University
Shihezi
China

Editorial Office
MDPI AG
Grosspeteranlage 5
4052 Basel, Switzerland

This is a reprint of the Special Issue, published open access by the journal *Water* (ISSN 2073-4441), freely accessible at: https://www.mdpi.com/journal/water/special_issues/H8J832955G.

For citation purposes, cite each article independently as indicated on the article page online and as indicated below:

Lastname, A.A.; Lastname, B.B. Article Title. *Journal Name* **Year**, *Volume Number*, Page Range.

ISBN 978-3-7258-5409-7 (Hbk)
ISBN 978-3-7258-5410-3 (PDF)
https://doi.org/10.3390/books978-3-7258-5410-3

Contents

water

MDPI

Editorial

Advances in Ecohydrology in Arid Inland River Basins

Lianqing Xue [1,*] and Guang Yang [2,*]

1 College of Hydrology and Water Resources, Hohai University, Nanjing 210098, China
2 College of Water Conservancy & Architectural Engineering, Shihezi University, Shihezi 832000, China
* Correspondence: lqxue@hhu.edu.cn (L.X.); mikeyork@163.com (G.Y.)

1. Introduction

Water is the foundation of life, ecosystems, and socioeconomic development [1]. Hydrological mechanisms, including flood and drought events, pollution, and water scarcity, underlie climate–soil–vegetation dynamics and control the most basic ecological patterns and processes, both for biota and materials [2]. Nowhere is this more evident than in arid and semi-arid regions, where the complex interplay between climate variability, human activities, and ecological systems shapes both the challenges and opportunities for sustainable water management [3–6]. Over the past few decades, rapid climatic shifts, intensified land use, and increasing anthropogenic pressures have exacerbated water scarcity, threatened ecological security, and heightened the need for innovative management strategies in these vulnerable landscapes.

The area of arid and semi-arid regions has expanded further with global climate warming, which has had profound impacts on the functional structure of ecosystems in these regions. This, in turn, has significantly affected humans' ability to access water resources [7]. Moreover, human activities, such as the development of water infrastructure and changes in land use, have caused profound disturbances to the ecological hydrological processes in arid and semi-arid regions [8,9]. Therefore, the ecological hydrological mechanisms in arid inland river basins require further investigation [10,11].

This Special Issue of *Water* brings together twelve cutting-edge studies that advance our understanding of water resources, hydrological processes, ecological risk, and sustainable management practices in arid and semi-arid environments. Collectively, these contributions offer novel insights into soil infiltration, hydrological modeling, drought assessment, water quality, vegetation dynamics, landscape ecological risk, water governance, and adaptive management strategies under the influence of both natural and human-induced changes.

This editorial aims to highlight emerging trends in the sustainable management of ecohydrological systems in arid regions. The studies in this Special Issue showcase interdisciplinary research approaches, provide diverse perspectives for uncovering the mechanisms of ecohydrology, and clarify how integrated scientific research can inform policy-making, support sustainable development in arid inland river basins, and enhance the resilience of ecosystems in arid inland river regions under the background of global climate change.

2. Review and Synthesis of Contributions

1. Soil Infiltration, Vegetation, and Land Cover Change

Soil infiltration is a key factor influencing surface runoff, groundwater recharge, and ecosystem functioning, particularly in regions where water resources are limited. The work in [Contribution 1] compares the soil infiltration capacity between native grasslands and thicketed oak woodlands across different soil types. Their findings indicate that converting oak savannahs to thicket-dominated woodlands can notably increase soil infiltration, especially in clay-rich soils. The study also evaluates several infiltration measurement methods and highlights the Simplified Steady Beerkan Infiltration (SSBI) technique as a rapid and practical alternative to traditional rainfall simulation for field assessments.

Changes in land use and land cover (LUCC) further shape hydrological processes and ecological risks. In the Kriya River Basin, the authors of [Contribution 9] use spatial modeling to show that both natural factors and human activities affect landscape ecological risk. The conversion of grasslands and unused land into croplands and built-up areas reflects the highly dynamic nature of land systems in arid regions. The research emphasizes that human activities have become the dominant force behind landscape risk and that different scenario-based projections—such as natural development, cropland protection, and ecological priority—lead to very distinct risk patterns. These results stress the importance of flexible land management and integrated risk assessment approaches to maintain ecological security in a changing environment.

2. Hydrological Modeling and Flood/Drought Risk

Hydrological processes in arid and semi-arid regions are often complex, with both episodic flooding and prolonged drought posing significant challenges for water resource management. The study of the Jingjiang–Dongting Lake system in [Contribution 2] highlights the intricate relationships between rivers, lakes, and flood control infrastructure. By building and validating a hydrological model with nearly 60 years of flood data, the research offers insights into the limitations of current flood mitigation strategies. Even with the operation of upstream infrastructure such as the Three Gorges Dam, it remains difficult to eliminate flood risks in the Dongting Lake area, suggesting the ongoing need for embankment reinforcement and policy innovation.

Concerns about drought are evident in the Manas River Basin [Contribution 5] and Huangshui River Basin [Contribution 3]. In the Manas River Basin, long-term analysis using the Standardized Precipitation Evapotranspiration Index (SPEI) and Standardized Runoff Index (SRI) shows that, while both meteorological and hydrological droughts have generally eased in recent decades, meteorological droughts during spring and autumn persist, and severe droughts are becoming more frequent. This highlights the need for scale-specific drought risk assessments and targeted management strategies. In the Huangshui River Basin, declining runoff and reduced ecological flow assurance since the early 1970s are closely related to intensified human activity. Still, ecological flows remain sufficient in most months, indicating the effectiveness of flow management but also the increasing vulnerability of river ecosystems to human pressures.

The authors of [Contribution 11] examine the impacts of water conservation projects and reservoir construction in the Weigan River estuary. Using a SWAT model integrated with glacier modules, the study quantifies how climate change, water engineering, and runoff have interacted over the past fifty years. The findings show that artificial regulation, particularly in spring, has significantly altered natural flow regimes, helping to ease seasonal shortages but requiring continuous evaluation to balance ecological and human needs.

3. Water Quality, Agricultural Practices, and Resource Allocation

Sustainable water management involves not only securing water quantity but also addressing water quality, especially in areas where agriculture, urban growth, and ecological conservation intersect. In south-central Ningxia, the authors of [Contribution 4] report that although water supply is generally adequate, water quality—specifically, persistently high total nitrogen—remains an issue. The study proposes and tests an optimized scheduling scheme, demonstrating that it is possible to improve water quality while meeting quantity demands, with reductions in total nitrogen exceeding 78%. These insights offer valuable guidance for other regions facing similar challenges.

Agricultural water use is closely tied to weed management and pesticide application, as discussed in [Contribution 10]. Limiting herbicide use in sensitive water zones can benefit aquatic biodiversity and environmental health but may also encourage the spread of particular weed species, complicating crop management. The research calls for innovative and integrated weed control approaches that balance agricultural productivity and water resource protection, underlining the need for cross-sectoral management.

Resource allocation is particularly complex in regions such as northwest China, where socioeconomic development depends on rational water distribution between urban and rural areas. These studies underscore the importance of considering both quantity and quality, optimizing allocation schedules, and involving local stakeholders to ensure sustainable outcomes.

4. Vegetation Dynamics and Ecological Restoration

Vegetation is central to ecosystem resilience, land–atmosphere interactions, and the provision of ecological services in drylands. The lower reaches of the Tarim River [Contribution 6] provide a telling example of vegetation recovery under changing climatic and human pressures. Using trend analysis, Hurst exponent projections, and GeoDetector modeling, the research documents significant improvements in the NDVI (Normalized Difference Vegetation Index) over the past twenty years, primarily driven by human intervention and proximity to river channels. However, the risk of future degradation persists, especially in areas far from water sources, as multiple factors combine to influence vegetation dynamics. These results are crucial for guiding targeted restoration and monitoring.

Research on the Great Lakes Depression [Contribution 12] further highlights the sensitivity of vegetation cover to hydro-climatic changes in Central Asia. In this region, rising temperatures, decreased precipitation, and lower river discharge have reduced lake water levels and altered vegetation patterns across diverse microenvironments. The study draws attention to the seasonal nature of vegetation responses to climate extremes and underscores the importance of integrated hydrometeorological and remote sensing analyses for monitoring and managing ecological change.

5. Climate Change, LUCC, and Streamflow Projections

Understanding how climate change and LUCC interact to affect hydrological processes is critical for future water security. The upper Tarim River [Contribution 7] is especially vulnerable, with projections suggesting that increased precipitation and temperature (under three socioeconomic scenarios) would raise streamflow, but ongoing LUCC—mainly the expansion of farmland at the expense of grasslands—would decrease it. Notably, LUCC is expected to have a greater impact on streamflow than climate change alone, with streamflow reductions projected under all future land use patterns. These findings reinforce the need for scenario-based integrated planning that accounts for multiple interacting drivers over time.

6. Water Governance and Socio-Institutional Innovation

Effective water governance is the foundation of sustainable management, particularly in regions that regularly experience both floods and droughts, such as Cambodia [Contribution 8]. The study identifies several barriers to improved governance, including fragmented sectoral responsibilities, weak coordination among agencies, and limited community participation. Through case studies of farmer water user groups and fisheries, the research shows that current, often top-down, management strategies do not always address local realities and can unintentionally worsen competition and conflict over resources. The study recommends integrating water governance at the district level to promote both sustainability and social equity.

3. Conclusions

This Special Issue of *Water* not only presents recent advances in water resource assessment but also offers an extensive view of water resource management issues, ecological risk analysis, and sustainable management in drylands. By drawing on diverse case studies from Asia and beyond, the collected studies provide a foundation for integrated solutions that address competing demands and foster resilience in vulnerable dry environments. As global water challenges intensify, ongoing collaboration among scientists, policymakers, and communities will be essential to advancing both understanding and actions. We sincerely thank all the authors, reviewers, and editorial staff for their contributions to this Special Issue. We hope these findings and perspectives will encourage further research, innovation, and dialogue on the path to sustainable water and ecological management.

Funding: This study was supported by the National Key Research and Development Program of China (2023YFC3206804).

Acknowledgments: As a guest editor of this Special Issue, the author would like to thank the journal editors and all authors who submitted their manuscripts to this Special Issue. Special thanks go to the reviewers for their careful review of all submitted papers and for providing professional suggestions, which have enhanced our understanding of the ecohydrological mechanisms in arid inland river basins. All authors have read and agree to the published version of the manuscript.

Conflicts of Interest: The authors declare no conflicts of interest.

List of Contributions:

1. Atalar, F.; Leite, P.A.M.; Wilcox, B.P. A Comparison of Three Methodologies for Determining Soil Infiltration Capacity in Thicketized Oak Woodlands and Adjacent Grasslands. *Water* **2025**, *17*, 518. https://doi.org/10.3390/w17040518.
2. Zhao, W.; Ji, W.; Wang, J.; Jiang, J.; Song, W.; Wang, Z.; Lv, H.; Lu, H.; Liu, X. Research on Flood Storage and Disaster Mitigation Countermeasures for Floods in China's Dongting Lake Area Based on Hydrological Model of Jingjiang–Dongting Lake. *Water* **2025**, *17*, 1. https://doi.org/10.3390/w17010001.
3. Liu, L.; Fan, L.; Hu, J.; Li, C. Human Activities Impacts on Runoff and Ecological Flow in the Huangshui River of the Yellow River Basin, China. *Water* **2024**, *16*, 2331. https://doi.org/10.3390/w16162331.
4. Zhang, Y.; Hu, T.; Xue, H.; Liu, X. Simulation and Optimal Scheduling of Water Quality in Urban and Rural Water Supply Systems: A Case Study in the Northwest Arid Region of China. *Water* **2024**, *16*, 2181. https://doi.org/10.3390/w16152181.
5. Dan, Y.; Tian, H.; Farid, M.A.; Yang, G.; Li, X.; Li, P.; Gao, Y.; He, X.; Li, F.; Liu, B.; et al. Evolution Characteristics of Meteorological and Hydrological Drought in an Arid Oasis of Northwest China. *Water* **2024**, *16*, 2088. https://doi.org/10.3390/w16152088.
6. Han, Q.; Xue, L.; Qi, T.; Liu, Y.; Yang, M.; Chu, X.; Liu, S. Dynamic Spatiotemporal Evolution and Driving Mechanisms of Vegetation in the Lower Reaches of the Tarim River, China. *Water* **2024**, *16*, 2157. https://doi.org/10.3390/w16152157.
7. Han, Q.; Xue, L.; Qi, T.; Liu, Y.; Yang, M.; Chu, X.; Liu, S. Assessing the Impacts of Future Climate and Land-Use Changes on Streamflow under Multiple Scenarios: A Case Study of the Upper Reaches of the Tarim River in Northwest China. *Water* **2024**, *16*, 100. https://doi.org/10.3390/w16010100.
8. Sithirith, M.; Sao, S.; de Silva, S.; Kong, H.; Kongkroy, C.; Thavrin, T.; Sarun, H. Water Governance in the Cambodian Mekong Delta: The Nexus of Farmer Water User Communities (FWUCs), Community Fisheries (CFis), and Community Fish Refuges (CFRs) in the Context of Climate Change. *Water* **2024**, *16*, 242. https://doi.org/10.3390/w16020242.
9. Li, J.; He, X.; Huang, P.; Wang, Z.; Wang, R. Landscape Ecological Risk Assessment of Kriya River Basin in Xinjiang and Its Multi-Scenario Simulation Analysis. *Water* **2023**, *15*, 4256. https://doi.org/10.3390/w15244256.

10. Winkler, J.; Říčica, T.; Hubačíková, V.; Koda, E.; Vaverková, M.D.; Havel, L.; Żółtowski, M. Water Protection Zones—Impacts on Weed Vegetation of Arable Soil. *Water* **2023**, *15*, 3161. https://doi.org/10.3390/w15173161.
11. Su, J.; Long, A.; Chen, F.; Ren, C.; Zhang, P.; Zhang, J.; Gu, X.; Deng, X. Impact of the Construction of Water Conservation Projects on Runoff from the Weigan River. *Water* **2023**, *15*, 2431. https://doi.org/10.3390/w15132431.
12. Dorjsuren, B.; Zemtsov, V.A.; Batsaikhan, N.; Yan, D.; Zhou, H.; Dorligjav, S. Hydro-Climatic and Vegetation Dynamics Spatial-Temporal Changes in the Great Lakes Depression Region of Mongolia. *Water* **2023**, *15*, 3748. https://doi.org/10.3390/w15213748.

References

1. Mishra, B.K.; Kumar, P.; Saraswat, C.; Chakraborty, S.; Gautam, A. Water Security in a Changing Environment: Concept, Challenges and Solutions. *Water* **2021**, *13*, 490. [CrossRef]
2. Yu, Z.; Lu, C.; Cai, J.; Yu, D.; Mahe, G.; Mishra, A.; Cudennec, C.; Van Lanen, H.A.J.; Orange, D.; Amani, A. Preface: Hydrological processes and water security in a changing world. *Proc. Int. Assoc. Hydrol. Sci.* **2020**, *383*, 3–4. [CrossRef]
3. Ahlström, A.; Raupach, M.R.; Schurgers, G.; Smith, B.; Arneth, A.; Jung, M.; Reichstein, M.; Canadell, J.G.; Friedlingstein, P.; Jain, A.K.; et al. The dominant role of semi-arid ecosystems in the trend and variability of the land CO_2 sink. *Science* **2015**, *348*, 895–899. [CrossRef] [PubMed]
4. Simpson, I.R.; McKinnon, K.A.; Kennedy, D.; Lawrence, D.M.; Lehner, F.; Seager, R. Observed humidity trends in dry regions contradict climate models. *Proc. Natl. Acad. Sci. USA* **2023**, *121*, e2302480120. [CrossRef] [PubMed]
5. Li, W.; Migliavacca, M.; Forkel, M.; Denissen, J.M.C.; Reichstein, M.; Yang, H.; Duveiller, G.; Weber, U.; Orth, R. Widespread increasing vegetation sensitivity to soil moisture. *Nat. Commun.* **2022**, *13*, 1–9. [CrossRef] [PubMed]
6. Shen, X.; Liu, Y.; Liu, B.; Zhang, J.; Wang, L.; Lu, X.; Jiang, M. Effect of shrub encroachment on land surface temperature in semi-arid areas of temperate regions of the Northern Hemisphere. *Agric. For. Meteorol.* **2022**, *320*, 108943. [CrossRef]
7. Tariq, A.; Sardans, J.; Zeng, F.; Graciano, C.; Hughes, A.C.; Farré-Armengol, G.; Peñuelas, J. Impact of aridity rise and arid lands expansion on carbon-storing capacity, biodiversity loss, and ecosystem services. *Glob. Change Biol.* **2024**, *30*, e17292. [CrossRef] [PubMed]
8. Huang, J.; Ji, M.; Xie, Y.; Wang, S.; He, Y.; Ran, J. Global semi-arid climate change over last 60 years. *Clim. Dyn.* **2015**, *46*, 1131–1150. [CrossRef]
9. Zhang, L.; Xiao, J.; Zheng, Y.; Li, S.; Zhou, Y. Increased carbon uptake and water use efficiency in global semi-arid ecosystems. *Environ. Res. Lett.* **2020**, *15*, 034022. [CrossRef]
10. Guo, R.; Guan, X.; He, Y.; Gan, Z.; Jin, H. Different roles of dynamic and thermodynamic effects in enhanced semi-arid warming. *Int. J. Clim.* **2017**, *38*, 13–22. [CrossRef]
11. Hamarash, H.; Hamad, R.; Rasul, A. Meteorological drought in semi-arid regions: A case study of Iran. *J. Arid. Land* **2022**, *14*, 1212–1233. [CrossRef]

water

MDPI

Article

A Comparison of Three Methodologies for Determining Soil Infiltration Capacity in Thicketized Oak Woodlands and Adjacent Grasslands

Furkan Atalar [1,2,†], Pedro A. M. Leite [2,*,†] and Bradford P. Wilcox [2]

1 Institute of Graduate Studies, Istanbul University-Cerrahpasa, 34320 Istanbul, Turkey; furkanatalar061@gmail.com

2 Department of Ecology and Conservation Biology, Texas A&M University, College Station, TX 77843, USA; bwilcox@tamu.edu

* Correspondence: pedroleite@tamu.edu

† These authors contributed equally to this work.

Abstract: This study had two primary objectives: (1) to determine relative differences in soil infiltration capacity between native grasslands and thicketized oak woodlands and (2) to compare the effectiveness of three infiltration measurement techniques—rainfall simulation, an automated Simplified Steady Beerkan Infiltration (SSBI) method, and the Saturo dual-head infiltrometer. The study was conducted at three sites with clay, loamy sand, and sandy soils. Rainfall simulation captured significant infiltration differences between vegetation covers at all three sites, while SSBI did so at two sites, and Saturo failed to detect significant differences. Consistent with past studies, rainfall simulation results showed significantly higher infiltration capacity in thicketized woodlands compared to adjacent grasslands, with mean infiltration capacity an order of magnitude greater in clay soils (67 mm h^{-1} vs. 7.5 mm h^{-1}) and more than twice as high in sandy (144.5 mm h^{-1} vs. 69 mm h^{-1}) and loamy sand (106 mm h^{-1} vs. 49 mm h^{-1}) soils. Across sites, rainfall simulation and SSBI showed strong positive correlations between infiltration capacity and dead biomass (R^2 = 0.74 and 0.46, respectively; $p < 0.001$ for both), as well as significant negative correlations with live biomass and bulk density. In contrast, the Saturo method exhibited higher variability, overestimating infiltration capacity by an average of 34.3 mm h^{-1} compared to rainfall simulation, and did not capture significant relationships with biomass or bulk density. Our findings have twofold importance: first, they demonstrate that thicketization of oak savannahs results in higher soil infiltration capacity; and second, they show that for determining soil infiltration capacity, the SSBI methodology is an accurate and practical alternative to the labor-intensive rainfall simulation.

Keywords: woody plant encroachment; shrub encroachment; soil hydraulic properties; saturated hydraulic conductivity; ecohydrological connectivity; bulk density; preferential flow pathways; surface runoff; *Juniperus virginiana*; *Ilex vomitoria*

1. Introduction

Thicketization is a form of woody plant encroachment (WPE) that, over time, leads to the closing of woody plant canopies. It is a common process in oak woodlands and savannahs in the central and eastern United States [1] that has been facilitated and accelerated by a combination of fire suppression and agricultural land abandonment [2,3]. It is especially prevalent in the Post Oak Savannah ecoregion of Texas, with the expansion of understory shrubs, such as Yaupon (*Ilex decidua, Ilex vomitoria*), as well as of Eastern red cedar trees (*Juniperus virginiana*) [4].

The impact of thicketization on the hydrological aspects of oak woodlands has received relatively little attention. The work that has been carried out suggests that the impact is significant. For example, in the Cross Timbers region of Oklahoma, field and modeling work has demonstrated that thicketization will lead to significant decreases in both groundwater recharge [5] and streamflows [6,7]. Similarly, Basant et al. [3] found that in sites overlying the regionally important Carrizo–Wilcox Aquifer, groundwater recharge was effectively eliminated in thicketized woodlands, whereas in open areas annual recharge rates ranging from 3 to 18 cm were recorded.

One of the most important factors that regulate the terrestrial water cycle is the infiltration capacity of soils. Commonly termed soil infiltrability, it is directly related to the soil's hydraulic conductivity and is influenced by soil properties, such as texture, bulk density, pore structure, aggregate stability, and organic matter content [8]. Changes in vegetation cover can alter these properties, consequently altering infiltration rates and important hydrological processes, such as soil water storage, groundwater recharge, and streamflow [9]. For this reason, soil infiltration capacity is a key parameter in many hydrological and Earth system models [7].

With respect to how WPE may influence water infiltration into the soil, it has been broadly reported that, in general, infiltration capacity and soil porosity (especially macroporosity) are higher under trees or shrubs than in adjacent open areas [10–13]—most likely owing to the formation of channels by decayed roots and the incorporation of leaf litter and organic matter [13,14]. To date, little, if any, work has evaluated the influence of thicketization on soil infiltration capacity in the Post Oak Savannah. However, in the Cross Timbers region of Oklahoma, which is ecologically similar, Zou et al. [6] found that soil infiltration capacity was three times higher under Eastern red cedar canopies than in open grasslands.

Various instruments have been developed to determine soil infiltration capacity in a variety of settings—the most commonly employed being rainfall simulators and ring infiltrometers [15]. Both have advantages and limitations in measuring infiltration rates and providing data useful for understanding the underlying processes. Rainfall simulators, for example, can provide information not only on soil infiltration capacity at the plot scale [16–18], but also on runoff and soil erosion rates [19–21]. Because rainfall simulation is generally applied to areas larger than those for which ponding methods are used, it is better able to capture the spatial variability of surface and subsurface conditions [22,23]. At the same time, rainfall simulation often requires heavy equipment, large amounts of water, and more than one operator—factors that can make multiple measurements difficult.

A more practical alternative for estimating infiltration capacity is the ring infiltrometer. Double-ring infiltrometers are well established and widely used for this purpose; however, they can be labor-intensive and consume more water than single-ring devices, without necessarily providing greater accuracy [24]. A relatively simple and inexpensive single-ring methodology is the Beerkan Estimation of Soil Transfer parameters (BEST) [25,26]. This technique involves inserting a ring into the soil at a shallow depth and sequentially applying small water volumes (ensuring negligible head within the ring) until infiltration rates stabilize, indicating the establishment of a steady-state condition. To reduce manual effort and human error, constant-head infiltrometers can be employed to automate the water application and data collection processes [27,28]. Various algorithms can then be applied to the cumulative infiltration data to estimate field saturated hydraulic conductivity (K_{fs}, mm h^{-1}), a widely used measure for assessing soil infiltration capacity for different vegetation covers [13,29,30]. Most of these algorithms also require soil texture, moisture, and bulk density data [26]. However, the Simplified Steady Beerkan Infiltration (SSBI) method proposed by Bagarello et al. [31] requires only an estimation of the α^* value, which represents capillary length and takes into account the three-dimensional aspects of infiltration. The α^* value of 0.012 mm was considered a good approximation for most field soils, but uncertainties in α^* estimates can propagate into the resulting K_{fs} value. Another potential limitation of this method is that an incorrect assumption of steady-state conditions could result in an overestimation of the soil's infiltration capacity [32].

Only a few commercially available infiltrometers do not require constant monitoring or post-processing of data. One notable example is the Saturo device, produced by Meter Group, which employs the two-ponding head technique developed by Reynolds and Elrick [33] to calculate K_{fs}. While this instrument does not require the estimation of other parameters, such as bulk density or α^* value, the user must set the configurations—including pressure heads, soak and hold times, and number of pressure cycles—which can also introduce bias. Additionally, the two-ponding head method often generates invalid (negative) K_{fs} results, especially in more heterogeneous soils with high microporosity [34]. Although fully automated and user-friendly, the Saturo device can be costly and might not be the most feasible option for many users.

Different techniques for measuring soil infiltration capacity can yield varying results. Many studies have compared different methods, but the conclusions drawn are often contrasting and highly site-specific [15]. Rainfall simulators, for instance, have shown results both similar to [22,35,36] and differing from [23,37] those of ring infiltrometers. While such studies provide important insights into the differences between methods, comprehensive comparisons across diverse soil textures and plant communities remain limited. For example, to our knowledge, no studies have compared multiple measurement techniques in woodland and grassland areas with different soil textures. This leaves a substantial gap in understanding how contrasting natural or unmanaged vegetation covers influence the performance, accuracy, and reliability of these methods. Comparative studies addressing this knowledge gap are essential for improving our ability to select the most appropriate techniques for specific environmental conditions.

Our study had two primary objectives: (1) to test the hypothesis that thicketization of oak savannahs increases soil infiltration capacity, driven by changes in soil porosity and addition of organic matter; and (2) to compare three different methodologies for estimating soil infiltration capacity, each of which has distinct advantages and disadvantages. The three methodologies we compared are rainfall simulation with a drip rainfall simulator [38], the SSBI method employing a constant-head infiltrometer [28,39], and measurement via a Saturo dual-head infiltrometer (Meter Group Inc., Pullman, WA, USA). Rainfall simulation was selected because it is a well-established and widely accepted method for accurately measuring soil infiltration capacity. The SSBI method was included for its simplicity, affordability, and positive comparisons with more complex single-ring methods. Finally, the Saturo infiltrometer was chosen as it is likely the only fully automated commercially available infiltrometer, though its performance in field soils has not been extensively evaluated. We hypothesize that these methods will yield significantly different results due to differences in their measurement scale, underlying assumptions, and calculation approaches.

2. Study Sites

This research was conducted at three sites, each representing a different soil textural class (clay, loamy sand, and sandy) and each consisting of paired grassland- and woodland-cover zones. The clay site was located at the Texas A&M University Beef Cattle Center in College Station, Texas (30°32′49″ N, 96° 25′03″ W, 68 m above sea level); the loamy sand site was located at the Texas A&M University Ecology and Natural Resources Teaching Area in College Station, Texas (30°34′39″ N, 96°21′04″ W, 88 m above sea level); and the sandy site was located at the Gus Engeling Wildlife Management Area, 32 km northwest of Palestine, Texas (31°56′21″ N, 95°53′45″ W, 104 m above sea level). The average annual temperature and precipitation are 21 °C and 1011 mm, respectively, at the clay and loamy sand sites, and 19.6 °C and 1083 mm, respectively, at the sandy site. All three sites lie within the Post Oak Savannah ecoregion of Texas and have a humid, subtropical climate.

According to the USDA/NRCS Web Soil Survey, the clay site is a Chromic Hapluderts (Ships series), the loamy sand site is a Chromic Vertic Albaqualfs (Boonville series), and the sandy site is an Arenic Plinthic Paleudults (Lilbert series). These soil types are characteristic of the Post Oak Savanna region of Texas, and our site selection aimed to capture some of this natural soil variability. While not exhaustive, selecting three distinct soil types and contrasting vegetation covers improves the study representativeness within this landscape.

The overstory component of the woodland zones at all three sites is dominated by post oak (*Quercus stellata*), blackjack oak (*Quercus marilandica*), winged elm (*Ulmus alata*), and a few large individuals of Eastern red cedar (*Juniperus virginiana*). The dense shrub understory consists mainly of encroaching species—Yaupon holly (*Ilex vomitoria*), the exotic Chinese privet (*Ligustrum sinense*), and young individuals of Eastern red cedar. The little herbaceous cover in the woodland zones is mostly C3 grasses such as Inland sea oats (*Chasmanthium latifolium*) and a few forb species.

The grassland zones at the loamy sand and sandy sites are dominated by native C4 grasses, including little bluestem (*Schizachyrium scoparium*) and switchgrass (*Panicum virgatum*), and many forb species. Neither of these sites has any recent history of grazing. At the clay site, the grassland zones are composed mainly of the introduced bermudagrass (*Cynodon dactylon*). This site experiences rotational grazing annually at relatively low stocking densities (approximately 1 animal unit/ha).

3. Materials and Methods

3.1. Experimental Design

At each site and within each cover type, we selected five measurement locations, spaced three meters apart along a linear transect. The transects for grassland and woodland zones were drawn parallel to each other and approximately 10 m from the boundary separating the two cover types. All measurements were conducted between April and May during the early growing season to minimize potential effects of seasonal variability. At each measurement location, a 65 cm × 65 cm runoff plot was established, enclosed by galvanized steel sheets inserted to a shallow depth (approximately 5 cm), for the rainfall simulation experiments. After the simulations had been completed, two corners of the plot were randomly selected for infiltration capacity measurements with the SSBI method, and a third corner was chosen for measurements with the Saturo infiltrometer (Figure 1). Figure 2 shows photographs of the three methodologies deployed in two different vegetation cover settings.

3.2. Infiltration Capacity Measurement Methods and Instruments

3.2.1. Rainfall Simulation Method

We utilized a drip-type rainfall simulator that has been widely employed by previous researchers [12,40–42] and is described in detail by Blackburn et al. [38]. The simulator (see Figure 2a) is placed approximately 40 cm above the soil surface and delivers droplets of water to a 90 cm × 90 cm surface area at a maximum rate of 154 mm h^{-1}. At each plot, we conducted a rainfall simulation over the 65 cm × 65 cm surface area. Because we expected high infiltration capacity, especially in the woodland zones, we applied the maximum rainfall intensity of 154 mm h^{-1} in order to maximize the chances of obtaining runoff. Runoff was collected for 30 s immediately after it commenced, followed by 30 s collections every 5 min until four consecutive runoff volumes were approximately the same. These four volumes were averaged to calculate a terminal runoff value, which was then subtracted from the rainfall intensity to obtain the steady-state infiltration rate [43,44]. The steady-state infiltration rate can be considered a good approximation of the soil's infiltration capacity, or K_{fs} [43], particularly under the assumption that lateral matrix flow is negligible owing to the buffering effect created by water falling outside the plot [45].

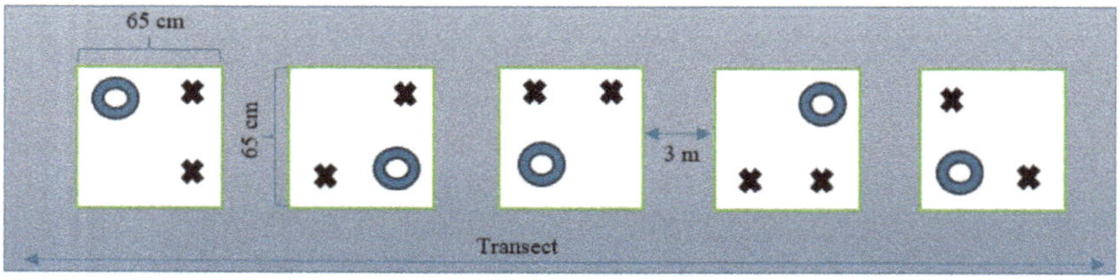

Figure 1. A schematic diagram showing the layout of the five measurement locations along each transect and the infiltration tests performed with the three methods. Green-outlined squares = rainfall simulation plots; black X = measurement via automated SSBI; blue circle = measurement via Saturo infiltrometer.

Figure 2. Soil infiltration capacity was measured via three different types of instruments under two vegetation cover conditions: (**a**) rainfall simulator in a thicketized woodland; (**b**) automated SSBI system in a grassland; and (**c**) Saturo infiltrometer in a grassland.

3.2.2. Automated Simplified Steady Beerkan Infiltration (SSBI) Method

The automated infiltrometers described in Leite et al. [28] were used to determine the K_{fs} values, which were calculated from steady-state infiltration rates obtained via the SSBI method [31]. A 10 cm diameter ring was inserted to a depth of 1 cm and the infiltrometers (see Figure 2b), filled with 2.2 L of water, were used to sustain a hydraulic head of about 1 cm within the ring. The test was conducted for a period of 60 min or until all water drained from the infiltrometer. Pressure readings were recorded every five seconds by a HOBO U20 datalogger (Onset, Bourne, MA, USA) and were subsequently used to generate cumulative infiltration curves via the calibration function provided in Leite et al. [28]. Steady-state infiltration rates (i_s; mm h^{-1}) were then derived from the linear (stable) segments of the infiltration curves and used to calculate K_{fs} (mm h^{-1}) according to Bagarello et al. [31], as follows

$$K_{fs} = \frac{i_s}{\frac{\gamma \gamma_w}{r\alpha^*} + 1}, \quad (1)$$

where r denotes the radius of the ring in millimeters, y and y_w are dimensionless constants, and α^* is an empirical parameter that captures the effects of gravitational and capillary forces. We adopted an α^* value of 0.012 mm, which is considered a good approximation for most field soils [31]. We performed two tests per plot one week after the rainfall simulations. The K_{fs} values for each plot represent the average of these two measurements.

3.2.3. Saturo Method

A Saturo dual-head infiltrometer (Meter Group Inc., Pullman, WA, USA) with a 14.4 cm (inner diameter) ring and a 5 cm insertion depth was employed to measure K_{fs} at one point per plot (see Figure 2c). An increasing number of studies have been adopting this instrument for evaluating soil infiltration capacity in engineered soils [46,47] and agricultural settings [48,49]. Because it is a relatively new instrument, there are few studies comparing the Saturo with other methods; however, one recent laboratory study on homogenous soil beds concluded that it yields results comparable with those obtained from rainfall simulators [36].

The instrument is fully automated and allows for setting different configurations—such as pressure heads, soak and hold times, and number of pressure cycles. We used pressure heads of 10 and 100 mm H_2O, a soak time and hold time of 10 min, and three pressure cycles, for a total test duration of 70 min. These configurations were selected to strike a balance between water consumption and test duration (higher pressure heads and longer hold times would have increased both water usage and the total duration of each test). Additionally, we sought to align the test duration with those of the other two methods to improve comparability across methodologies.

The final K_{fs} value is automatically calculated by the instrument by means of the simplified version of the two-ponding head technique developed by Reynolds and Elrick [33], as follows:

$$K_{fs} = \frac{\Delta(i_1 - i_2)}{D_1 - D_2}, \tag{2}$$

where D_1 represents the higher-pressure head (mm H_2O), D_2 denotes the lower-pressure head (mm H_2O), and Δ is calculated as 0.993 times the ring insertion depth (mm) plus 0.578 times the ring radius (mm). The variables i_1 and i_2 are the steady-state infiltration rates (mm h^{-1}) at D_1 and D_2, respectively.

3.3. Additional Data

For each test, we also collected 5 cm diameter soil cores at the 0 to 5 cm and 5 to 10 cm depth intervals for measuring bulk density via a standard oven-drying technique (105 °C for 48 h). After the rainfall simulations and prior to the other infiltration tests, we collected all the live (clipped at ground level) and dead plant tissues from the surface of each plot to determine live and dead plant biomass. The samples were taken to the laboratory and oven-dried at 65 °C for 48 h.

3.4. Data Analysis

Infiltration capacity differences between vegetation cover types (woodland vs. grassland) for the three sites (clay soil, loamy sand soil, and sandy soil) were tested by means of t-tests, and comparisons of the three measurement methodologies were analyzed with one-way ANOVA and Tukey's HSD post-hoc test. Prior to these tests, skewness and heteroscedasticity were taken into account by the application of log transformations that met parametric assumptions. We then conducted simple linear regressions to evaluate how vegetation biomass (live and dead) and bulk density correlated with infiltration capacity values obtained with each method. Finally, we conducted paired t-tests to assess whether, across all sites and treatments, the infiltration capacity values obtained from the automated

SSBI and Saturo methods differed significantly from those obtained via rainfall simulation. All analyses were performed using R version 4.4.1.

4. Results

4.1. Rainfall Simulation Method

The results obtained with the rainfall simulator showed significantly higher infiltration rates for the thicketized woodland zones than for the grassland zones at all three sites (clay, loamy sand, and sandy) (Figure 3, Table 1). Specifically, at the clay soil sites, mean infiltration capacity values in the WPE-affected zones were an order of magnitude higher than those in the grassland zones; and at the loamy sand and sandy soil sites they were more than twice as high as those in grassland zones. At the sandy site, infiltration capacity for the thicketized woodland exceeded rainfall intensity (154 mm h^{-1}) in four of the five plots. The mean from the rainfall simulation is likely underestimated, as there was no runoff from those four plots, even at the maximum simulated rainfall intensity of 154 mm h^{-1}. For this reason, statistical comparisons between methodologies were not performed for the woodland tests at this site.

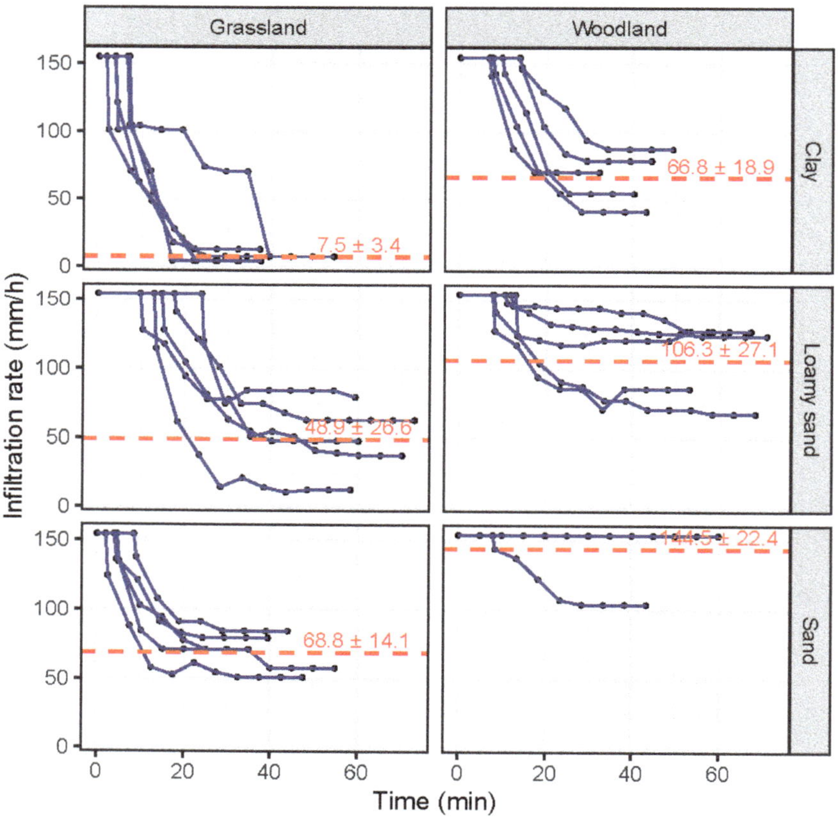

Figure 3. Infiltration rate curves from tests conducted with the rainfall simulator. The red dashed line represents the average steady-state infiltration rate (infiltration capacity) for each site. For the woodland zone in the sandy soil site, from which there was no runoff in four of the five plots, the infiltration rates shown (measured at the maximum simulated rainfall intensity of 154 mm h^{-1}) are likely underestimations.

Table 1. Infiltration capacity values (mean ± standard deviation, mm h^{-1}) obtained via the three measurement methodologies for each soil textural class and cover type. Capital *A* or *B* represents a significant difference (revealed by *t*-tests) in the mean values obtained by each methodology for the two cover types (grassland vs. woodland) within each site, while lowercase *a* or *b* represents a significant difference (revealed by Tukey's HSD test) between the mean values obtained by different methodologies within the same site and cover type.

Site/ Soil Texture	Cover Type	Methodology		
		Rainfall Simulation	Automated SSBI	Saturo
Clay	Grassland	7.5 ± 3.4 *A b*	4.3 ± 3 *A b*	47.4 ± 18.6 *A a*
	Woodland	66.8 ± 18.9 *B a*	71.7 ± 92.4 *B a*	153.3 ± 138.2 *A a*
Loamy sand	Grassland	48.9 ± 26.6 *A a*	101.6 ± 94.3 A *a*	113 ± 63.6 *A a*
	Woodland	106.3 ± 27.1 *B a*	79.2 ± 57.4 *A a*	56.3 ± 85.8 *A a*
Sandy	Grassland	68.8 ± 14.1 *A a*	13.4 ± 2.1 *A b*	108.4 ± 81.6 *A a*
	Woodland	144.5 ± 22.4 *B*	120.9 ± 28.6 *B*	183.4 ± 84.3 *A*

4.2. Automated SSBI Method

With the automated SSBI method, we found significantly higher infiltration capacity for the woodland-cover zones than for the grassland-cover zones at the clay and sandy soil sites, but no significant difference at the loamy sand site (Figure 4, Table 1). For the clay and sandy soils, mean infiltration capacity in woodland zones was an order of magnitude higher than in grassland zones.

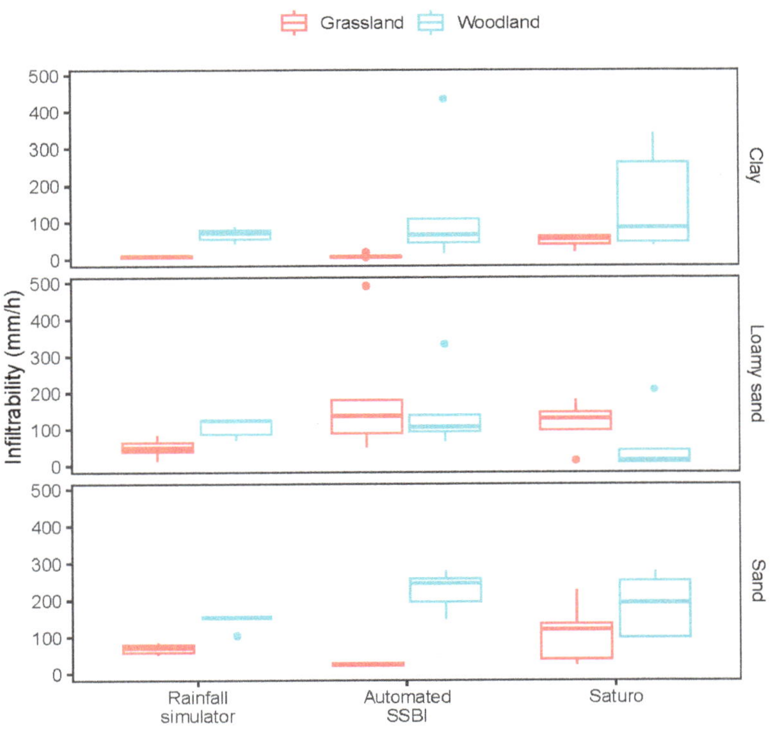

Figure 4. Box plots of infiltration capacity values for grassland and woodland zones obtained at the three sites via the three measurement methodologies.

4.3. Saturo Method

Measurements with the Saturo infiltrometer indicated higher median infiltration capacity in thicketized woodlands for the clay and sandy soil sites (Figure 4), but in both cases the difference was not significant (Table 1) because of the higher variability of this instrument. Similarly, for the grassland zone at the clay soil site, the Saturo-obtained mean infiltration capacity was an order of magnitude higher than those obtained with the other two methods; and for the grassland zone at the sandy soil site, the Saturo-obtained mean value was 1.6 times higher than that obtained via rainfall simulation and 8 times higher than that obtained with the automated SSBI method (Table 1).

4.4. Influence of Biomass and Bulk Density on Infiltration Capacity

With respect to the influence of biomass on infiltration capacity, we found that the rainfall simulation and automated SSBI methods both indicated a significant negative correlation (adj R^2 = 0.32, p < 0.001 and adj R^2 = 0.28, p = 0.002, respectively) between live biomass and infiltration capacity (Figure 5)—which suggests that higher live biomass was associated with reduced infiltration rates. The Saturo method, however, showed an insignificant correlation (adj R^2 = -0.03, p = 0.874), indicating that this method was not sensitive to differences in live biomass. Additionally, both the rainfall simulation (adj. R^2 = 0.74, p = 0.001) and the automated SSBI (adj. R^2 = 0.46, p < 0.001) methods showed strong positive correlations between dead biomass and infiltration capacity, suggesting that an increase in dead biomass significantly enhances soil infiltration. This effect is likely due to decomposed organic matter and aggregate breakdown, which slow runoff and facilitate infiltration. The Saturo method, on the other hand, showed a result similar to that for live biomass, i.e., no significant relationship between dead biomass and infiltration capacity (adj. R^2 = 0.02, p = 0.213).

Finally, with respect to the effects of bulk density on infiltration, both rainfall simulation and the automated SSBI method yielded significant negative correlations, on the basis of measurements at two depth intervals (0–5 cm and 5–10 cm). For the 0 to 5 cm depth, the adjusted R^2 values for rainfall simulation and for automated SSBI were, respectively, 0.25 (p = 0.003) and 0.65 (p < 0.001); and in the 5 to 10 cm depth, they were 0.37 (p < 0.001) and 0.65 (p < 0.001)—implying that higher bulk density reduces infiltration capacity. In contrast, the Saturo method showed non-significant relationships at both the 0 to 5 cm and 5 to 10 cm depths, with adjusted R^2 values of 0.08 (p = 0.067) and 0.03 (p = 0.166), respectively.

4.5. Differences Among Methodologies

With respect to variability, rainfall simulation generally showed the least of the three methods. Coefficients of variability ranged from 15.5% to 54.3% for the rainfall simulator, from 15.6% to 128.9% for the automated SSBI, and from 39.3% to 152.5% for the Saturo. When we compared the rainfall simulation and automated SSBI infiltration capacity values for all sites and both cover types, the paired t-test yielded a t-value of -0.49, a p-value of 0.627, and a mean difference of -5.42 mm h^{-1}—that is, there was no significant difference between the two methods. Conversely, when we compared the rainfall simulation and the Saturo methods in the same way, the paired t-test revealed a significant difference (t = 2.15, p = 0.041), with the Saturo method producing higher infiltration capacity values on average (mean difference = 34.28 mm h^{-1}). In other words, the Saturo method may overestimate infiltration capacity relative to rainfall simulation.

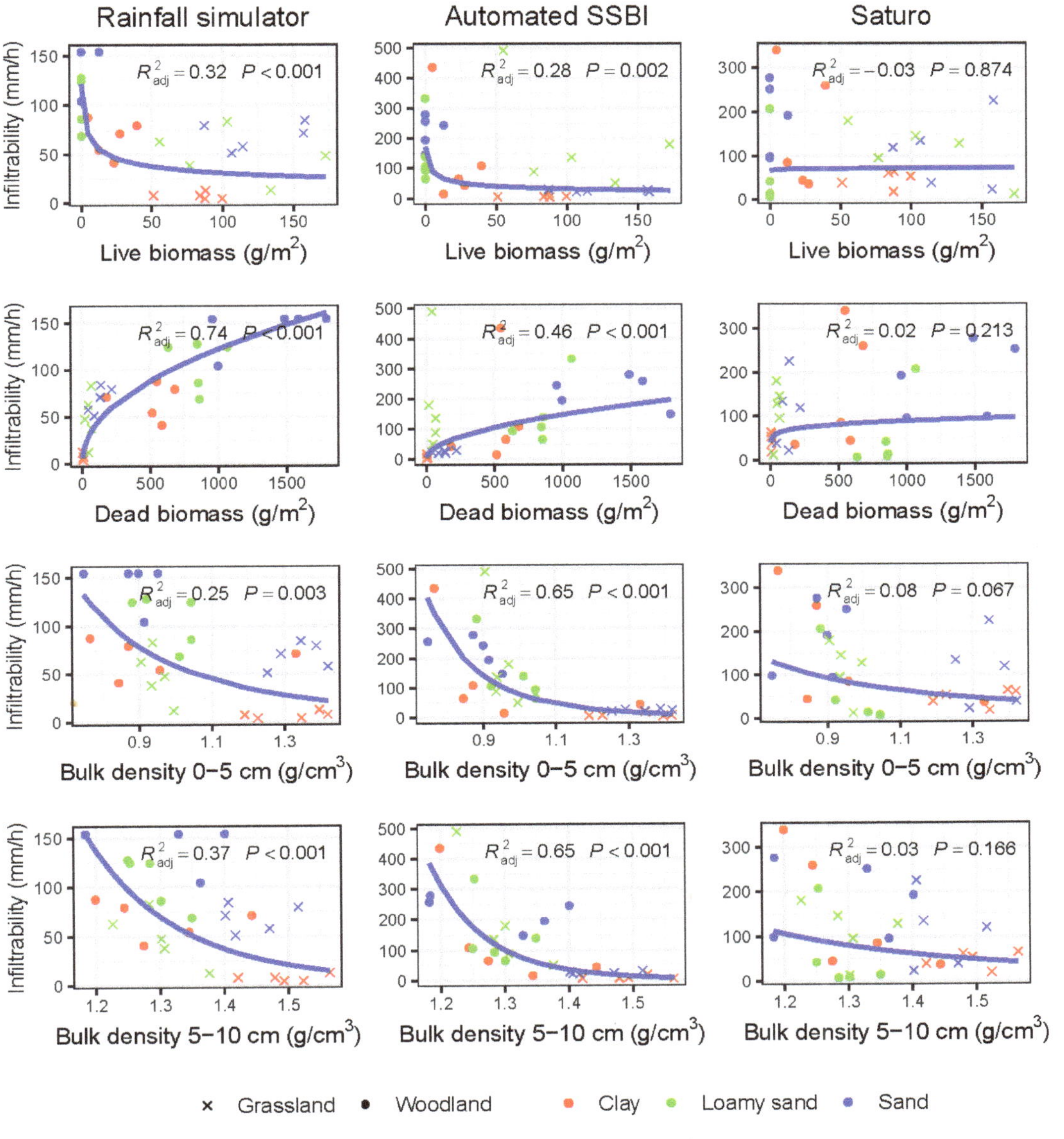

Figure 5. Correlations between infiltration capacity values (mm h^{-1}) obtained with each method (rainfall simulation, automated SSBI, and Saturo) and influencing variables: live biomass and dead biomass (g m^{-2}), and soil bulk density (g cm^{-3}) in the 0 to 5 cm and 5 to 10 cm depth intervals. Note that the regression line, adjusted R^2, and *p*-values are based on a log–log relationship, but the scales on both axes are presented in their original form for better visualization.

5. Discussion

We designed this study with two objectives in mind. The first was to evaluate differences in soil infiltration capacity in thicketized woodlands vs. more open grasslands in the Post Oak Savannah, and the second was to compare three field-based methods for determining soil infiltration capacity.

5.1. Effects of Thicketization on Soil Infiltration Capacity

We found that, regardless of soil texture, soil infiltration capacity was higher in thicketized woodlands than in open grasslands. This finding is based primarily on the results obtained via rainfall simulation, which (as we explain below) we consider the gold standard for determining soil infiltration capacity. In addition, this finding is generally supported by the other methodologies we tested.

Higher infiltration capacity in the thicketized woodland zones is broadly consistent with what has been demonstrated elsewhere [6,12,13,20,30]. The closest comparison to our site would be the Cross Timbers region, where juniper is encroaching upon grasslands and open woodlands. As noted previously, Zou et al. [6] found that infiltration capacity was three times higher in juniper-encroached areas than in open grasslands—most likely as a result of the addition of organic matter from leaf litter.

Notably, we found that the highest contrast in infiltration capacity between encroached woodlands and grasslands occurred at the clay site, which was also the only site subject to grazing. This aligns with previous studies showing that grazing reduces infiltration capacity [12,40] and that the contrast between woody and grass cover is most pronounced in the presence of grazing [50].

While increased infiltration capacity may benefit soil water retention and aboveground biomass productivity, it does not necessarily translate into higher streamflow or groundwater recharge, particularly in thicketized landscapes. For example, in regions like the Cross Timbers, thicketization has been linked to lower streamflows due to enhanced evapotranspiration and reduced surface runoff, which is the primary contributor to streamflow in the region [5,6]. Similarly, Basant et al. [3] found that in a Post Oak Savannah site overlying deep sands, thicketization was associated with greater rooting depths and effectively eliminated groundwater recharge. Nonetheless, higher infiltration under woody plants can have positive aspects. At our clay site, for instance, the extremely low infiltration capacity of the grassland (<10 mm/h) can lead to substantial runoff generation during storms, contributing to high erosion rates, nutrient and contaminant transport, and flash floods. Therefore, maintaining areas of thicketized woodlands within grazed grasslands and pastures could be beneficial for capturing runoff, particularly near riparian corridors, swales, and drainage depressions prone to gully erosion.

As shown in Figure 5, soil infiltration capacity at our sites was highly correlated with dead biomass or leaf litter, a finding that also aligns with those of numerous other studies reporting that litter enhances soil infiltration capacity [51–53]. The lower bulk densities for soils in thicketized areas (Figure 5) are also likely a reflection of additions of organic matter. The negative relationship between bulk density, which is an expression of soil compaction, and soil infiltration capacity has been shown repeatedly [13,50,54,55].

One counter-intuitive result that we found was a negative correlation between infiltration capacity and live biomass (Figure 5). For the thicketized zones, the fact that there was relatively little live biomass on the ground surface compared with the grassland zones explains the negative correlation. Wilcox et al. [16] found grass biomass to be positively correlated with soil infiltration capacity, whereas a meta-analysis by Thompson et al. [15] reported both positive and non-significant relationships, depending on climate. In contrast, findings from other studies have suggested that the dense root systems associated with high levels of herbaceous cover may contribute to reduced free pore spaces, which impedes the movement of water into the soil [56,57].

5.2. Comparison of Methodologies

Rainfall simulators have been widely used to measure surface runoff and infiltration rates across areas that differ with respect to land use and vegetation cover. As noted above, rainfall simulation is often considered the gold standard for determining soil infiltration capacity [22,58–60] because it is more comparable to natural rainfall than ponding methods and its gradual wetting of the soil avoids problems such as air entrapment and rapid soil slaking [58]. Another advantage over ring infiltration methods is that rainfall simulations are normally performed over a larger soil surface, enabling them to capture greater spatial variability of factors that influence infiltration capacity (e.g., microtopography and infiltration hotspots such as soil macropores). The net result is less variability in infiltration capacity values [22,23,61]. In the case of our study, the larger measurement area of the rainfall simulations offers a plausible explanation for the generally higher values than those obtained with the SSBI method, as well as the lower variability in values than observed with both the SSBI and Saturo methods.

Some types of rainfall simulators attempt to mimic natural rainfall events to take into account the effect of raindrops' kinetic energy on aggregate breakdown and soil sealing. In our study, we opted to perform simulations from a short distance above the ground surface to make results more comparable with those obtained by the two ring infiltrometers. Nonetheless, one potential reason for differences between the rainfall simulation results and those of the other methods could be the protective effect of leaf litter and herbaceous cover. Because leaf litter and herbaceous cover—which are known to protect the soil against surface sealing by soil aggregate breakdown [52,53,62]—were removed for the infiltration tests with the automated SSBI and Saturo instruments, only the rainfall simulations would have benefited from this protective effect. This could partially explain why the automated SSBI infiltrometers yielded lower infiltration capacity values for the sandy soils. Previous studies have demonstrated that surface sealing can lead to major reductions in infiltration capacity [44], and that the effect can be accentuated for coarser-textured soils (which tend to have lower aggregate stability than finer-textured soils) [63].

One drawback of rainfall simulators is that their maximum rainfall intensity imposes an upper limit on measurable infiltration capacity values—in our case, 154 mm h^{-1}. As a result, soils with exceptionally high infiltration capacity, such as those observed in the woodland zone at our sandy site, may exceed this limit, leading to imprecise measurements and potentially undermining meaningful statistical analysis. Additionally, they are labor-intensive and often require large volumes of water, making them less suitable for use in areas with difficult access and limiting the number of replicates.

The Saturo infiltrometer, while fully automated and capable of providing K_{fs} values without the need for additional data processing, shares some of these drawbacks. Like a rainfall simulator, it is relatively bulky and water-demanding, limiting its applicability in certain field situations. Further, with the Saturo device it is crucial to monitor the equipment carefully; if the water runs out, the test can be interrupted and the results lost. Additionally, the Saturo's complexity and reliance on interconnected components make it prone to malfunctions requiring specialized servicing, further increasing costs and downtime.

Consistent with what has been found elsewhere, the Saturo infiltrometer generally yielded higher infiltration capacity values than the other methods [64]. This difference could be explained by the methodological approach of the Saturo. In contrast to rainfall simulation and the automated SSBI method, with which ponding and water-pressure heads are mostly negligible, the Saturo requires at least one relatively high-pressure head (10 cm in our study)—which could exaggerate the role of preferential flow. Higher-pressure heads could cause water to flow to a disproportionately greater extent through preferential pathways, such as macropores or cracks in the soil, than through the soil matrix [65]. And the fact that these pathways are more common in finer-textured soils [66] would explain why the difference in values was particularly striking at the clay soil site.

Compared with the SSBI method, the Saturo method involves the insertion of heavier-gauge rings deeper into the soil, which causes greater disturbance to the soil structure, potentially masking effects of vegetation cover on surface hydraulic properties. This may also explain the lack of correlation between Saturo data and variables such as vegetation biomass or bulk density. In a study conducted across 124 sites in North America, [49] used the Saturo to measure K_{fs} and found that it was not significantly affected by land management or soil properties—a finding that contradicts a substantial body of research [12,39,67–70]. While the authors acknowledged that the limited sampling size (one point per experimental unit) and the inherently high variability of K_{fs} likely influenced their results, we further propose that their use of the Saturo may also have contributed to these unexpected findings.

For the automated SSBI method, a potential source of uncertainty in the K_{fs} values obtained is the choice of the α^* value. However, for most applications, large variations in the α^* value would have relatively small impacts on K_{fs} estimates [71]. For example, even a 10-fold overestimation of the α^* value would result in just a 50% error in K_{fs} estimates [72]. Within each site, the relative differences in K_{fs} between vegetation covers should be mostly unaffected by the choice of α^* value, as the soil class remains the same. For our clay and loamy sand sites, the K_{fs} values obtained with the SSBI method were within the same range as those obtained by other methods, suggesting that the recommended first-approximation α^* value of 0.012 mm was appropriate. However, for the sandy site, the SSBI method yielded K_{fs} values lower than those of the other methods, and this may have been caused in part by underestimation of the α^* value. Thus, with the SSBI method, caution is warranted in using a fixed α^* value to compare K_{fs} across sites having different soil types.

Despite potential uncertainties, the automated SSBI method's ability to collect more replicates in a shorter time offers a clear advantage over both rainfall simulation and the Saturo method. Being lightweight, requiring less water, and producing results within a range comparable to those of the other methods—while generally aligning more closely with rainfall simulation than with Saturo—it offers a practical compromise between the two.

As expected from a single-ring method, the infiltration capacity data obtained with the SSBI method exhibit greater variability than those from rainfall simulation. However, the automated SSBI remains more affordable and significantly lighter than the other two instruments. During our study, one operator using 10 automated SSBI infiltrometers (at the cost of a single Saturo unit) was able to collect all 20 data points per site in less than four hours. In contrast, collection of 10 data points with either a Saturo or a rainfall simulator would take two days because of the time required for setup and testing. Additionally, the automated SSBI required significantly less water per test (2.2 L or less) than either the Saturo (5–30 L) or the rainfall simulator (60–80 L). This reduced water demand is especially advantageous in remote or water-scarce environments, where transporting large volumes of water may be impractical.

6. Summary and Conclusions

This study compared infiltration capacity determined through the use of three different measurement methodologies: rainfall simulation, automated SSBI, and the Saturo dual-head infiltrometer method. The tests were carried out at three experimental sites having soils of different textures (clay, loamy sand, and sandy) and each having both a grassland and a woodland component. Measurements done with the rainfall simulator revealed significantly higher infiltration capacity for the woodland-cover zones than for the grassland zones across all three soil textures. For the woodland zones of the sites with loamy sand and sandy soils, mean infiltration capacity was more than double that found in the grassland zones, and for the woodland zone of the clay soil site it was nearly ten times higher than for the grassland zone.

Paired t-tests using data from all three sites showed no significant difference in infiltration capacity between the rainfall simulation and automated SSBI methods ($t = -0.49$, $p = 0.627$). In contrast, the test results showed a significant difference for the Saturo method ($t = 2.15$, $p = 0.041$), indicating a slight overestimation compared with rainfall simulation. For both the rainfall simulation and automated SSBI methods, regressions analyses revealed significant correlations between infiltration capacity and influencing factors (e.g., live and dead biomass, bulk density), whereas no such correlations were observed for the Saturo.

Each of the three methods tested in this study has its advantages and disadvantages and was able to generate infiltration capacity values within a comparable range. However, with respect to ability to capture differences between woodland and grassland covers, the three differed substantially. The Saturo infiltrometer, while fully automated, showed high variability and failed to capture significant differences between covers. Rainfall simulation yielded measurements with lower variability and showing significant differences between grassland and woodland cover on all three sites—but its labor-intensive and water-demanding nature limits its practicality. The automated SSBI method, despite also showing high variability, captured differences in infiltration capacity between cover types at two of the three sites. This method also offers several advantages over the other two, including the ability to collect a larger number of replicates in less time, portability, and reduced water requirements. For soils with natural vegetation cover and high spatial heterogeneity, the SSBI method may be a practical alternative to rainfall simulation.

Author Contributions: Conceptualization, P.A.M.L. and B.P.W.; Methodology, P.A.M.L.; Formal analysis, F.A. and P.A.M.L.; Investigation, F.A. and P.A.M.L.; Writing—original draft, F.A. and P.A.M.L.; Writing—review & editing, P.A.M.L. and B.P.W.; Supervision, B.P.W.; Project administration, B.P.W.; Funding acquisition, B.P.W. All authors have read and agreed to the published version of the manuscript.

Funding: This research was funded by National Institute of Food and Agriculture, grant number 13386861.

Data Availability Statement: The original contributions presented in this study are included in the article. Further inquiries can be directed to the corresponding author.

Conflicts of Interest: The authors declare no conflicts of interest.

References

1. Scifres, C.J.; Kelly, D. Range Vegetation Response to Burning Thicketized Live Oak Savannah. In *Texas FARMER Collection*; Texas Agricultural Experiment Station: Weslaco, TX, USA, 1979.
2. Dyksterhuis, E. The Savannah Concept and Its Use. *Ecology* **1957**, *38*, 435–442. [CrossRef]
3. Basant, S.; Wilcox, B.P.; Parada, C.; Wyatt, B.M.; Newman, B.D. Thicketized Oak Woodlands Reduce Groundwater Recharge. *Sci. Total Environ.* **2023**, *862*, 160811. [CrossRef] [PubMed]
4. Olariu, H.G.; Wilcox, B.P.; Popescu, S.C. Examining Changes in Woody Vegetation Cover in a Human-Modified Temperate Savanna in Central Texas between 1996 and 2022 Using Remote Sensing. *Front. For. Glob. Change* **2024**, *7*, 1396999. [CrossRef]
5. Acharya, B.S.; Hao, Y.; Ochsner, T.E.; Zou, C.B. Woody Plant Encroachment Alters Soil Hydrological Properties and Reduces Downward Flux of Water in Tallgrass Prairie. *Plant Soil* **2017**, *414*, 379–391. [CrossRef]
6. Zou, C.B.; Turton, D.J.; Will, R.E.; Engle, D.M.; Fuhlendorf, S.D. Alteration of Hydrological Processes and Streamflow with Juniper (*Juniperus virginiana*) Encroachment in a Mesic Grassland Catchment. *Hydrol. Process.* **2014**, *28*, 6173–6182. [CrossRef]
7. Qiao, L.; Zou, C.B.; Will, R.E.; Stebler, E. Calibration of SWAT Model for Woody Plant Encroachment Using Paired Experimental Watershed Data. *J. Hydrol.* **2015**, *523*, 231–239. [CrossRef]
8. Hillel, D. *Soil and Water: Physical Principles and Processes*; Elsevier: Amsterdam, The Netherlands, 2012; ISBN 0-323-15670-3.
9. Ding, J.; Eldridge, D.J. Contrasting Global Effects of Woody Plant Removal on Ecosystem Structure, Function and Composition. *Perspect. Plant Ecol. Evol. Syst.* **2019**, *39*, 125460. [CrossRef]

10. Eldridge, D.J.; Maestre, F.T.; Maltez-Mouro, S.; Bowker, M.A. A Global Database of Shrub Encroachment Effects on Ecosystem Structure and Functioning: *Ecological Archives* E093-234. *Ecology* **2012**, *93*, 2499. [CrossRef]
11. Bargués Tobella, A.; Reese, H.; Almaw, A.; Bayala, J.; Malmer, A.; Laudon, H.; Ilstedt, U. The Effect of Trees on Preferential Flow and Soil Infiltrability in an Agroforestry Parkland in Semiarid Burkina Faso. *Water Resour. Res.* **2014**, *50*, 3342–3354. [CrossRef]
12. Basant, S.; Wilcox, B.P.; Leite, P.M.; Morgan, C.L. When Savannas Recover from Overgrazing, Ecohydrological Connectivity Collapses. *Environ. Res. Lett.* **2020**, *15*, 054001. [CrossRef]
13. Leite, P.A.M.; Wilcox, B.P.; McInnes, K.J. Woody Plant Encroachment Enhances Soil Infiltrability of a Semiarid Karst Savanna. *Environ. Res. Commun.* **2020**, *2*, 115005. [CrossRef]
14. Wu, G.-L.; Cui, Z.; Huang, Z. Contribution of Root Decay Process on Soil Infiltration Capacity and Soil Water Replenishment of Planted Forestland in Semi-Arid Regions. *Geoderma* **2021**, *404*, 115289. [CrossRef]
15. Thompson, S.E.; Harman, C.J.; Heine, P.; Katul, G.G. Vegetation-infiltration Relationships across Climatic and Soil Type Gradients. *J. Geophys. Res.* **2010**, *115*, 2009JG001134. [CrossRef]
16. Wilcox, B.P.; Wood, M.K.; Tromble, J.M. Factors Influencing Infiltrability of Semiarid Mountain Slopes. *J. Range Manag.* **1988**, *41*, 197. [CrossRef]
17. Cerda, A.; Schnabel, S.; Ceballos, A.; Gomez-Amelia, D. Soil Hydrological Response under Simulated Rainfall in the Dehesa Land System (Extremadura, SW Spain) under Drought Conditions. *Earth Surf. Process. Landf.* **1998**, *23*, 195–209. [CrossRef]
18. Leite, P.A.M.; de Souza, E.S.; dos Santos, E.S.; Gomes, R.J.; Cantalice, J.R.; Wilcox, B.P. The Influence of Forest Regrowth on Soil Hydraulic Properties and Erosion in a Semiarid Region of Brazil. *Ecohydrology* **2018**, *11*, e1910. [CrossRef]
19. Seyfried, M.S. Infiltration Patterns from Simulated Rainfall on a Semiarid Rangeland Soil. *Soil Sci. Soc. Am. J.* **1991**, *55*, 1726–1734. [CrossRef]
20. Taucer, P.I.; Munster, C.L.; Wilcox, B.P.; Owens, M.K.; Mohanty, B.P. Large-Scale Rainfall Simulation Experiments on Juniper Rangelands. *Trans. ASABE* **2008**, *51*, 1951–1961. [CrossRef]
21. Williams, C.J.; Pierson, F.B.; Kormos, P.R.; Al-Hamdan, O.Z.; Nouwakpo, S.K.; Weltz, M.A. Vegetation, Hydrologic, and Erosion Responses of Sagebrush Steppe 9 Yr Following Mechanical Tree Removal. *Rangel. Ecol. Manag.* **2019**, *72*, 47–68. [CrossRef]
22. Gupta, R.K.; Rudra, R.P.; Dickinson, W.T.; Patni, N.K.; Wall, G.J. Comparison of Saturated Hydraulic Conductivity Measured by Various Field Methods. *Trans. ASAE* **1993**, *36*, 51–55. [CrossRef]
23. Vlček, L.; Šípek, V.; Zelíková, N.; Čáp, P.; Kincl, D.; Vopravil, J. Water Retention and Infiltration Affected by Conventional and Conservational Tillage on a Maize Plot; Rainfall Simulator and Infiltrometer Comparison Study. *Agric. Water Manag.* **2022**, *271*, 107800. [CrossRef]
24. Verbist, K.; Torfs, S.; Cornelis, W.M.; Oyarzún, R.; Soto, G.; Gabriels, D. Comparison of Single- and Double-Ring Infiltrometer Methods on Stony Soils. *Vadose Zone J.* **2010**, *9*, 462–475. [CrossRef]
25. Lassabatère, L.; Angulo-Jaramillo, R.; Soria Ugalde, J.M.; Cuenca, R.; Braud, I.; Haverkamp, R. Beerkan Estimation of Soil Transfer Parameters through Infiltration Experiments—BEST. *Soil Sci. Soc. Am. J.* **2006**, *70*, 521. [CrossRef]
26. Angulo-Jaramillo, R.; Bagarello, V.; Di Prima, S.; Gosset, A.; Iovino, M.; Lassabatere, L. Beerkan Estimation of Soil Transfer Parameters (BEST) across Soils and Scales. *J. Hydrol.* **2019**, *576*, 239–261. [CrossRef]
27. Di Prima, S. Automated Single Ring Infiltrometer with a Low-Cost Microcontroller Circuit. *Comput. Electron. Agric.* **2015**, *118*, 390–395. [CrossRef]
28. Leite, P.A.M.; Di Prima, S.; Schmidt, L.M.; Wilcox, B.P. A Simple Infiltrometer Automated with a User-friendly Pressure Datalogger. *Vadose Zone J.* **2024**, *23*, e20366. [CrossRef]
29. Wilcox, B.P.; Breshears, D.D.; Turin, H.J. Hydraulic Conductivity in a Piñon-Juniper Woodland: Influence of Vegetation. *Soil Sci. Soc. Am. J.* **2003**, *67*, 1243–1249. [CrossRef]
30. Eldridge, D.J.; Wang, L.; Ruiz-Colmenero, M. Shrub Encroachment Alters the Spatial Patterns of Infiltration. *Ecohydrology* **2015**, *8*, 83–93. [CrossRef]
31. Bagarello, V.; Di Prima, S.; Iovino, M. Estimating Saturated Soil Hydraulic Conductivity by the near Steady-State Phase of a Beerkan Infiltration Test. *Geoderma* **2017**, *303*, 70–77. [CrossRef]
32. Bagarello, V.; David, S.M. Run Duration Effects on the Hydrodynamic Properties of a Loam Soil Estimated by Steady-State Infiltration Methods. *J Agric. Eng.* **2020**, *51*, 229–238. [CrossRef]
33. Reynolds, W.D.; Elrick, D.E. Ponded Infiltration From a Single Ring: I. Analysis of Steady Flow. *Soil Sci. Soc. Am. J.* **1990**, *54*, 1233–1241. [CrossRef]
34. Bagarello, V.; Iovino, M.; Lai, J.-B. Field and Numerical Tests of the Two-Ponding Depth Procedure for Analysis of Single-Ring Pressure Infiltrometer Data. *Pedosphere* **2013**, *23*, 779–789. [CrossRef]
35. Lewis, J.; Amoozegar, A.; McLaughlin, R.A.; Heitman, J.L. Comparison of Cornell Sprinkle Infiltrometer and Double-ring Infiltrometer Methods for Measuring Steady Infiltration Rate. *Soil Sci. Soc. Am. J* **2021**, *85*, 1977–1984. [CrossRef]
36. Naik, A.P.; Norbu, T.; Pekkat, S. Comparison of Flux-Based and Head-Based Methods for Determination of near-Surface Saturated Hydraulic Conductivity. *Hydrol. Sci. J.* **2024**, *69*, 275–293. [CrossRef]

37. Di Prima, S.; Bagarello, V.; Lassabatere, L.; Angulo-Jaramillo, R.; Bautista, I.; Burguet, M.; Cerdà, A.; Iovino, M.; Prosdocimi, M. Comparing Beerkan Infiltration Tests with Rainfall Simulation Experiments for Hydraulic Characterization of a Sandy-Loam Soil. *Hydrol. Process.* **2017**, *31*, 3520–3532. [CrossRef]
38. Blackburn, W.H.; Meeuwig, R.O.; Skau, C.M. A Mobile Infiltrometer for Use on Rangeland. *Rangel. Ecol. Manag./J. Range Manag. Arch.* **1974**, *27*, 322–323. [CrossRef]
39. Leite, P.A.M.; Castellanos, A.E.; Wilcox, B.P.; Vega-Puga, M.; Martínez, E.; Dennis, S.; Choza, S.; Acuña-Acosta, D.M. Contrasting Effects of Native and Exotic Vegetation on Soil Infiltrability in the Sonoran Desert. *Sci. Total Environ.* **2022**, *852*, 158544. [CrossRef] [PubMed]
40. McCalla, G.R.; Blackburn, W.H.; Merrill, L.B. Effects of Livestock Grazing on Infiltration Rates, Edwards Plateau of Texas. *J. Range Manag.* **1984**, *37*, 265. [CrossRef]
41. Knight, R.W.; Blackburn, W.H.; Merrill, L.B. Characteristics of Oak Mottes, Edwards Plateau, Texas. *J. Range Manag.* **1984**, *37*, 534. [CrossRef]
42. Wilcox, B.P.; Wilding, L.P.; Woodruff, C.M. Soil and Topographic Controls on Runoff Generation from Stepped Landforms in the Edwards Plateau of Central Texas. *Geophys. Res. Lett.* **2007**, *34*, L24S24. [CrossRef]
43. White, I.; Sully, M.J.; Melville, M.D. Use and Hydrological Robustness of Time-to-Incipient-Ponding. *Soil Sci. Soc. Am. J.* **1989**, *53*, 1343–1346. [CrossRef]
44. Di Prima, S.; Concialdi, P.; Lassabatere, L.; Angulo-Jaramillo, R.; Pirastru, M.; Cerdà, A.; Keesstra, S. Laboratory Testing of Beerkan Infiltration Experiments for Assessing the Role of Soil Sealing on Water Infiltration. *CATENA* **2018**, *167*, 373–384. [CrossRef]
45. Bowyer-Bower, T.A.S.; Burt, T.P. Rainfall Simulators for Investigating Soil Response to Rainfall. *Soil Technol.* **1989**, *2*, 1–16. [CrossRef]
46. Ebrahimian, A.; Sample-Lord, K.; Wadzuk, B.; Traver, R. Temporal and Spatial Variation of Infiltration in Urban Green Infrastructure. *Hydrol. Process.* **2020**, *34*, 1016–1034. [CrossRef]
47. Tecca, N.P.; Nieber, J.; Gulliver, J. Bias of Stormwater Infiltration Measurement Methods Evaluated Using Numerical Experiments. *Vadose Zone J.* **2022**, *21*, e20210. [CrossRef]
48. Norris, C.E.; Bean, G.M.; Cappellazzi, S.B.; Cope, M.; Greub, K.L.H.; Liptzin, D.; Rieke, E.L.; Tracy, P.W.; Morgan, C.L.S.; Honeycutt, C.W. Introducing the North American Project to Evaluate Soil Health Measurements. *Agron. J.* **2020**, *112*, 3195–3215. [CrossRef]
49. Bagnall, D.K.; Morgan, C.L.S.; Bean, G.M.; Liptzin, D.; Cappellazzi, S.B.; Cope, M.; Greub, K.L.H.; Rieke, E.L.; Norris, C.E.; Tracy, P.W.; et al. Selecting Soil Hydraulic Properties as Indicators of Soil Health: Measurement Response to Management and Site Characteristics. *Soil Sci. Soc. Am. J.* **2022**, *86*, 1206–1226. [CrossRef]
50. Wilcox, B.P.; Caldeira, M.C.; Leite, P.A.M.; Lobo-do-Vale, R.; Bugalho, M.N. Understory Shrubs Improve Soil Infiltrability in Overgrazed Mediterranean Oak Woodlands, but Have Little Impact on Ungrazed Woodlands. *For. Ecol. Manag.* **2024**, *569*, 122186. [CrossRef]
51. Marín-Castro, B.E.; Negrete-Yankelevich, S.; Geissert, D. Litter Thickness, but Not Root Biomass, Explains the Average and Spatial Structure of Soil Hydraulic Conductivity in Secondary Forests and Coffee Agroecosystems in Veracruz, Mexico. *Sci. Total Environ.* **2017**, *607–608*, 1357–1366. [CrossRef]
52. Xia, L.; Song, X.; Fu, N.; Cui, S.; Li, L.; Li, H.; Li, Y. Effects of Forest Litter Cover on Hydrological Response of Hillslopes in the Loess Plateau of China. *CATENA* **2019**, *181*, 104076. [CrossRef]
53. Wang, L.; Zhang, G.; Zhu, P.; Wang, X. Comparison of the Effects of Litter Covering and Incorporation on Infiltration and Soil Erosion under Simulated Rainfall. *Hydrol. Process.* **2020**, *34*, 2911–2922. [CrossRef]
54. Jabro, J.D. Estimation of Saturated Hydraulic Conductivity of Soils From Particle Size Distribution and Bulk Density Data. *Trans. ASAE* **1992**, *35*, 557–560. [CrossRef]
55. Hu, W.; Shao, M.A.; Si, B.C. Seasonal Changes in Surface Bulk Density and Saturated Hydraulic Conductivity of Natural Landscapes. *Eur. J. Soil Sci.* **2012**, *63*, 820–830. [CrossRef]
56. Liu, Y.; Guo, L.; Huang, Z.; López-Vicente, M.; Wu, G.-L. Root Morphological Characteristics and Soil Water Infiltration Capacity in Semi-Arid Artificial Grassland Soils. *Agric. Water Manag.* **2020**, *235*, 106153. [CrossRef]
57. Liu, Y.-F.; Zhang, Z.; Liu, Y.; Cui, Z.; Leite, P.A.M.; Shi, J.; Wang, Y.; Wu, G.-L. Shrub Encroachment Enhances the Infiltration Capacity of Alpine Meadows by Changing the Community Composition and Soil Conditions. *CATENA* **2022**, *213*, 106222. [CrossRef]
58. Ogden, C.B.; Van Es, H.M.; Schindelbeck, R.R. Miniature Rain Simulator for Field Measurement of Soil Infiltration. *Soil Sci. Soc. Am. J.* **1997**, *61*, 1041–1043. [CrossRef]
59. Szabó, J.A.; Centeri, C.; Keller, B.; Hatvani, I.G.; Szalai, Z.; Dobos, E.; Jakab, G. The Use of Various Rainfall Simulators in the Determination of the Driving Forces of Changes in Sediment Concentration and Clay Enrichment. *Water* **2020**, *12*, 2856. [CrossRef]
60. Simelane, M.P.Z.; Soundy, P.; Maboko, M.M. Effects of Rainfall Intensity and Slope on Infiltration Rate, Soil Losses, Runoff and Nitrogen Leaching from Different Nitrogen Sources with a Rainfall Simulator. *Sustainability* **2024**, *16*, 4477. [CrossRef]

61. Lai, J.; Ren, L. Assessing the Size Dependency of Measured Hydraulic Conductivity Using Double-Ring Infiltrometers and Numerical Simulation. *Soil Sci. Soc. Am. J.* **2007**, *71*, 1667–1675. [CrossRef]
62. Lado, M.; Paz, A.; Ben-Hur, M. Organic Matter and Aggregate Size Interactions in Infiltration, Seal Formation, and Soil Loss. *Soil Sci. Soc. Am. J.* **2004**, *68*, 935–942. [CrossRef]
63. Lado, M.; Ben-Hur, M.; Shainberg, I. Soil Wetting and Texture Effects on Aggregate Stability, Seal Formation, and Erosion. *Soil Sci. Soc. Am. J.* **2004**, *68*, 1992–1999. [CrossRef]
64. Radinja, M.; Vidmar, I.; Atanasova, N.; Mikoš, M.; Šraj, M. Determination of Spatial and Temporal Variability of Soil Hydraulic Conductivity for Urban Runoff Modelling. *Water* **2019**, *11*, 941. [CrossRef]
65. Zhang, J.; Lei, T.; Chen, T. Impact of Preferential and Lateral Flows of Water on Single-Ring Measured Infiltration Process and Its Analysis. *Soil Sci. Soc. Am. J.* **2016**, *80*, 859–869. [CrossRef]
66. Grant, K.N.; Macrae, M.L.; Ali, G.A. Differences in Preferential Flow with Antecedent Moisture Conditions and Soil Texture: Implications for Subsurface P Transport. *Hydrol. Process.* **2019**, *33*, 2068–2079. [CrossRef]
67. Zimmermann, B.; Elsenbeer, H.; De Moraes, J.M. The Influence of Land-Use Changes on Soil Hydraulic Properties: Implications for Runoff Generation. *For. Ecol. Manag.* **2006**, *222*, 29–38. [CrossRef]
68. Singh, B.P.; Cowie, A.L.; Chan, K.Y. (Eds.) *Soil Health and Climate Change*; Soil Biology; Springer: Berlin/Heidelberg, Germany, 2011; Volume 29, ISBN 978-3-642-20255-1.
69. Jarvis, N.; Koestel, J.; Messing, I.; Moeys, J.; Lindahl, A. Influence of Soil, Land Use and Climatic Factors on the Hydraulic Conductivity of Soil. *Hydrol. Earth Syst. Sci.* **2013**, *17*, 5185–5195. [CrossRef]
70. Sun, D.; Yang, H.; Guan, D.; Yang, M.; Wu, J.; Yuan, F.; Jin, C.; Wang, A.; Zhang, Y. The Effects of Land Use Change on Soil Infiltration Capacity in China: A Meta-Analysis. *Sci. Total Environ.* **2018**, *626*, 1394–1401. [CrossRef] [PubMed]
71. Bagarello, V.; Di Prima, S.; Iovino, M.; Provenzano, G. Estimating Field-Saturated Soil Hydraulic Conductivity by a Simplified Beerkan Infiltration Experiment. *Hydrol. Process.* **2014**, *28*, 1095–1103. [CrossRef]
72. Stewart, R.D.; Abou Najm, M.R. A Comprehensive Model for Single Ring Infiltration II: Estimating Field-Saturated Hydraulic Conductivity. *Soil Sci. Soc. Am. J.* **2018**, *82*, 558–567. [CrossRef]

water

MDPI

Article

Research on Flood Storage and Disaster Mitigation Countermeasures for Floods in China's Dongting Lake Area Based on Hydrological Model of Jingjiang–Dongting Lake

Wengang Zhao [1], Weizhi Ji [1], Jiahu Wang [2], Jieyu Jiang [1], Wen Song [1], Zaiai Wang [1], Huizhu Lv [1], Hanyou Lu [1] and Xiaoqun Liu [1,*]

[1] Hunan Institute of Water Resources and Hydropower Research, Changsha 410007, China; zwg921153@163.com (W.Z.); jiweizhi1986@163.com (W.J.); jjy19950728@163.com (J.J.); songwen0879@163.com (W.S.); wangzaiai0457@163.com (Z.W.); lhz4554@163.com (H.L.); luhanyou666@163.com (H.L.)

[2] College of Hydrology and Water Resources, Hohai University, Nanjing 210098, China; tigerlly@hhu.edu.cn

* Correspondence: lxq1488@163.com

Abstract: China's Dongting Lake area is intertwined with rivers and lakes and possesses many water systems. As such, it is one of the most complicated areas in the Yangtze River Basin, in terms of the complexity of its flood control. Over time, siltation and reclamation in the lake area have greatly weakened the river discharge capacity of the lake area, and whether it can endure extreme floods remains an open question. As there is no effective scenario simulation model for the lake area, this study constructs a hydrological model for the Jingjiang–Dongting Lake system and verifies the model using data from 11 typical floods occurring from 1954 to 2020. The parameters derived from 2020 data reflect the latest hydrological relationship between the lake and the river, while meteorological data from 1954 and 1998 are used as inputs for various scenarios with the aim of evaluating the flood pressure of the lake area, using the water levels at the Chenggling ji and Luoshan stations as indicators. The preliminary results demonstrate that the operation of the upstream Three Gorges Dam and flood storage areas cannot completely offset the flood pressure faced by the lake area. Therefore, the reinforcement and raising of embankments should be carried out, in order to cope with potential extreme flood events. The methodology and results of this study have reference value for policy formation, flood control, and assessment and dispatching in similar areas.

Keywords: excess flood volume; Jingjiang–Dongting lake hydrological model; flood control and disaster mitigation countermeasures

1. Introduction

Floods and droughts are common global challenges at present [1,2]. In particular, major flood events (e.g., Pakistan in 2022, Libya in 2023, and the Midwest and south of the United States in 2024) have impacted tens of millions of people and caused property losses worth tens of billions [3]. Rapid and accurate hydrological forecasting, coupled with the effective planning and management of water conservancy projects—including reservoirs and dams—has become crucial in mitigating the impacts of flood and drought disasters [4–6]. Water conservancy projects play a vital role in reducing flood peaks, supplementing irrigation, and supporting ecological balance, all of which contribute to sustainable development [7,8]. However, they also significantly alter the morphology of rivers, leading to uncertainty in flood modeling and simulation endeavors [9]. Combined with the impacts of climate change, such changes present substantial challenges for accurately simulating floods within complex river and lake systems [10,11].

To date, three types of models have been developed for flood simulation and forecasting; namely, hydraulic, hydrological, and data-driven models [12,13]. Among them, hydraulic models (represented by MIKE and HEC) are widely used globally due to their clear physical meaning and relatively reliable simulation results [14]. However, for complex river–lake water networks, models with high efficiency and high-precision flood forecasting capabilities are still under development due to limitations associated with acquiring refined topography and boundary conditions [15,16]. Traditional hydrological models (represented by SWAT and the Muskingum method) have achieved remarkable results in the field of flood simulation due to their high computational efficiency and ease of data collection [17,18]. However, for river–lake water networks in plains heavily influenced by human activities and significant erosion and deposition changes, they can no longer meet the needs of flood simulation. As black-box models, data-driven models simplify the basic principles of hydraulics through seeking the non-linear relationships between hydrological elements such as water level and flow rate, playing a significant role in rapid flood element simulation [19]. However, the effectiveness of these models varies greatly with the scale of the training dataset and the selection of driving factors [20]. Additionally, when considering complex river–lake water networks, they ignore the interaction and hysteresis effects between upstream and downstream hydrological elements, and further improvements are needed for the simulation of hydrological processes under extreme flooding and drought events. To overcome the shortcomings of each model, some researchers have studied hydrological–hydrodynamic coupling models and hydrodynamic data-driven coupling models [21,22]. However, addressing the issue of low operational efficiency in large and complex river–lake systems without compromising simulation accuracy remains a key problem [23]. Therefore, it is necessary to build new flood forecasting hydrological models based on traditional hydrological models, incorporating refined flood simulation methods while considering erosion and deposition changes through hydrodynamic models and the operational efficiency of data-driven models, in order to achieve rapid and precise simulation and facilitate flood control and disaster reduction management.

The Yangtze River is the most flood-prone river in China [24,25]. As an important flood regulation and storage area of the Yangtze River, the Jingjiang–Dongting Lake reach has been described as follows: "while the Yangtze River stretches for thousands of miles, dangers may occur in Jingjiang River and challenges may be confronted by the Dongting Lake". To address the complex flood issues in this region, a flood control engineering system centered on the Three Gorges Reservoir, Jingjiang Embankment, and flood storage and detention areas has been established through long-term practice and exploration [26–31]. During the flood season, through ensuring that the water level at Shashi City on the Yangtze River mainstream does not exceed 45 m and the water level at Lianhuatang does not exceed 34.4 + 0.5 m, the system effectively controls floods below the 1954 standard (once-in-a-century) in the middle and lower reaches of the Yangtze River, safeguarding regional flood safety [32,33]. However, in recent years, due to the impacts of climate change and variations in river discharge capacity, floods have concentrated near Chenglingji, resulting in a flood control situation in the region that is still severe. In the event of a major flood, the limited storage capacity of 12.02 billion cubic meters near Chenglingji can be utilized; however, its activation would have significant economic and social impacts [34,35].

This study takes the Jingjiang–Dongting Lake reach as the research object and assumes that, under the periodic action of water cycle due to climate change, floods similar to those in 1954 and 1998 will recur in the Yangtze River Basin. Under the influence of river–lake erosion and deposition changes and the operation of the reservoir group, significant differences can be expected to emerge under the existing river–lake relationships. Through a predictive analysis using a self-constructed model, this hypothesis was verified and the flood risks faced by the region under the present river–lake relationships were further revealed, providing important technical support for rapid flood forecasting and flood control layout optimization in the basin.

2. Study Area and Data

2.1. Study Area

The Jingjiang–Dongting Lake river section (27.8840°–30.4383° N Latitude and 113.3275°–111.37742° E Longitude) is located in the middle of China, within the midstream of the Yangtze River basin. The Jingjiang River starts from Zhicheng Hydrological Station and ends at Chenglingji Lianhuatang Hydrological Station, with a total length of 339 km. The Yangtze River is diverted into Dongting Lake through the Songzi Estuary, Taiping Estuary, Ouchi Estuary, and Diaoxian Estuary (which have been gated and controlled since 1958). After converging with the waters of Xiangjiang, Zijiang, Yuanjiang, and Lishui rivers, the combined flow then rejoins the Yangtze River at Chenglingji and flows out through Luoshan Hydrological Station—a well-known bottleneck of the Yangtze River where the relationship between water level and flow rate remains stable [35,36]. Xiangtan Hydrological Station, Taojiang Hydrological Station, Taoyuan Hydrological Station, and Shimen Hydrological Station are located on the Xiangjiang River, Zijiang River, Yuanjiang River, and Lishui River, respectively, serving as control stations in their respective downstream regions. For a long time, in order to ensure flood control safety within the region as well as in the middle and lower reaches of the Yangtze River, a total of 27 flood retention basins have been established, with an effective flood storage capacity totaling 16.68 billion cubic meters, as shown in Figure 1.

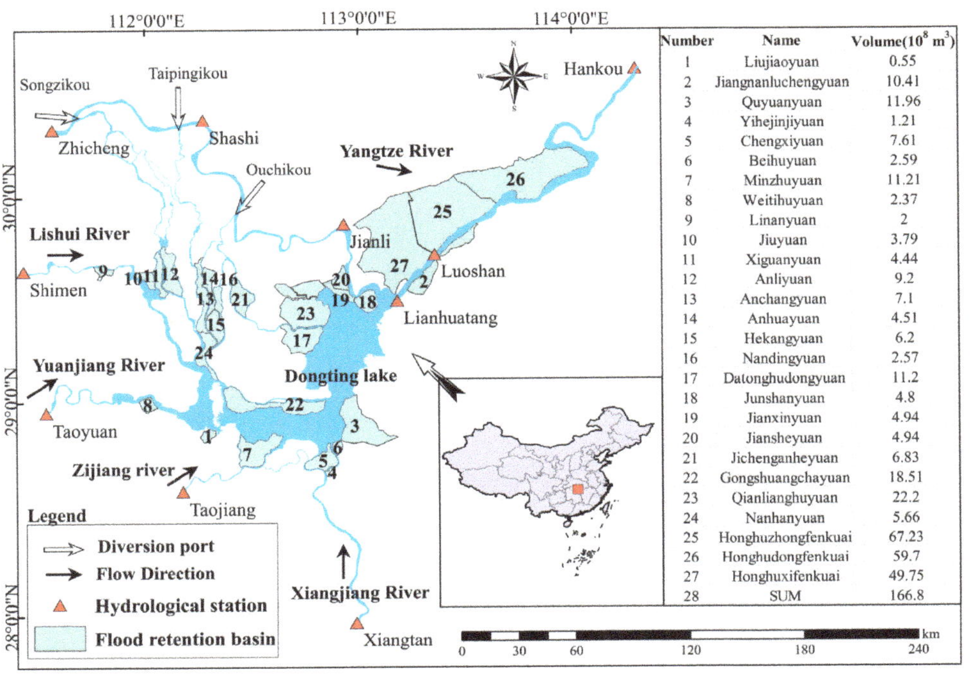

Number	Name	Volume(10^8 m^3)
1	Liujiaoyuan	0.55
2	Jiangnanluchengyuan	10.41
3	Quyuanyuan	11.96
4	Yihejinjiyuan	1.21
5	Chengxiyuan	7.61
6	Beihuyuan	2.59
7	Minzhuyuan	11.21
8	Weitihuyuan	2.37
9	Linanyuan	2
10	Jiuyuan	3.79
11	Xiguanyuan	4.44
12	Anliyuan	9.2
13	Anchangyuan	7.1
14	Anhuayuan	4.51
15	Hekangyuan	6.2
16	Nandingyuan	2.57
17	Datonghudongyuan	11.2
18	Junshanyuan	4.8
19	Jianxinyuan	4.94
20	Jiansheyuan	4.94
21	Jichenganheyuan	6.83
22	Gongshuangchayuan	18.51
23	Qianlianghuyuan	22.2
24	Nanhanyuan	5.66
25	Honghuzhongfenkuai	67.23
26	Honghudongfenkuai	59.7
27	Honghuxifenkuai	49.75
28	SUM	166.8

Figure 1. Map of the study area's location.

2.2. Data Sources

Considering that this study is primarily focused on the issue of floods, typical high-water years were selected from the measured hydrological sequences between 1954 and 2024 as the research objects, including 1954, 1983, 1995, 1996, 1998, 1999, 2002, 2003, 2016, 2017, and 2020. Through integrating the measured data, flood survey results, and reconstruction data from various high flood years, the model construction and flood control situation analysis were successfully completed. Table 1 outlines the primary sources of meteorological and hydrological elements used in this process. Specifically, the investigation and reconstruction data related to precipitation and floods in 1954 were sourced from the Hydrology Bureau of the Yangtze River Water Resources Bureau and relevant references. For other years, the measured hydrological data—including precipitation, water level, and discharge at Lianhuatang and Luoshan stations—were obtained from the Hydrology Bureau of the Yangtze River Water Resources Bureau and related references [37]. Additionally, the measured hydrological data for precipitation, water level, and discharge at Xiangtan, Taojiang, Taoyuan, and Shimen stations were sourced from the Hunan Hydrology and Water Resources Survey Center. Furthermore, precipitation data for Changde, Changsha, and Yueyang stations were retrieved from the China Meteorological Data Network (http://data.cma.cn/dataService/cdcindex/datacode/A.0012.0001.html, accessed on 22 September 2023).

Table 1. List of data sources for relevant meteorological and hydrological factors.

Factors	Stations	Years
Discharge	Xiangtan, Taojiang, Taoyuan, Shimen, Zhicheng, Lianhuatang, Luoshan	1954, 1983, 1995, 1996, 1998, 1999, 2002, 2003, 2016, 2017, 2020
Water level	Lianhuatang, Luoshan	
Precipitation	Xiangtan, Taojiang, Taoyuan, Shimen, Zhicheng, Changsha, Changde, Yueyang	

3. Methods

3.1. Jingjiang–Dongting Lake Hydrological Model

3.1.1. Flood Routing Model

The Jingjiang–Dongting Lake hydrological model, based on the Saint-Venant equations, simplifies the continuity equation into a water balance equation for the river segment and the dynamic equation into a storage equation using the empirical hydrological storage curve method, as follows:

$$Idt - Qdt = dW, \tag{1}$$

$$W = f(Q,\ I), \tag{2}$$

where I, Q, and W denote the inflow, outflow, and storage volume of the river segment, respectively.

For the calculation, the storage volume of the river segment and the stage–discharge relationship curve, which takes the daily stage increase rate, downstream backwater effect, and initial stage as parameters, are used to rewrite Equation (1) as:

$$\frac{I_1 + I_2}{2}\Delta t - \frac{Q_1 + Q_2}{2}\Delta t = W_2 - W_1, \tag{3}$$

where I_1 and I_2 represent the inflow at the beginning and end of the time period, respectively; Q_1 and Q_2 represent the outflow at the beginning and end of the time period, respectively; Δt represents the time duration (in days); and W_1 and W_2 represent the storage volume at the beginning and end of the time period, respectively.

When performing the specific calculations, Equation (3) is transformed into:

$$I_1 + I_2 - Q_1 + \frac{2W_1}{\Delta t} = Q_2 + \frac{2W_2}{\Delta t}. \tag{4}$$

Based on the inflow values I_1 and I_2 at the beginning and end of the time period, as well as the initial water level H_1, and considering that Q_1 W_1 is a function of H_1, the relevant calculation curves are utilized to perform the computations:

$$I_1 + I_2 - Q_1 + \frac{2W_1}{\Delta t} = \mathrm{M}. \tag{5}$$

Then, assuming a water level H_2 at the end of the time period and considering that Q_2 W_2 is a function of H_2, the relevant calculation curves are used to compute:

$$Q_2 + \frac{2W_2}{\Delta t} = \mathrm{M}^*. \tag{6}$$

If the calculated values meet the criteria (with ε representing the allowable error limit), Q_2 and H_2 are considered as the desired values at the end of the time period. Otherwise, the bisection method is used for iterative calculation [38].

$$\left|\mathrm{M}^* - \mathrm{M}\right| < \varepsilon. \tag{7}$$

3.1.2. Interval Runoff Generation and Concentration Model

For the interval runoff generation and concentration model, the Coupled Routing and Excess Storage (CREST) approach was adopted. With the aid of a Digital Elevation Model (DEM), the watershed was divided into numerous regular units. For each unit, the rainfall runoff generation is calculated using the storage capacity curve, and rapid and slow runoffs are distinguished based on the soil's steady infiltration rate [39]. The main structure of the model is illustrated in Figure 2.

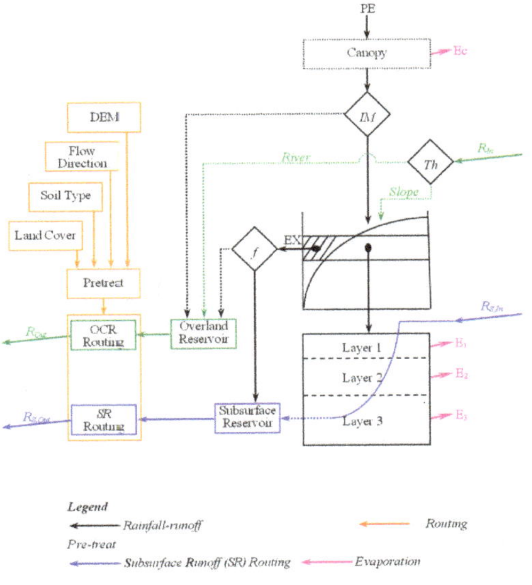

Figure 2. Flowchart of the CREST Model. (PE: rainfall minus evapotranspiration capacity; IM: impermeability factor; Ec: canopy evaporation; Th: thresholds for delineation of slopes and channels; other symbols are customary in hydrological modelling).

3.1.3. Parameter Processing

(1) The stage–discharge relationship at Luoshan

The relationship between water level and discharge generally exhibits a complex loop pattern, and is influenced by backwater from tributaries or fluctuations of flood levels. For the sake of convenience in calculation, the concept of fluctuation rate is introduced to simplify the water level–discharge relationship at Luoshan into a cluster of lines.

Assuming that, during the rise and fall of a flood, for the same water level Z within a time interval Δt, the increment in water level is Δh, the surface slope is ΔI, and the flood propagation time is U, then:

$$\Delta I = \frac{\Delta h}{\Delta L} = \frac{\Delta h}{U \cdot \Delta t} = \frac{1}{U}\frac{\Delta h}{\Delta t}, \tag{8}$$

According to the Chézy formula, there is a direct proportional relationship between the discharge at the same water level and the square root of the slope.

$$\frac{Q'}{Q} = \frac{\sqrt{I + \Delta I}}{\sqrt{I}} = \sqrt{1 + \frac{1}{UI}\frac{\Delta h}{\Delta t}}, \tag{9}$$

where Q' represents the flow rate considering flood fluctuations, Q represents the steady flow rate, and I represents the slope of the stable flow water surface.

During this process, the relationship between water level and discharge under steady flow conditions is known and, based on measured discharge data, a relationship curve between Z and UI can be plotted. Based on this, assuming that there is a power exponential relationship between water level fluctuations and discharge, the water level–discharge relationship at Luoshan station under simplified water level fluctuation conditions can be established as follows:

$$Z = k_1 Q^m, \tag{10}$$

where k_1 and m are respectively referred to as the ZQ coefficient and the ZQ exponent.

Meanwhile, S—namely, the storage capacity of the composite system with Luoshan Station as the outlet—exhibits a linear relationship with the outflow discharge Q [40]:

$$Q\Delta t = Sk_2, \tag{11}$$

where k_2 is the outflow coefficient, $k_2 \in (0, 1]$.

(2) Diversion of water flow through three outlets during high water levels

According to the study by Zhu et al. on the relationship between the diversion flows at Songzi, Taiping, and Ouchikou and the flow at Zhicheng from 1955 to 2020, the diversion ratios at gates with Zhicheng flow exceeding 30,000 m^3/s follow a linear relationship [41]. These ratios can be calculated using Equations (12)–(14).

$$Q_{\text{songzi}} = 0.154 \cdot Q_{\text{zhicheng}} - 850, \tag{12}$$

$$Q_{\text{taiping}} = 0.04 \cdot Q_{\text{zhicheng}} + 100, \tag{13}$$

$$Q_{\text{ouchi}} = 0.3025 \cdot Q_{\text{zhicheng}} - 5175. \tag{14}$$

(3) Calculation of excess flood volume

The excess flood volume is calculated by flattening the water level curve, which means that the cumulative flow remaining after deducting the discharge capacity corresponding to the Lianhuatang water level of 34.4 m is subtracted from the Luoshan discharge process, and the remaining cumulative discharge is regarded as the excess flood volume.

3.1.4. Parameter Optimization

In this study, the SCE-UA algorithm—which has been widely used in conceptual, semi-distributed, and distributed hydrological models—was selected for parameter optimization [42–44]. The main parameters optimized include k_1, m, and k_2.

3.2. Evaluation Indices for Model Effects

For evaluation of model performance, we primarily employed the Nash–Sutcliffe Efficiency Coefficient (NSE) and the Absolute Error (AE).

3.3. Scenario Design

The impact of controlled water levels on excess flood volumes is significant. This study explores the reasonable range of controlled water levels at Lianhuatang station from the perspective of smooth water surface profiles, combined with a fitting analysis of the correlation between the measured super-warning high flood levels at Lianhuatang station and those at upstream and downstream control stations from 1954 to 2020.

To avoid the influence of the Three Gorges Dam and upstream cascade reservoirs on the consistency of the hydrological series, the year 2003 was taken as the dividing point to separately plot the water levels at downstream Hankou station before and after the operation of the Three Gorges Dam, in comparison with the water levels at Lianhuatang station. Through linear fitting it was found that, under high flood conditions both before and after the operation of the Three Gorges Dam, the correlation between the water levels at Hankou and Lianhuatang was relatively good, with R^2 values exceeding 0.5 in both cases (Figure 3). When the controlled water level at Hankou is 29.73 m, based on the correlation before and after the operation of the Three Gorges Dam, the corresponding controlled water levels at Lianhuatang are 35.45 m and 35.18 m, respectively.

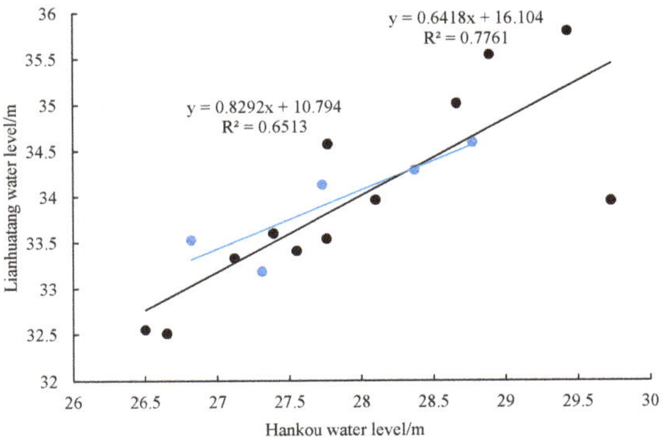

Figure 3. Correlations between water levels at Hankou and Lianhuatang stations before and after the operation of the Three Gorges Dam.

Based on the current controlled water level and warning water level at Lianhuatang, along with the calculated controlled water level, further plotting of the measured highest water levels at Shashi, Jianli, Lianhuatang, Luoshan, and Hankou revealed the following: the controlled water level surface profile from Shashi to Hankou tends to follow a concave curve, while the measured high water level surface profile exhibits a trend of first concave, then convex, and finally concave again, with the water level at Lianhuatang located at the inflection point of the curve.

In the observed high-water-level series, the water levels at various stations in 1954, 1980, and 1983 all fell within the ranges defined by the warning and control levels. In 1976 and 1989, despite Wuhan and Luoshan stations recording water levels below the warning level, Lianhuatang, Jianli, and Shashi stations experienced water levels exceeding the warning level. In 1988, 1991, 1995, 1996, 2002, 2003, 2010, 2016, 2017, and 2020, when Shashi station's water level was below the warning level and Hankou's was near the warning level, the water levels at Luoshan and Lianhuatang stations approached or exceeded the guaranteed level. During the years of 1998 and 1999, when Hankou's water level was below the guaranteed level and Shashi's was close to the guaranteed level, the water levels at Lianhuatang, Luoshan, and Jianli stations far exceeded the guaranteed level. A statistical analysis of the relationships between the water levels at various stations and the guaranteed level in high water years before and after the operation of the Three Gorges Dam, as well as over the long-term, revealed the following: at Shashi station, the average distance from the high water level to the guaranteed level was −1.48 m before the dam operation, −2.97 m after the dam operation, and −1.92 m over the long-term; at Lianhuatang station, these averages were −0.42 m before the dam operation, −0.45 m after the dam operation, and −0.43 m over the long-term; and, at Hankou station, the averages were −1.77 m before the dam operation, −1.93 m after the dam operation, and −1.82 m over the long-term. Overall, these findings indicate that, regardless of whether the Three Gorges Dam was operating, the water level near Lianhuatang presented a relatively small difference from the guaranteed level, leading to prominent flooding issues in the region. Considering the smoothness of the water surface profile and based on the calculated control levels of 35.18 m and 35.45 m, only the floods in 1998 and 1999 exceeded Lianhuatang's control level (Figure 4).

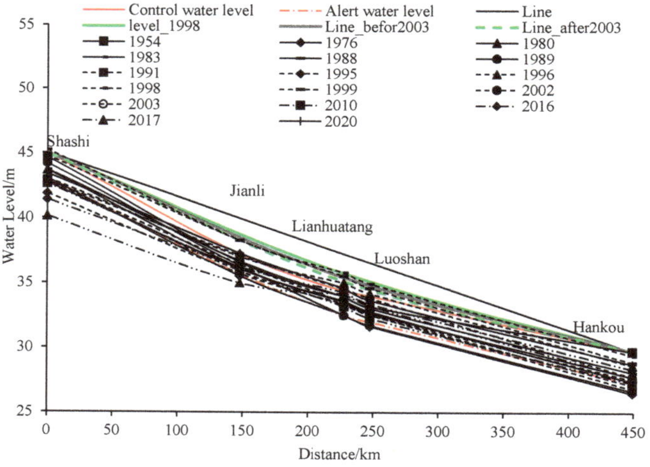

Figure 4. Relationships between the highest water levels at various control stations and the warning and control water levels during high flood years.

Therefore, combining the current controlled water levels, water levels analyzed through correlation analysis, and historical highest flood levels, ten scenarios were set for flood routing in the years of 1954 and 1998. Meanwhile, based on potential floodwater diversion and storage plans, an analysis of the regional flood control situation was conducted under the scenario where the control water level remained unchanged. The specific plan design is outlined in Table 2.

Table 2. Scenario design for typical flood years in Dongting Lake area.

Scenarios	Year	Control Water Level (m)	Note
1	1954	34.40	Current controlled water level
2	1954	34.90	Current operating water level
3	1954	35.18	Projected water level after the Three Gorges Dam operation
4	1954	35.45	Projected water level before the Three Gorges Dam operation
5	1954	35.80	Highest historical water level
6	1998	34.40	Current controlled water level
7	1998	34.90	Current operating water level
8	1998	35.18	Projected water level after the Three Gorges Dam operation
9	1998	35.45	Projected water level before the Three Gorges Dam operation
10	1998	35.80	Highest historical water level
11	1954	34.40	Activate Flood Storage and Detention Areas
12	1998	34.40	Activate Flood Storage and Detention Areas
13	1954	34.40	\
14	1998	34.40	\

4. Results

4.1. Training and Verification of Jingjiang–Dongting Lake Hydrological Model

To construct the Jingjiang–Dongting Lake hydrological model, daily hydrological data from high flood years spanning from 1954 to 2020 and precipitation data from corresponding stations were utilized. Figure 5 and Table 3 present the training and testing results of the model, from which it can be seen that the simulated flow rates and water levels were in high agreement with the observed values. Specifically, the Nash–Sutcliffe Efficiency (NSE) coefficients were no less than 0.86. The absolute error for the highest water level ranged from 0.02 to 0.26 m, with an average error of −0.06 m, accounting for 0.18% of the multi-year average highest water level. The absolute error for peak flood flow ranged from 47.7 to 2643.7 m^3/s, with an average error of 315.88 m^3/s, accounting for 0.53% of the multi-year average peak flood flow. This indicates that the model performed excellently in simulating flow rates and water levels. Meanwhile, all the values were within reasonable parameter ranges: k_1 ranged from 1 to 3.24, k_2 from 0.13 to 0.19, and m from 0.21 to 0.32.

Table 3. Evaluation of results and calibration parameters for the Jingjiang–Dongting Lake hydrological model (Luohshan station).

Year	Discharge			Water Level			
	NSE	Absolute Error of Peak Flood Discharge (m^3/s)	k_2	NSE	Absolute Error of Peak Water Level (m)	k_1	m
1954	0.86	2643.7	0.15	0.86	0.04	2.21	0.24
1983	0.95	−797.8	0.19	0.95	−0.08	2.03	0.25
1988	0.98	326.1	0.19	0.98	−0.18	2.00	0.25
1989	0.94	47.7	0.17	0.94	−0.1	3.24	0.21
1995	0.99	70	0.17	0.99	0.14	1.81	0.27
1996	0.99	71	0.13	0.99	0.26	1.93	0.26
1998	0.98	1712.5	0.18	0.99	−0.18	2.04	0.25
1999	0.99	10.8	0.14	0.99	0.03	2.04	0.25
2002	0.97	144.4	0.17	0.97	−0.20	2.23	0.24
2003	0.98	−379.3	0.15	0.98	−0.23	2.55	0.23
2016	0.98	−72.4	0.13	0.98	−0.02	1.00	0.32
2017	0.98	201.1	0.14	0.98	−0.14	1.51	0.28
2020	0.99	128.6	0.16	0.99	−0.08	1.17	0.31
Average	-	315.88	-	-	−0.06	-	-

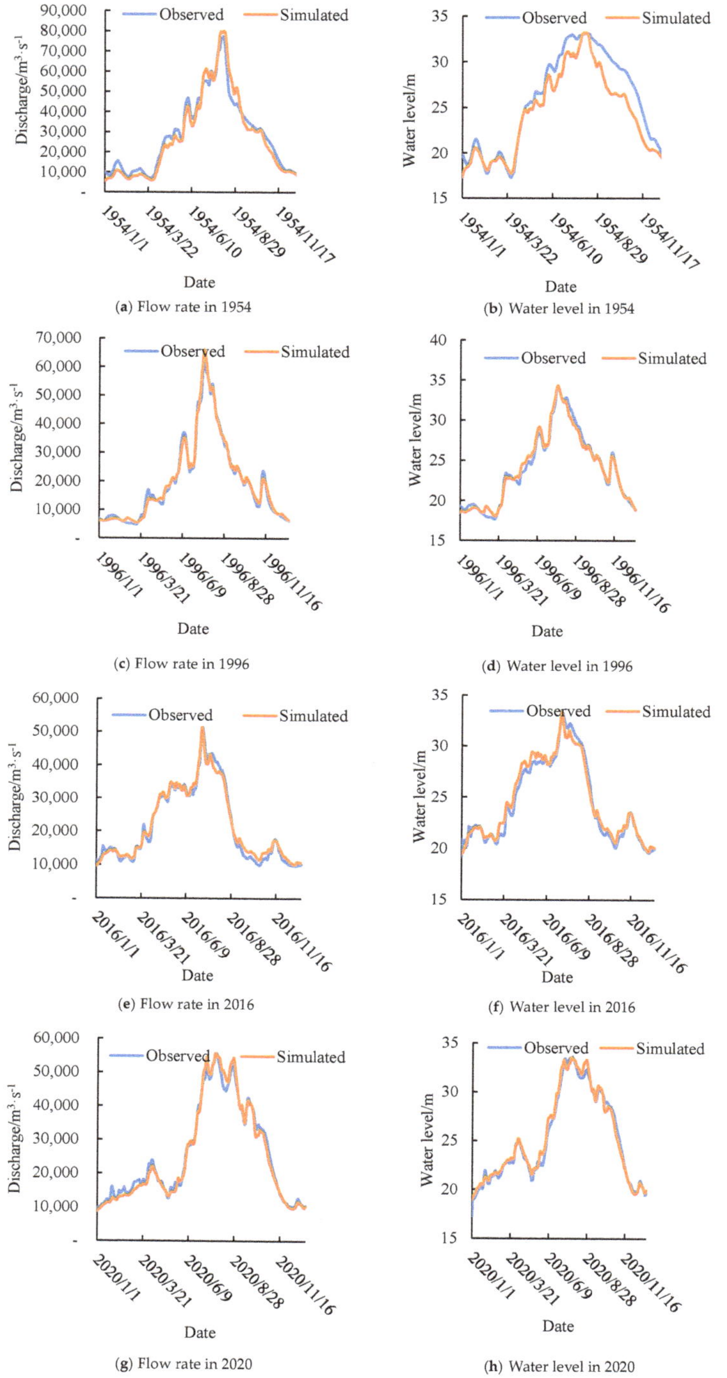

Figure 5. Comparisons of simulated and observed water levels and flow rates for typical years at Luoshan station.

4.2. Impact on Flood Control Situations in Dongting Lake Under Different Flood Scenarios

Relying on the constructed Jingjiang–Dongting Lake hydrological model, in order to invert the flood impacts on the Dongting Lake area caused by the floods in 1954 and 1998 under the current river–lake relationship conditions, the measured flood series at the Yangtze River's Zhicheng boundary was utilized, with the discharge from the Three Gorges and upstream cascaded reservoirs set to no more than 30,000 m^3/s. Given the limited flood regulation capabilities of the reservoirs on Xiangjiang, Zijiang, Yuanjiang, and Lishui rivers, the corresponding measured flood series at the boundaries of Xiangtan on the Xiangjiang River, Taojiang on the Zijiang River, Taoyuan on the Yuanjiang River, and Shimen on the Lishui River were adopted. The entire model operated under the river–lake relationship conditions of 2020.

4.2.1. Scenario of Elevated Flood Control Water Level

The overall flood control standard for the middle and lower reaches of the Yangtze River aims to defend against the largest flood that has occurred since 1949; namely, the catastrophic basin-wide flood in 1954. The flood control standard for the Jingjiang reach is for a flood with a 100-year return period [45]. The 1998 flood was another major basin-wide flood of the Yangtze River, characterized by its large magnitude, wide coverage, long duration, and severe flooding disasters [46].

Under the river–lake relationship conditions for 2020, the maximum water level at Lianhuatang station during the 1954 flood was 37.2 m and the maximum discharge at Luoshan station was 73,770 m^3/s. With water level controls at Lianhuatang station set at 34.4 m, 34.9 m, 35.18 m, 35.45 m, and 35.8 m, the excess flood volumes near Chenglingji were 17.864 billion m^3, 7.957 billion m^3, 6.926 billion m^3, 4.418 billion m^3, and 1.421 billion m^3, respectively. The excess flood volume in 1954, under the current engineering system, was basically consistent with the previously studied figure of 17.6 billion m^3 [47]. When the controlled water level at Lianhuatang was raised to 35.18 m, 35.45 m, and 35.80 m, respectively, compared to the controlled water level of 34.4 m, the excess flood volume near Chenglingji decreased by 10.938 billion m^3, 13.446 billion m^3, and 16.443 billion m^3, respectively, with reduction percentages of 61.23%, 75.27%, and 92.05%.

For the 1998 flood, when the controlled water level at Lianhuatang was raised to 35.18 m, 35.45 m, and 35.80 m, respectively, compared to the controlled water level of 34.4 m, the excess flood volume near Chenglingji decreased by 5.314 billion m^3, 7.359 billion m^3, and 7.556 billion m^3, respectively, with reduction percentages of 65.28%, 90.41%, and 92.75% (Table 4).

Table 4. Analysis of flood control situation near Chenglingji under different controlled water level conditions.

Scenarios	Year	Control Water Level (m)	Peak Discharge/34.4 m (m^3/s)	Peak Water Level (m)	Excess Flood Volume (10^8 m^3)		
					Before	After	Reduction
1	1954	34.40	73,770/56,300	37.2	178.64	-	-
2	1954	34.90	73,770/56,300	37.2	178.64	79.57	99.07
3	1954	35.18	73,770/56,300	37.2	178.64	69.26	109.38
4	1954	35.45	73,770/56,300	37.2	178.64	44.18	134.46
5	1954	35.80	73,770/56,300	37.2	178.64	14.21	164.43
6	1998	34.40	68,280/56,300	36.4	81.40	-	-
7	1998	34.90	68,280/56,300	36.4	81.40	39.66	41.74
8	1998	35.18	68,280/56,300	36.4	81.40	28.26	53.14
9	1998	35.45	68,280/56,300	36.4	81.40	7.81	73.59
10	1998	35.80	68,280/56,300	36.4	81.40	5.90	75.5

In summary, raising the controlled water level at Lianhuatang can basically address the excess flood volume near Chenglingji. Moreover, the highest water level during the historic flood in 1998 reached 35.8 m. Considering the current regional flood control layout, if the water level at Lianhuatang was raised to 35.8 m, it would only require raising of the 550 km main dike along the Yangtze River from Jianli to Hankou within the region, as well as reinforcement of the main dikes totaling 2377 km in length at 11 key polders and 24 flood storage polders in the Dongting Lake area [34]. Based on the unit investment cost for the first-phase risk removal and reinforcement of key polders in Dongting Lake, the estimated cost was CNY 20.376 billion.

4.2.2. Scenario of Using Flood Storage and Detention Areas

Taking into consideration the construction status of flood storage and detention areas, current operable areas include Chengxi polder (5), Weidihu polder (8), Linan polder (9), Datonghu dong polder (17), Gongshuangcha polder (22), Qianlianghu polder (23), and Honghudongfenkuai (26). As shown in Table 5, for the 1954 flood, the excess flood volume near Chenglingji was 17.864 billion cubic meters. Under current conditions, all completed flood storage and detention areas would need to be activated, with a cumulative floodwater storage and diversion of 12.36 billion cubic meters, reducing the peak flow at Luoshan by 19.07% and lowering the water level at Lianhuatang Station by 0.23 m. For the 1998 flood, considering economic losses, the effectiveness of floodwater storage and diversion by the polders, and the need for roughly equal floodwater distribution between Hunan and Hubei provinces, the activation of Weidihu polder (8), Linan polder (9), Gongshuangcha polder (22), and Honghudongfenkuai (26) would be required, which would lower the water level at Lianhuatang Station by 2 m and reduce the peak flow at Luoshan by 17.98%.

Table 5. Analysis of flood control situation near Chenglingji with the activation of different flood retention basins.

Scenarios	Year	Activated Flood Retention Basins	Peak Discharge (34.4 m)			Peak Water Level (m)			Excess Flood Volume (10^8 m^3)		
			Before (m^3/s)	After (m^3/s)	Decay Rate	Before	After	Reduction	Before	After	Reduction
11	1954	5, 8, 9, 17, 22, 23, 26	73,770 (56,300)	59,700 (53,900)	19.07% (4.03%)	37.2	34.9	2.30	178.64	55.04	123.6
12	1998	8, 9, 22, 26	68,280 (56,300)	56,000	17.98% (0.53%)	36.4	34.4	2.00	81.40	0	81.4

Combining previous studies, the estimated inundation losses from floodwater storage and detention areas during the 1954 flood were CNY 26.978 billion, while the estimated inundation losses for the 1998 flood were CNY 11.027 billion [48,49].

After comprehensively considering both the economic cost and the rationality of the water surface profile, raising the controlled water level at Lianhuatang to 35.8 m and implementing supporting measures to heighten and reinforce the dikes and polders is a feasible strategy for flood control and disaster reduction in the middle reaches of the Yangtze River, especially in the vicinity of Chenglingji.

5. Discussion

5.1. Analysis of the Impact of River–Lake Conditions on the Flood Control Situation near Chenglingji

Under the condition of raising the controlled water level at Lianhuatang, the impacts of changes in channel storage due to erosion and deposition, as well as variations in the discharge capacity of Luoshan, on the excess flood volume within the region cannot be ignored. Under the determined controlled water level at Lianhuatang, the excess flood volume near Chenglingji is greatly influenced by the state of the river–lake system, especially the diversions at the three outlets of Songzi, Taiping, and Ouchi; the discharging capacity of Luoshan; and the changes in the storage capacity of the river–lake system. Table 6 reflects the flood situations in 1954 under the river–lake states of 1983, 1988, 1989, 1995, 1996, 1998, 1999, 2002, 2003, 2016, 2017, and 2020. It is evident that the highest water level at Lianhuatang Station did not exceed the current controlled water level of 34.4 m in 1954 under the river–lake states of 1988, 1989, 1996, 2003, and 2017, with no excess flood volume. This is likely mainly due to the river–lake conditions in the years 1988, 1989, 2002, 2003, and 2017, which were characterized by the convergence of floods from the Yangtze and Yuanjiang Rivers, floods in the Yangtze River, floods from the Yangtze and Yuanjiang Rivers, floods from the Yangtze and Lishui Rivers, and floods from the Yangtze, Xiangjiang, and Zijiang Rivers. These floods mainly originated in the upstream areas, with combined inflow rates ranging from 54,900 to 66,000 m^3/s. When the downstream water level at Hankou was below 28 m and the drop between Luoshan and Hankou was maintained in the range of 5.08 to 5.75 m, the attenuation effect on the discharge capacity of Luoshan was limited, allowing it to remain around the theoretical value of 65,000 m^3/s. This is generally consistent with the discharge capacity of 65,800 m^3/s at Luoshan, as estimated by An Shenyi et al. in 2003, when the water levels at Chenglingji and Hankou were 34.4 m and 27.5 m, respectively [50]. Adequate discharge capacity ensured that the floods were transmitted downstream without significant accumulation in the region, thus not generating excessive flood volumes.

Table 6. Flood control situation near Chenglingji under different river–lake conditions.

Serial Number	Conditions of River–Lake Interactions	Peak Water Level of Lianhuatang Station (m)	Peak Discharge of Luoshan Station (m^3/s)	Excess Flood Volume (10^8 m^3)	
				34.4	35.8
1	1983	34.6	65,580	3.4	-
2	1988	33.3	60,610	-	-
3	1989	33.3	60,399	-	-
4	1995	34.6	59,502	3.2	-
5	1996	34.2	59,944	-	-
6	1998	34.6	60,736	2.3	-
7	1999	34.5	61,570	0.6	-
8	2002	33.9	63,840	-	-
9	2003	33.5	60,961	-	-
10	2016	35.7	59,914	57.2	-
11	2017	34.0	60,772	-	-
12	2020	37.2	73,770	178.64	14.21

In 1996, the river–lake conditions were characterized by major floods in the middle and lower reaches of the Yangtze River. With an upstream combined inflow of 64,800 m^3/s, a downstream water level at Hankou of 28.66 m, and a drop of 5.52 m, the backwater effect was significant, reducing the discharge capacity of Luoshan to 60,000 m^3/s. However, with the initial water level at Lianhuatang at 27.18 m and limited base water in the rivers and lakes, no excessive flood volumes were generated after centralized storage and regulation.

The river–lake conditions in 1995, 1998, and 1999 were all characterized by major floods in the middle and lower reaches of the Yangtze River. With upstream combined inflow rates ranging from 50,700 to 66,900 m^3/s, downstream water levels at Hankou of 27.79 to 29.43 m, and drops of 4.79 to 5.71 m, the backwater effect at Hankou was significant, reducing the discharge capacity of Luoshan to 60,000 m^3/s. Coupled with the initial high water levels at Lianhuatang (ranging from 28.05 to 29.72 m) and limited storage and regulation space in the rivers and lakes, the region generated excessive flood volumes of 0.06 to 0.32 billion m^3.

In 1983, the river–lake relationship had not been reshaped by the multiple floods occurring in the 1990s and still maintained the discharge capacity of 65,000 m^3/s determined in the flood control planning of the Yangtze River Basin in the 1960s. However, due to the impact of early-stage reclamation, the storage and regulation space in the rivers and lakes has reduced, resulting in an excessive flood volume of 0.33 billion m^3 in the region.

In 2016 and 2020, the river–lake conditions were characterized by high water levels in the region due to the backwater effect of floods in the Poyang Lake water system or Hanjiang River water system downstream. When the water level at Hankou was above 28.3 m and the drop between Hankou and Luoshan was below 5 m, the discharge capacity of Luoshan further decreased to 53,000–56,000 m^3/s. With high initial water levels at Lianhuatang and reduced storage and regulation space in the rivers and lakes, the region generated excessive flood volumes of 5.72 to 17.864 billion m^3.

5.2. The Limitation of the Study

This study primarily utilized actual flood data recorded in the Yangtze River basin, considering an ideal regulation scenario where the combined discharge from the Three Gorges Dam and upstream cascade reservoirs does not surpass 30,000 m^3/s. These data were integrated with the actual flood processes observed in the Xiangjiang, Zijiang, Yuanjiang, and Lishui rivers, employing a specially constructed hydrological model of the Jingjiang–Dongting Lake system in order to delve into the flood control challenges faced by typical high flood water level regions within the current river–lake system (i.e., in 2020). This analysis holds immense significance for flood control and disaster mitigation initiatives.

However, it is crucial to acknowledge that the ongoing operation of upstream cascade reservoirs, led by the Three Gorges Dam, alongside the 53 control reservoirs that operate under the unified management of the Yangtze River—which collectively possess a flood storage capacity of 70.6 billion m^3—will further modify downstream discharges during flood seasons [51,52]. Subsequently, refined reservoir operations will lead to corresponding changes in the inflow into tributaries within the Jingjiang–Dongting Lake area. Taking the 2024 flood season as an illustrative example, the Three Gorges Dam maintained a minimum discharge of 14,000 m^3/s, well below the 30,000 m^3/s threshold, thereby notably enhancing the region's flood control capabilities.

Moreover, the activation of flood detention basins exerts an interactive influence on the flood control dynamics of external rivers—an aspect that was simplified in this study by considering its impact solely on the flood volume of external rivers. Therefore, future research endeavors must strive to further refine regional inflow optimization approaches in tandem with control objectives, and incorporate hydrodynamic models of local flood detention basins to conduct a detailed re-assessment of the region's future flood control landscape.

6. Conclusions

A hydrological model of the Jingjiang–Dongting Lake system was constructed using data from 12 typical high flood years between 1954 and 2020. The Nash–Sutcliffe Efficiency (NSE) coefficients of the model exceeded 0.86, with an average absolute error in peak flow simulation of 315.88 m^3/s, accounting for 0.53% of the multi-year average peak flow. The average absolute error in the highest water level was 0.06 m, representing 0.18% of the average annual highest water level. This indicates that the constructed Jingjiang–Dongting Lake hydrological model performs well in simulating both flows and water levels.

Based on the river–lake conditions in 2020, the Jingjiang–Dongting Lake hydrological model was used to study the floods of 1954 and 1998. The findings revealed that, without activating flood storage and detention areas, raising the control water level at Lianhuatang to 35.8 m could reduce the excess flood volume near Chenglingji by 92.05% and 92.75%, respectively. Alternatively, without raising the control water level at Lianhuatang, activating the flood storage and detention areas could reduce the peak flow at Luoshan Station by 19.07% and 17.98%, decreasing the excess flood volume near Chenglingji by 69.19% and 100%, respectively. From the perspectives of reducing disaster losses and hydraulic rationality, raising the controlled water level at Lianhuatang to 35.8 m and fully utilizing the storage capacity of rivers and lakes, while reinforcing and raising the 2377 km of main levees, can both address the issue of excessive flood volumes and avoid direct losses of approximately 2.6978 billion Yuan due to flood diversion and storage. This has significant practical implications for ensuring flood control safety, thus promoting the economic and social development of the Yangtze River Economic Belt.

Author Contributions: Conceptualization, X.L. and W.Z.; methodology, W.J. and J.W.; Software J.W. and W.S.; formal analysis, J.J. and Z.W.; investigation, H.L. (Huizhu Lv) and H.L. (Hanyou Lu); writing—original draft preparation, W.Z., W.J. and J.W.; funding acquisition, X.L. All authors have read and agreed to the published version of the manuscript.

Funding: This research was funded by the National Key Research and Development Program Project (Grant No. 2022YFC3201804-03), Hunan Provincial Water Conservancy Science and Technology Planning Project (Grant No. XSKJ2022068-13, XSKJ2022068-12, XSKJ2023059-05), and Hunan Xiaohe science and technology talent project (Grant No. 2024TJ-X76).

Data Availability Statement: The datasets generated during and/or analyzed during the current study can be made available upon request.

Acknowledgments: The research team acknowledges data support from Changjiang Water Resources Bureau Hydrology Bureau and the Hunan Provincial Hydrology and Water Resources Survey Center.

Conflicts of Interest: The authors declare no conflicts of interest.

References

1. Masson-Delmotte, V.; Zhai, P.; Pirani, S.; Connors, C.; Péan, S.; Berger, N.; Caud, Y.; Chen, L.; Goldfarb, M.; Scheel Monteiro, P.M. Ipcc, 2021: Summary for policymakers. In *Proceedings of the Climate Change 2021: The Physical Science Basis. Contribution of Working Group I to the Sixth Assessment Report of the Intergovernmental Panel on Climate Change*; Cambridge University Press: Cambridge, UK; New York, NY, USA, 2021.
2. Mirza, M.M.Q. Global warming and changes in the probability of occurrence of floods in Bangladesh and implications. *Glob. Environ. Chang.* **2002**, *12*, 127–138. [CrossRef]
3. Bates, P.D. Flood Inundation Prediction. *Annu. Rev. Fluid Mech.* **2022**, *54*, 287–315. [CrossRef]
4. Brunner, M.I.; Slater, L.; Tallaksen, L.M.; Clark, M. Challenges in modeling and predicting floods and droughts: A review. *WIREs Water* **2021**, *8*, e1520. [CrossRef]
5. Patel, D.P.; Ramirez, J.A.; Srivastava, P.K.; Bray, M.; Han, D. Assessment of flood inundation mapping of Surat city by coupled 1D/2D hydrodynamic modeling: A case application of the new HEC-RAS 5. *Nat. Hazards* **2017**, *89*, 93–130. [CrossRef]
6. Yang, T.-H.; Liu, W.-C. A General Overview of the Risk-Reduction Strategies for Floods and Droughts. *Sustainability* **2020**, *12*, 2687. [CrossRef]
7. Fekete, B.M.; Bogárdi, J.J. Role of engineering in sustainable water management. *Earth Perspect.* **2015**, *2*, 2. [CrossRef]
8. Lebu, S.; Lee, A.; Salzberg, A.; Bauza, V. Adaptive strategies to enhance water security and resilience in low-and middle-income countries: A critical review. *Sci. Total Environ.* **2024**, *925*, 171520. [CrossRef]
9. Teraguchi, H.; Nakagawa, H.; Kawaike, K.; Yasuyuki, B.; Zhang, H. Effects of hydraulic structures on river morphological processes. *Int. J. Sediment Res.* **2011**, *26*, 283–303. [CrossRef]
10. Vionnet, V.; Fortin, V.; Gaborit, E.; Roy, G.; Abrahamowicz, M.; Gasset, N.; Pomeroy, J.W. Assessing the factors governing the ability to predict late-spring flooding in cold-region mountain basins. *Hydrol. Earth Syst. Sci.* **2020**, *24*, 2141–2165. [CrossRef]
11. Contreras, M.T.; Gironás, J.; Escauriaza, C. Forecasting flood hazards in real-time: A surrogate model for hydrometeorological events in an Andean watershed. *Nat. Hazards Earth Syst. Sci. Discuss.* **2020**, *20*, 3261–3277. [CrossRef]
12. Kumar, V.; Sharma, K.; Caloiero, T.; Mehta, D.; Singh, K. Comprehensive Overview of Flood Modeling Approaches: A Review of Recent Advances. *Hydrology* **2023**, *10*, 141. [CrossRef]

13. Zou, J.; Huang, F.; Yu, F.; Shen, X.; Han, S.; Qian, Z.; Jiang, H. Differentiating the Effects of Streamflow and Topographic Changes on the Water Level of Dongting Lake, China, Using the LSTM Network and Scenario Analysis. *Water* **2023**, *15*, 3742. [CrossRef]
14. Ansarifard, S.; Eyvazi, M.; Kalantari, M.; Mohseni, B.; Ghorbanifard, M.; Moghaddam, H.J.; Nouri, M. Simulation of floods under the influence of effective factors in hydraulic and hydrological models using HEC-RAS and MIKE 21. *Discov. Water* **2024**, *4*, 92. [CrossRef]
15. Long, Y.; Chen, W.; Jiang, C.; Huang, Z.; Yan, S.; Wen, X. Improving streamflow simulation in Dongting Lake Basin by coupling hydrological and hydrodynamic models and considering water yields in data-scarce areas. *J. Hydrol. Reg. Stud.* **2023**, *47*, 101420. [CrossRef]
16. Chen, Y.; Wang, Y.; Zhang, Y.; Luan, Q.; Chen, X. Flash floods, land-use change, and risk dynamics in mountainous tourist areas: A case study of the Yesanpo Scenic Area, Beijing, China. *Int. J. Disaster Risk Reduct.* **2020**, *50*, 101873. [CrossRef]
17. Suwannachai, L.; Sriworamas, K.; Sivanpheng, O.; Kangrang, A. Application of SWAT Model for Assessment of Surface Runoff in Flash Flood Areas. *Water* **2024**, *16*, 495. [CrossRef]
18. Salvati, A.; Moghaddam Nia, A.; Salajegheh, A.; Shirzadi, A.; Shahabi, H.; Ahmadisharaf, E.; Han, D.; Clague, J.J. A systematic review of Muskingum flood routing techniques. *Hydrol. Sci. J.* **2024**, *69*, 810–831. [CrossRef]
19. Zhou, Q.; Teng, S.; Situ, Z.; Liao, X.; Feng, J.; Chen, G.; Zhang, J.; Lu, Z. A deep-learning-technique-based data-driven model for accurate and rapid flood predictions in temporal and spatial dimensions. *Hydrol. Earth Syst. Sci.* **2023**, *27*, 1791–1808. [CrossRef]
20. Chafjiri, A.S.; Gheibi, M.; Chahkandi, B.; Eghbalian, H.; Waclawek, S.; Fathollahi-Fard, A.M.; Behzadian, K. Enhancing flood risk mitigation by advanced data-driven approach. *Heliyon* **2024**, *10*, e37758. [CrossRef]
21. Zhang, K.; Ji, Z.; Luo, X.; Liu, Z.; Zhong, H. Flood Simulation in the Complex River Basin Affected by Hydraulic Structures Using a Coupled Hydrological and Hydrodynamic Model. *Water* **2024**, *16*, 2383. [CrossRef]
22. Huang, H.; Lei, X.; Liao, W.; Liu, D.; Wang, H. A hydrodynamic-machine learning coupled (HMC) model of real-time urban flood in a seasonal river basin using mechanism-assisted temporal cross-correlation (MTC) for space decoupling. *J. Hydrol.* **2023**, *624*, 129826. [CrossRef]
23. Pang, C.; Zhou, Z.; Liu, J.; Tianyu, S.; Du, C.; Wang, K.; Yu, X. Parallel computing performance of distributed hydrological model accelerated by GPU. *South-North Water Transf. Water Sci. Technol.* **2024**, *22*, 33–38.
24. Zhang, D.; Shi, X.; Xu, H.; Jing, Q.; Pan, X.; Liu, T.; Wang, H.; Hou, H. A GIS-based spatial multi-index model for flood risk assessment in the Yangtze River Basin, China. *Environ. Impact Assess. Rev.* **2020**, *83*, 106397. [CrossRef]
25. Zhang, C.; Sun, F.; Sharma, S.; Zeng, P.; Mejia, A.; Lyu, Y.; Gao, J.; Zhou, R.; Che, Y. Projecting multi-attribute flood regime changes for the Yangtze River basin. *J. Hydrol.* **2023**, *617*, 128846. [CrossRef]
26. Duan, G.-L.; Long, H.; Guo, M.-J. Evolvement of Riverbed Near Bank of Dangerous Reach along Jingjiang River Levee after the Impoundment of Three Gorges Reservoir. *J. Yangtze River Sci. Res. Inst.* **2014**, *31*, 117.
27. Jiang, Y.-H.; Cheng, H.-Q.; Zhou, Q.-P.; Li, Y.; Yang, G.-Q.; Jin, Y.; Mei, S.-J.; Gu, X.; Zhang, H. Influence of major water conservation projects on river channels and shorelines in the middle and lower reaches of the Yangtze River. *Arab. J. Geosci.* **2021**, *14*, 884. [CrossRef]
28. Li, S.; Li, Y.; Yuan, J.; Zhang, W.; Chai, Y.; Ren, J. The impacts of the Three Gorges Dam upon dynamic adjustment mode alterations in the Jingjiang reach of the Yangtze River, China. *Geomorphology* **2018**, *318*, 230–239. [CrossRef]
29. Chang, J.; Li, J.; Lu, D.; Zhu, X.; Lu, C.; Zhou, Y.; Deng, C. The hydrological effect between Jingjiang River and Dongting Lake during the initial period of Three Gorges Project operation. *J. Geogr. Sci.* **2010**, *20*, 771–786. [CrossRef]
30. Liao, S.; Wang, C.; Ji, R.; Zhang, X.; Wang, Z.; Wang, W.; Chen, N. Balancing Flood Control and Economic Development in Flood Detention Areas of the Yangtze River Basin. *ISPRS Int. J. Geo-Inf.* **2024**, *13*, 122. [CrossRef]
31. Xia, J.; Chen, J. A new era of flood control strategies from the perspective of managing the 2020 Yangtze River flood. *Sci. China Earth Sci.* **2021**, *64*, 1–9. [CrossRef]
32. Xing, L.; Zhang, D.-D.; Shuai, L. Influence of Three Gorges Reservoir on Flood Regime of Mid-lower Reaches of Yangtze River. *Water Resour. Power* **2019**, *37*, 44–46.
33. Hu, C.; Zhang, S. Discussion on development and protection strategy of Yangtze River. *Yangtze River* **2020**, *51*, 1–5.
34. Shen, X.; Liu, X.; Zhao, W. Reflections on Flood Control in the Middle Reaches of Yangtze River. *Technol. Econ. Change* **2023**, *7*, 43–48.
35. Lai, X.; Jiang, J.; Yang, G.; Lu, X.X. Should the Three Gorges Dam be blamed for the extremely low water levels in the middle–lower Yangtze River? *Hydrol. Process.* **2013**, *28*, 150–160. [CrossRef]
36. Wang, H.; Sun, F.; Liu, W. Characteristics of streamflow in the main stream of Changjiang River and the impact of the Three Gorges Dam. *Catena* **2020**, *189*, 104498. [CrossRef]
37. Bureau of Hydrology, Changjiang Water Resources Commission. *The 1954 Flood of the Yangtze River*; Changjiang Publishing House: Wuhan, China, 2004.
38. Zhang, Y.; Liu, X.; Lu, X. Study on Flood Simulation in the Middle and Lower Reaches of the Yangtze River. *Hydraul. Hydropower Eng.* **2003**, *5*, 17–19.

39. Wang, J.; Hong, Y.; Li, L.; Gourley, J.J.; Khan, S.I.; Yilmaz, K.K.; Adler, R.F.; Policelli, F.S.; Habib, S.; Irwn, D.; et al. The coupled routing and excess storage (CREST) distributed hydrological model. *Hydrol. Sci. J.* **2011**, *56*, 84–98. [CrossRef]
40. Zou, B.; Shiqiang, L. Applicability of Single-valued Lake Routing Model at the Luoshan Station. *J. China Hydrol.* **2011**, *31*, 140–142+147.
41. Zhu, Y.; Guo, X.; Li, L. Changes of flood diversion capacity of three outlets on Jingjiang reach and outflow of Dongting Lake during 2020 flood season. *Yangtze River* **2020**, *51*, 210–215.
42. Yan, B.; Gu, Y.; Li, E.; Xu, Y.; Ni, L. Runoff Prediction of Tunxi Basin under Projected Climate Changes Based on Lumped Hydrological Models with Various Model Parameter Optimization Strategies. *Sustainability* **2024**, *16*, 6897. [CrossRef]
43. Dariane, A.B.; Bagheri, R.; Ghasemi, M.; Asadi, R. Comparative implementation of melody search in auto-calibrating SWAT. *Arab. J. Geosci.* **2024**, *17*, 168. [CrossRef]
44. Sohi, H.Y.; Zahraie, B.; Dolatabadi, N.; Zebarjadian, F. Application of VIC-WUR model for assessing the spatiotemporal distribution of water availability in anthropogenically-impacted basins. *J. Hydrol.* **2024**, *637*, 131365. [CrossRef]
45. Zhang, X.; Li, Y.; Niu, W. Strategy of water projects joint operation program on upper-middle reaches of Changjiang River for controlling 1954 flood under current conditions. *Yangtze River* **2020**, *51*, 141–148.
46. Hu, X.; Ding, Y.; Zou, Q.; Li, A. Study and application of collaborative operation model of reservoir groups in upper reaches of Changjiang River for multi-regional flood control. *Yangtze River* **2020**, *51*, 56–63+79.
47. Zou, Q.; Yu, B.; Ding, Y.; He, X. Study on the Flood Control Role of Reservoir Group in the Upper Reaches of the Yangtze River during the 1998 Flood. In Proceedings of the 2023 China Water Resources Academic Conference, Zhengzhou, China, 7–8 July 2023; p. 9.
48. Cai, L. An Evaluation of the Honghu East Block Flood Project'Social Impact—Based on an Survey of the Social and Economic Development at the Honghu Est Block Storage Area. *J. Hubei Univ. Technol.* **2016**, *31*, 26–30.
49. Shen, X.; Shi, Y.; Liu, X.; Song, P.; Luan, Z.; Zhao, W. *Study on Countermeasures for Flood Storage and Disaster Mitigation during Extraordinary Floods in Dongting Lake Region*; Changjiang Publishing House: Wuhan, China, 2022.
50. An, S.; Guo, L. Discussion on the Flood Discharge Capacity of Chenglingji (Luoshan Station). *Hog Shui River* **2005**, *24*, 1–4.
51. Liu, Z.; Chen, X.; Liu, F.; Lin, K.; He, Y.; Cai, H. Joint Dependence Between River Water Temperature, Air Temperature, and Discharge in the Yangtze River: The Role of the Three Gorges Dam. *J. Geophys. Res. Atmos.* **2018**, *123*, 11938–11951. [CrossRef]
52. Tan, G.; Chen, P.; Deng, J.; Xu, Q.; Tang, R.; Feng, Z.; Xiong, Y. Estimations and changes of the dominant discharge in Three Gorges Reservoir channel. *Arab. J. Geosci.* **2019**, *12*, 82. [CrossRef]

water

MDPI

Article

Human Activities Impacts on Runoff and Ecological Flow in the Huangshui River of the Yellow River Basin, China

Lanxin Liu [1], Lijuan Fan [2,*], Jing Hu [1] and Chunhui Li [1,*]

[1] Key Laboratory for Water and Sediment Sciences of Ministry of Education, School of Environment, Beijing Normal University, Beijing 100086, China; 202111180043@mail.bnu.edu.cn (L.L.); 202321180028@mail.bnu.edu.cn (J.H.)

[2] Haidong Municipal Water Resources Bureau of Qinghai Province, Haidong 810600, China

* Correspondence: fanlj87@hotmail.com (L.F.); chunhuili@bnu.edu.cn (C.L.)

Abstract: This study analyzed 61 years of hydrological data from the Minhe and Xiangtang Hydrological Stations (1956–2016) to examine hydrological changes and ecological flow assurance rates in the Huangshui River Basin, China. Using the Mann–Kendall trend test, IHA/RVA method, and ecological flow calculation methods, the study revealed the following results: (1) After 1994, increased human activity in the Datong River led to a measured runoff decrease compared to natural runoff. Although human activities in the Huangshui River's main stream were present before 1972, after 1972, these activities intensified, resulting in a more pronounced decrease in the measured runoff. (2) Ecological flow analysis indicated that the main stream of the Huangshui River and the Datong River have ecological flow assurance rates of 100% for all but a few months, where the rates are 98%. The water volume is sufficiently abundant to meet ecological water demands.

Keywords: human activities; ecological flow; IHA/RVA; ecological flow assurance rates; Huangshui River Basin

1. Introduction

Climate change and the intensification of human activities are causing a decrease in river runoff worldwide, posing significant challenges to water resource management and ecological development. Activities such as agricultural irrigation, water conservancy projects, and land-use changes can lead to substantial hydrological alterations. Scholars globally are conducting quantitative research on how these factors affect river runoff. Climate change is expected to modify the behavior of droughts within watersheds, thereby threatening drought monitoring and early warning systems [1]. For instance, Sadio et al. [2] utilized the GR2M model to assess the impact of climate change on runoff characteristics in watersheds in Senegal and Guinea. The interplay between climate change and watershed characteristics can influence runoff evolution to varying extents. Based on the Budyko hypothesis, Lv et al. [3] analyzed runoff and meteorological data from the Huangshui River Basin, concluding that while runoff showed a non-significant decreasing trend, changes in watershed characteristics were the primary contributors to the reduction in runoff. Previous research indicates that human activities are the predominant factor affecting changes in watershed runoff [4]. Li et al. [5] investigated the impact of human activities on hydrological and drought dynamics in the Xilin River Basin, revealing that these activities were the main cause of the significant reduction in runoff, accounting for 68% of the change.

Ensuring ecological flow is essential for sustaining a healthy aquatic ecosystem, especially amid changes in river hydrological conditions. Hydrological rhythms serve as a vital reference for establishing reasonable ecological flow rates. Among the various methods for calculating ecological flow, the hydrological method is the most widely used, including techniques such as the Tennant method [6] and the 7Q10 method [7]. The data needed for the Tennant method can be derived from hydrological monitoring stations without the need for on-site measurement. In cases where no monitoring station exists for the river under study, hydrological technology can be employed to obtain the necessary data quickly and efficiently. However, this method has its limitations, as it does not account for rivers with high sediment content, pronounced seasonality, significant flow variability, or the substantial influence of geometric shapes on flow.

Hydrological research encompasses a range of methods, such as wavelet analysis (WA) [8,9], hydrological change indicators, the hydrological change range method [10,11], and various hydrological models [12–15]. The Indicators of Hydrologic Alteration (IHA), introduced by Richter et al. [10] in 1996, represent the most widely adopted indicator system. To identify indicators that more precisely describe the extent of hydrological changes, Richter et al. [11] proposed the Range of Variability Approach (RVA) in 1997, building upon the IHA system. The combination of IHA and RVA offers a more systematic and comprehensive assessment of river hydrological condition changes through multi-indicator analysis. IHA is predominantly utilized in hydrological research to evaluate temporal changes in hydrological conditions due to factors such as human activities, climate change, and land transformation [16–18], as well as to assess the ecological response to current or past hydrological conditions in relation to specific ecological variables [19]. Kannan et al. [20] also explored the use of IHA to assess the overall health status of rivers.

The Huangshui River is the largest and most important tributary of the upper reaches of the Yellow River, and also a concentrated area for economic activities. In recent years, the level of human activity in the basin has increased, and multiple water conservancy projects have been constructed, such as the Datong River to Huangshui River Water Diversion Project and the Yellow River to Xining Water Diversion Project. In this situation, it is necessary to increase attention to the hydrological changes and ecological flow guarantee of the long sequence of the Huangshui River and Datong River to provide scientific theoretical guidance for the construction of water conservancy projects.

Therefore, this study primarily employs the IHA/RVA method and related ecological water demand methods to analyze the hydrological condition changes in the Huangshui River Basin and the ecological flow guarantee rates of the Datong River and the Huangshui River. The aim is to identify the influencing factors and offer targeted recommendations.

2. Study Area

The Huangshui River and the Datong River both originate within Qinghai Province. The Datong River, originating on Muli Mountain, flows from northwest to southeast and merges with the Huangshui River at Xiangtangxia, on the border of Gansu and Qinghai provinces. The Huangshui River, the largest tributary of the upper Yellow River, originates on Daban Mountain. Its upper reaches, known as the Bawutu River, flow from north to south before Haiyan, then gradually turn from west to east. Exiting Minhe in Qinghai, the Huangshui River meets the Datong River in Haishiwan, Gansu Province, and is collectively referred to as the Huangshui River. It eventually flows into the Yellow River near the Bapan Gorge in Gansu Province.

The Huangshui River basin has a total controlled area of 32,863 km^2. The watershed area monitored by the Minhe Hydrological Station is 15,342 km^2, and that of the Datong River at the Xiangtang Hydrological Station is 15,126 km^2. Together, the Minhe and Xiangtang stations control a watershed area of 30,468 km^2. Established by the Hydrological and Water Resources Survey Bureau of the Yellow River Commission, these stations observe various parameters including water level, flow rate, suspended sediment transport rate, sediment concentration, water temperature, and ice conditions. Since their establishment in 1950, the stations have undergone reorganization and now offer excellent control conditions for observation. The site of Huangshui River Basin is shown in Figure 1.

For this study, hydrological data from the Minhe and Xiangtang Hydrological Stations spanning 61 years, from 1956 to 2016, were selected to analyze the hydrological changes and the ecological flow assurance rate in the Huangshui River Basin. These data include measured runoff, natural runoff, and precipitation.

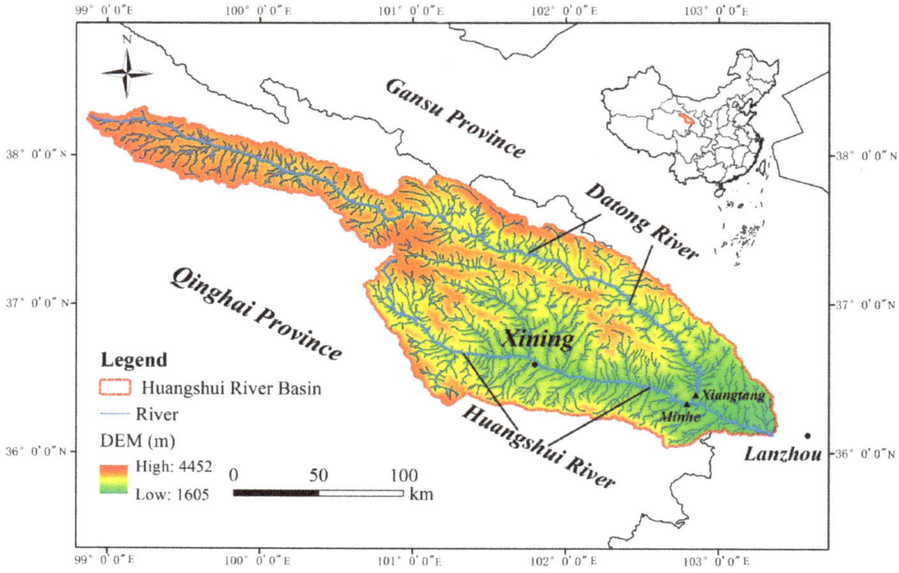

Figure 1. Site of Huangshui River Basin.

3. Study Methods

3.1. Mann–Kendall Mutation Test Method

The Mann–Kendall mutation test is a non-parametric hypothesis testing method used to examine trend changes in time series. This testing method detects monotonic trends (upward, downward, or no trend) in time series by comparing the size of each data point with its previous data points.

In the Mann–Kendall test, the null hypothesis H_0 is a time series $(x_1, \ldots, x_n)$ consisting of n independent, identically distributed samples of random variables, without a certain upward or downward trend, the alternative hypothesis H_1 is a bilateral test, and for all $k, j \leq n$, and $k \neq j$, the distributions of x_k and x_j are not the same. The test statistic S is calculated as follows:

$$S = \sum_{k=1}^{n-1} \sum_{j=k+1}^{n} \mathrm{sgn}(x_j - x_k) \tag{1}$$

where, S represents a statistical quantity that follows a normal distribution.

$$\mathrm{sgn}(x_j - x_k) = \begin{cases} 1, & x_j - x_k > 0 \\ 0, & x_j - x_k = 0 \\ -1, & x_j - x_k < 0 \end{cases} \tag{2}$$

where, sgn(·) represents the sign function.

When n > 10, the standard normal system variable is calculated by the following formula:

$$Z = \begin{cases} \frac{S-1}{\sqrt{\mathrm{Var}(S)}} & S > 0 \\ 0 & S = 0 \\ \frac{S+1}{\sqrt{\mathrm{Var}(S)}} & S < 0 \end{cases} \tag{3}$$

where, Z represents the standard normal distribution statistic and Var(S) represents variance.

$$\mathrm{Var}(S) = \frac{n(n-1)(2n+5)}{18} \tag{4}$$

In this way, in the bilateral trend test, if $|Z| \geq Z_{1-\alpha/2}$ at a given alpha confidence level, the null hypothesis is unacceptable. This shows that there is a significant upward or downward trend in the time series data at the α confidence level. The trend type is determined by the sign of the statistic Z. $Z > 0$ indicates an upward trend and $Z < 0$ indicates a downward trend. Whether this trend is significant is then determined by the magnitude of the Z value.

The steps for using the M-K mutation algorithm for mutation testing are as follows:

① For the time series x_i, construct a rank sequence r_i to represent the cumulative number of samples where $x_i > x_j$ $(1 \leq j \leq i)$, and define s_k as:

$$s_k = \sum_{i=1}^{k} r_i r_i = \begin{cases} 1\, x_i > x_j \\ 0 \quad \text{else} \end{cases} j = 1, 2, \ldots, i \tag{5}$$

② Assuming the time series is randomly independent, the statistical variable UF_k is defined as [21]:

$$UF_k = \frac{s_k - E(s_k)}{\sqrt{Var(s_k)}}\ (k = 2, 3, \ldots, n) \tag{6}$$

where, UF_k—standard normal distribution statistic; $E(s_k)$—the mean value of s_k; and $\sqrt{Var(s_k)}$—the variance of s_k.

Among them, $UF_1 = 0$, and at a given significance level α, if $|UF_k| \geq U_\alpha$, this indicates a significant trend change in the sequence.

③ Arrange the time series x in reverse order, repeat ① and ②, and make:

$$UB_k = -UF_k\ (k = n, n-1, \ldots, 1) \tag{7}$$

Among them, $UB_1 = 0$. By analyzing the statistical sequences UB_k and UF_k, we can further analyze the time nodes of sequence x mutations and display the mutation regions. If $UF_k > 0$, it indicates that the sequence is showing an upward trend; conversely, UF_k shows a downward trend. When they exceed any critical line, this indicates a significant upward or downward trend. If the UB_k and UF_k lines intersect and the intersection point is between two critical lines, then the moment corresponding to the intersection point is the moment when the mutation begins.

3.2. Indicators of Hydrologic Alteration (IHA)

To quantitatively assess the effects of human activities on hydrological conditions, Richter et al. (1996) introduced the Indicators of Hydrologic Alteration (IHA). This comprehensive framework comprises 33 distinct indicators categorized into five ecologically significant components. These components reflect various aspects of hydrological change, including variability in flow magnitude, frequency of occurrence, timing of events, duration of hydrologic events, and the rate of change in these conditions (Table 1).

Table 1. Indicators of Hydrologic Alteration and their characteristics.

IHA Parameter Groups	Characteristics	Indicators
Average monthly runoff	Runoff Time	Average monthly runoff value
Annual extreme runoff	Runoff Duration	Annual maximum and minimum average runoff (1, 3, 7, 30, 90 d) Period of river interruption Base flow coefficient

Table 1. *Cont.*

IHA Parameter Groups	Characteristics	Indicators
Extreme runoff occurrence time	Time	Annual maximum and minimum 1-day runoff occurrence time
Frequency and duration of high and low runoff	Runoff Frequency Duration	Number of occurrences of high and low runoff per year Average duration of high and low runoff per year
Rate and frequency of runoff changes	Rate of change Frequency	Average rate of runoff increase and decrease Number of runoff reversals

3.3. *Range of Variability Approach (RVA)*

In 1997, Richter proposed the Range of Variability Approach (RVA) based on the IHA index system. The core of this method is to construct the RVA target by adding or subtracting the standard deviation of the average value of each index before the impact, or the values of 75% and 25% of the frequency of occurrence of each index as the upper and lower limits of each index. The degree of hydrological change is evaluated by the proportion of each index value falling within the RVA target range after the impact and the difference before the impact. The calculation formula for hydrological change degree is:

$$D_i = \frac{N_{io} - N_{ie}}{N_{ie}} \times 100\% \tag{8}$$

where, $N_{ie} = rN_T$,D_i is the hydrological change degree of the *i*-th IHA indicator. When D_i is a positive value, this indicates that the *i*-th indicator falls within the RVA target after being affected by human activities, and when D_i is a negative value, this indicates that the *i*-th indicator falls outside the RVA target after being affected by human activities. N_{io} is the number of years in which the *i*-th indicator still falls within the RVA target after being affected by human activities. N_{ie} is the expected number of years for the *i*-th indicator to fall within the RVA target after being affected by human activities. r is the proportion of the *i*-th indicator falling within the RVA target before being affected by human activities. N_T is the total number of years of hydrological series affected by human activities. In order to objectively evaluate the degree of hydrological change in IHA indicators, this method divides hydrological change into three levels: $0 \leq |D_i| < 33\%$ is a low degree change, $33\% \leq |D_i| < 67\%$ is a moderate change, and $67\% \leq |D_i| \leq 100\%$ is a high degree change.

3.4. *Ecological Water Demand Method*

3.4.1. Tennant Method

This method considers 10% of the annual average flow as the minimum ecological water demand of the river, 30% of the annual average flow as the optimal flow to meet the survival of aquatic organisms, and 60% to 100% as the ecological flow to maintain the original natural river ecosystem. The calculation formula is:

$$Q_T = \sum_{i}^{12} Q_i \times Z_i \tag{9}$$

where, Q_T is the ecological water demand of the river channel (m^3), Q_i is the average annual flow rate for the *i*-th month within a year (m^3), and Z_i corresponds to the recommended base flow percentage (%) for the *i*-th month.

The Tennant method has 8 levels, and the recommended ecological environment water demand in the river is divided into the general water use period (from October to March of the following year) and the fish spawning and juvenile period (from April to September).

The recommended value is based on the percentage of runoff. The recommended runoff of the Tennant method is shown in Table 2.

Table 2. Recommended ecological flow by Tennant method.

Qualitative Description of Habitats	Recommended Base Flow Standard (Percentage of Annual Average Flow)	
	General Water Use Period (from October to March of the Following Year)	Fish Spawning and Juvenile Period (from April to September)
Maximum	200	200
Optimum flow	60~100	60~100
Excellent	40	60
Very good	30	50
Good	20	40
Become vestigial	10	30
Poor or minimum	10	10
Extremely poor	<10	<10

3.4.2. Ecological Flow Assurance Rate and Evaluation Standard

The ecological flow of rivers and lakes is evaluated based on the monthly average satisfaction level. The degree of ecological flow satisfaction is defined as the ratio of the number of months where the monthly flow value is greater than the monthly ecological flow value to the total number of months corresponding to the long-term runoff year. The calculation formula is:

$$D_i = (T_{bi}/T_i) \times 100\%, 1 \leq i \leq 12 \tag{10}$$

where, D_i represents the ecological flow guarantee rate for the *i*-th month, T_{bi} is the number of months that satisfy the ecological flow in the *i*-th month of the calculation year, and T_i is the total number of months in the *i*-th month of the calculation year.

The evaluation criteria for the ecological flow satisfaction index are shown in Table 3.

Table 3. Evaluation criteria for ecological flow satisfaction index.

Index	Evaluation Criteria (%)				
	Excellent	Good	Medium	Poor	Inferior
Ecological flow assurance rate	100	95~100	90~95	80~90	<80

4. Result Analysis

4.1. Hydrological Change Trends and Impact Analysis of Huangshui River and Datong River

Measured runoff refers to the amount of water that passes through a certain cross-section of a river during a certain period of time. Natural runoff refers to the amount of water that has been reverted from the measured river runoff, which generally refers to the measured runoff plus the utilization of water above the measured cross-section. A Mann–Kendall trend test analysis was conducted on the measured annual runoff data of the Huangshui River's Minhe section and the Datong River's Xiangtang section from 1956 to 2016 (Figure 2). At the Xiangtang section, there were four intersections between the UF and UB curves during this period, occurring in the years 2000, 2007, 2011, and 2013. These intersections indicate significant shifts in the measured runoff of the Datong River. Prior to 1973, positive UF values suggested an increasing trend in runoff. Between 1973 and 1987, negative UF values indicated a decreasing trend. From 1987 to 1995, positive UF values again pointed to an increasing trend. Post-1995, negative UF values with a strengthening decreasing trend marked a pronounced reduction in runoff.

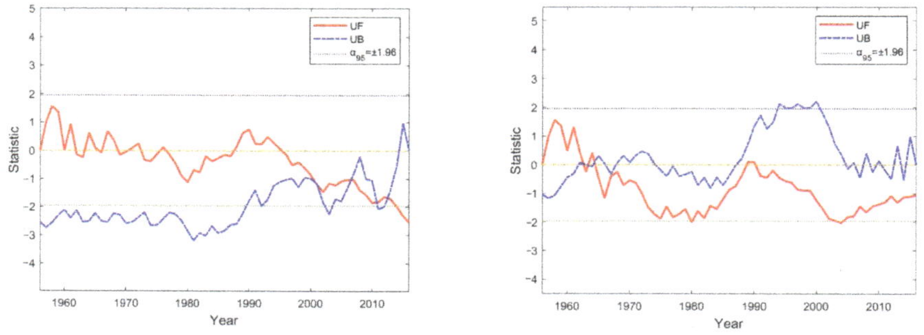

(a) (b)

Figure 2. Annual measured runoff, M-K mutation test. (**a**) Xiangtang; (**b**) Minhe.

In the Minhe section, three intersections between the UF and UB curves were observed in 1963, 1964, and 1965, indicating abrupt changes in the main stream of the Huangshui River during those years. Before 1965, positive UF values reflected an increasing trend in runoff. After 1965, negative UF values indicated a decreasing trend that intensified from 1965 to 1980 and then moderated around 1980. The decreasing trend picked up again after 1989, and by around 2004, it began to subside.

Figure 3 shows the natural and measured runoff variations for the Xiangtang and Minhe sections from 1956 to 2016. The deviation rate of measured runoff, denoted as α, is calculated using the formula α = (measured runoff − natural runoff)/natural runoff. Figure 4 shows the variations in the deviation rate.

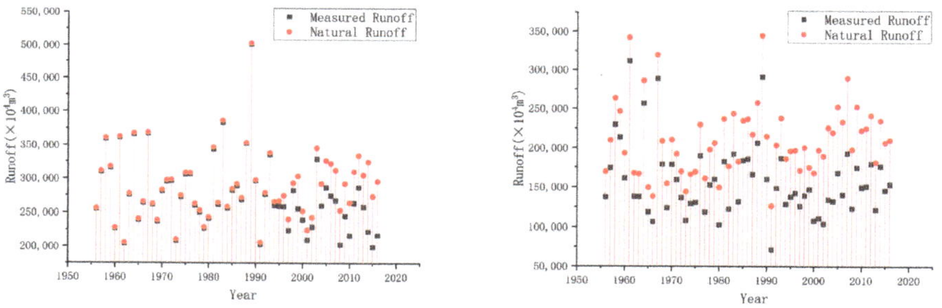

(a) (b)

Figure 3. Changes in natural runoff and measured runoff of Xiangtang Station and Minhe Station from 1956 to 2016. (**a**) Xiangtang Station; (**b**) Minhe Station.

At the Xiangtang section, prior to 1994, the absolute value of the measured runoff deviation rate remained under 1%, indicating that the river was predominantly in a natural state with minimal human influence. Between 1994 and 2016, the absolute deviation rate increased significantly, ranging from 2% to approximately 30%.

For the Minhe section, before 1972, the absolute deviation rate of measured runoff was generally under 20%. The linear regression equation y = −0.0014x + 2.6807 reflected minimal human impact during this period. However, between 1973 and 2016, the absolute deviation rate consistently increased, typically exceeding 20%, with a peak reaching 45%. The linear regression equation for this period was y = −0.0029x + 5.4149, featuring a slope twice that of the pre-1972 period, indicating a significant increase in human activity levels.

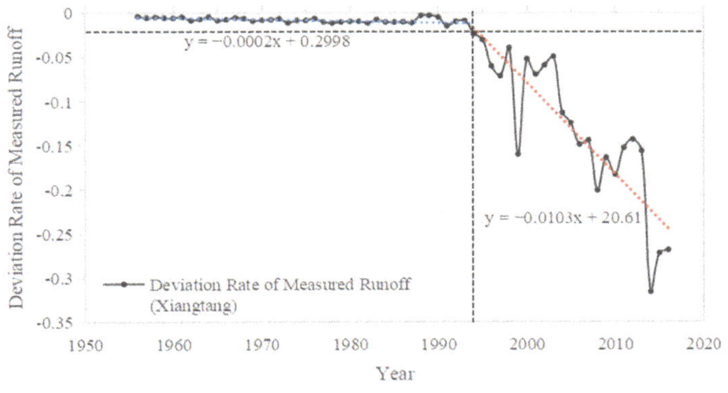

(a)

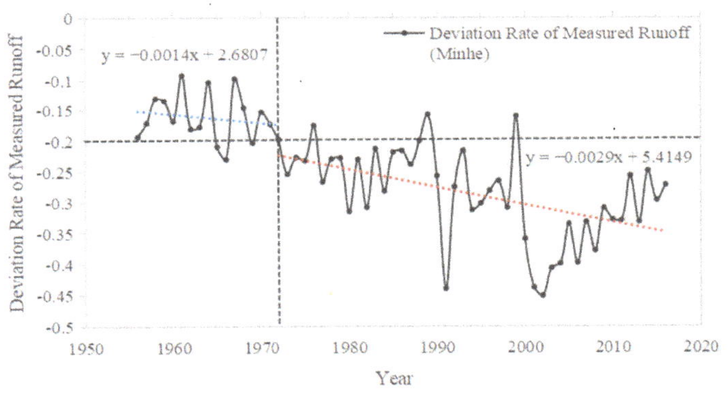

(b)

Figure 4. Changes in deviation rate of measured runoff of Xiangtang Station and Minhe Station from 1956 to 2016. (**a**) Xiangtang Station; (**b**) Minhe Station.

Both the Minhe and Xiangtang sections exhibited a rising trend in the absolute deviation rate of measured runoff annually, suggesting an escalating impact of human activities on the hydrological conditions of these rivers.

Therefore, the Xiangtang section was considered to be in a natural state before 1994, and after 1994, it transitioned into a phase influenced by both human activities and climate change. For the Minhe section, it can be approximated as being in a natural state stage before 1972, and after 1972, it entered a stage influenced by human activities and climate change.

Regarding the Xiangtang section, the runoff deviation rates from March to November showed a decreasing trend under the influence of human activities and climate change, with rates less than zero, compared to the natural state. However, from December to February of the following year, the deviation rate was positive, indicating an increasing trend. Except for February, the runoff in other months displayed moderate to high variability due to human activities and climate change. The degree of change in hydrological indicators in Xiangtang is shown in Table 4.

For the Minhe section, the runoff deviation rates from February to December were negative, reflecting a decreasing trend following an increase in human activity levels, compared to earlier stages of lower human activity. June and September showed low variability, while other months exhibited moderate variability. Consequently, it is evident that there was a significant increase in human activity levels in the main stream of the

Huangshui River from January to May, July to August, and October to December after 1972. The degree of change in hydrological indicators in Minhe is shown in Table 5.

Table 4. Degree of change in hydrological indicators in Xiangtang.

Indicator (Average Runoff)	1956—1993 (Approximate Natural State)			1994—2016 (Human Activities + Climate Change)		
	Mean Value	RVA Target Value (m^3/s)		Mean Value	Deviation Rate (%)	D_i (%)
		Lower Limit	Upper Limit			
January	19.52	16.95	22.30	21.29	9.07	−52.17 (M)
February	19.26	17.15	22.55	21.14	9.75	−30.43 (L)
March	25.97	22.80	28.45	25.66	−1.17	−43.48 (M)
April	54.00	45.85	61.90	46.50	−13.89	−65.22 (M)
May	99.19	75.79	118.00	62.65	−36.84	−73.91 (H)
June	131.19	90.91	168.00	91.16	−30.52	−52.17 (M)
July	213.18	174.51	241.00	172.20	−19.23	−60.87 (M)
August	200.42	152.50	258.49	182.89	−8.75	−34.78 (M)
September	171.60	116.51	199.50	167.51	−2.38	−43.48 (M)
October	92.98	71.85	107.00	80.83	−13.07	−43.48 (M)
November	45.25	38.51	52.41	43.43	−4.00	−65.22 (M)
December	26.98	23.15	29.90	30.82	14.23	−69.57 (H)

Table 5. Degree of change in hydrological indicators in Minhe.

Indicator (Average Runoff)	1956—1972 (Approximate Natural State)			1973—2016 (Human Activities + Climate Change)		
	Mean Value	RVA Target Value (m^3/s)		Mean Value	Deviation Rate (%)	D_i (%)
		Lower Limit	Upper Limit			
January	24.02	19.00	28.80	26.47	10.19	−43.18 (M)
February	26.92	21.70	32.16	26.25	−2.48	−36.36 (M)
March	28.16	22.95	32.85	20.65	−26.66	−65.90 (M)
April	33.51	18.30	44.10	28.96	−13.55	−36.36 (M)
May	49.62	22.80	57.55	28.53	−42.51	−43.18 (M)
June	46.76	19.00	73.36	42.28	−9.58	−25.00 (L)
July	85.97	54.51	116.51	71.52	−16.80	−36.36 (M)
August	115.88	63.84	171.50	87.39	−24.59	−38.64 (M)
September	110.31	63.25	171.49	94.74	−14.11	−15.91 (L)
October	81.52	58.36	97.24	72.74	−10.77	−40.91 (M)
November	44.36	33.20	49.34	37.47	−15.53	−52.27 (M)
December	32.48	25.80	37.04	31.98	−1.54	−34.09 (M)

The observed decrease in measured runoff compared to the natural state is attributed to the combined effects of human activities and climate change. Interestingly, this analysis also noted an increase in runoff during specific months: from December to February at the Xiangtang Station and from January at the Minhe Station. The following analysis explores the reasons for these increases.

The primary source of runoff in the Huangshui River basin is atmospheric precipitation, with rainwater as the predominant source and snow as the secondary source. The year can be divided into distinct hydrological periods: the spring flood season from May to June, replenished by the melting of ice and snow in the upstream areas and rainfall; the summer and autumn flood season from July to early September, mainly supplemented by large-scale precipitation; the autumn normal water period from October to December, primarily supported by groundwater recharge and river channel storage; and the winter dry season from January to April of the following year, mainly sustained by groundwater with a small and stable water volume.

After increases in human activity levels, the months when runoff at both Xiangtang and Minhe Stations increases correspond to the river's dry season. While not entirely excluding the influence of climate change, considering the 'flood storage and dry discharge' operation mode of reservoirs, it can be inferred that the increase in runoff during the dry season is likely related to the discharge practices of the upstream reservoirs relative to the hydrological stations.

The difference between the annual natural runoff and the annual measured runoff at Xiangtang Station is used to estimate the volume of water extracted by human activities in the Datong River. Similarly, the difference at Minhe Station can be used to estimate water extraction in the main stream of the Huangshui River. The changes in these values from 1956 to 2015 are depicted in Figures 5 and 6.

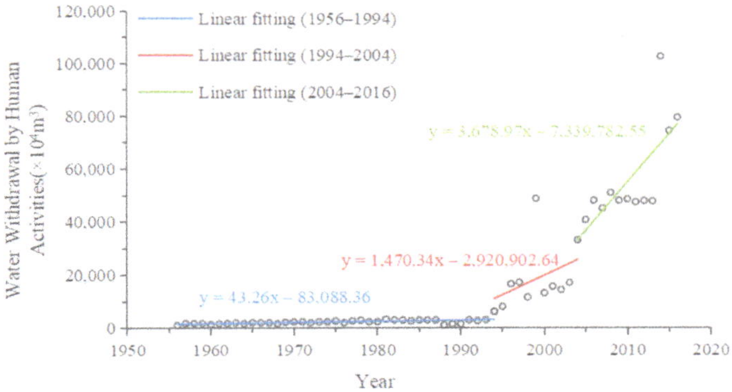

Figure 5. Changes in water withdrawal from human activities at Xiangtang Station.

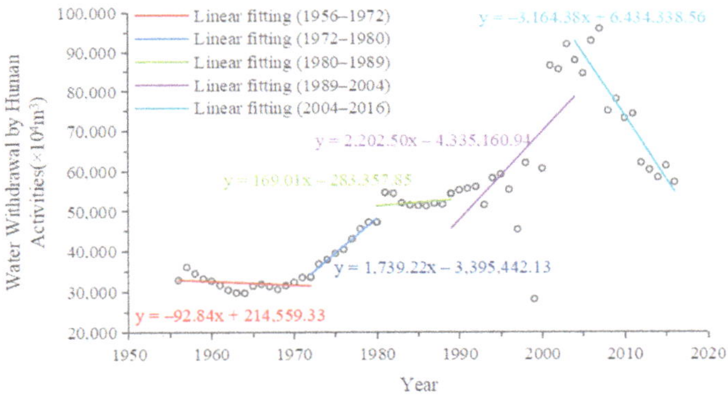

Figure 6. Changes in water withdrawal from human activities at Minhe Station.

Before 1994, human activity levels in the Datong River Basin were relatively low, resulting in small volumes of water being taken from the river, with natural runoff closely aligning with measured runoff. Since 1994, the volume of water extracted by human activities in the Datong River Basin has gradually increased, with a notable spike around 2004. Concurrently, there were significant increases in human activity levels in 1999 and 2014. The Mann–Kendall (M-K) mutation test identified breakpoints that suggest human activities as the primary cause of the sudden changes in the Datong River runoff.

The significant increases in water withdrawal at Xiangtang Station in 1999 and 2014 are attributed to specific developments. The first phase of the Datong-to-Huangshui River Diversion Project, which includes the Heiquan Reservoir and the first phase of the Huangshui North Main Canal, began construction in 1996. The Heiquan Reservoir started intercepting the river flow in 1997 and completed its main project, allowing for water storage, in 2001. The second phase of the project, featuring the Shitouxia Reservoir and the second phase of the Huangshui North Main Canal, concluded at the end of 2014 with the completion of the Shitouxia Hydropower Station. The construction and operation of these projects have significantly impacted water withdrawal, as evidenced by the abrupt increases in 1999 and 2014.

Although human activities in the main stream of the Huangshui River were present before 1972, water withdrawal from the river exhibited only slight fluctuations, remaining relatively stable. However, since 1972, there has been a gradual increase in the level of human activity. Between 1972 and 1989, the volume of water extracted through human activities saw a gradual rise. From 1989 to 2004, human activity levels escalated further, but there was a notable decrease around 1999. Post-2004, water intake has gradually decreased.

The Mann–Kendall (M-K) mutation test indicates that the changes in human activities within the Huangshui River Basin generally correspond to various breakpoints and trends identified in the analysis. These findings suggest that human activities have had a significant and evolving impact on water withdrawal patterns over the years.

The significant decrease in human activities in the Huangshui River Basin around 1999 can be attributed to several factors. The Hehuang Valley, primarily shaped by the Huangshui River, is a region of intense human activity on the Qinghai–Tibet Plateau and an early center of human activity within the Yellow River Basin. According to the Qinghai Provincial Water Resources Bulletin, irrigation water consumption for farmland in the Huangshui River Basin constitutes the largest portion of total water consumption, ranging from 45% to 55%. This is followed by water consumption for forestry, animal husbandry, fishing, and other livestock-related activities. The primary sources of irrigation water in the basin are river withdrawal and natural precipitation, with climate factors exerting a significant influence on irrigation needs.

Analyzing precipitation data, the Huangshui River Basin experienced relatively high levels of rainfall from 1997 to 1999, contrasting with both the preceding and subsequent three-year periods. In other years, higher precipitation was often an isolated event. The sustained high precipitation over three consecutive years likely led to a reduction in the amount of water extracted from rivers for agricultural irrigation in 1999. Additionally, in 1998, nationwide heavy rainfall and catastrophic floods in major river basins severely impacted farmland, which in turn affected agricultural production and led to a decrease in irrigation water usage in the following year.

The annual variation in precipitation at the Minhe Station, which provides insight into these climatic influences, is illustrated in Figure 7.

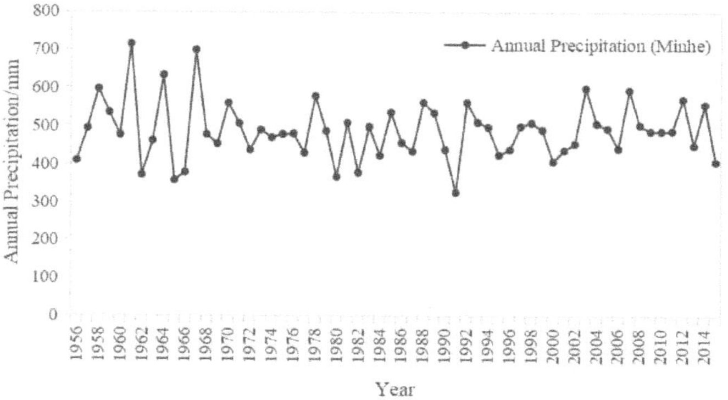

Figure 7. Annual variation of precipitation at Minhe Station from 1956 to 2015.

As depicted in Figure 8, the total amount of human water withdrawal at the Xiangtang and Minhe Stations indicates a gradual increase in human activity levels in the Huangshui River Basin since 1972. The activity level has remained relatively stable since 1980. However, a significant surge occurred in 1994, followed by a decline after 2007. Notably, there was a sharp increase in 2014, after which the withdrawal rate gradually decreased.

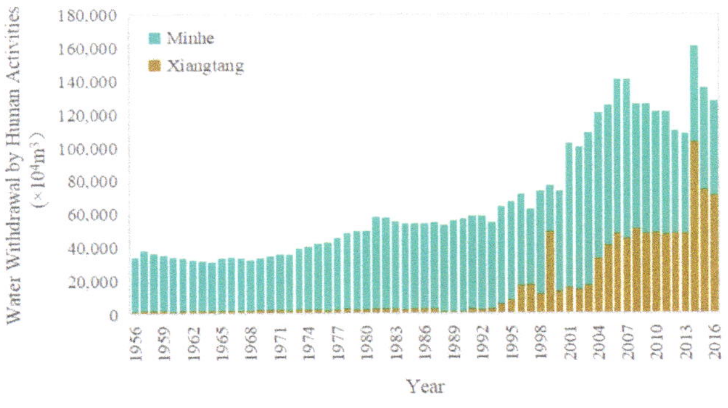

Figure 8. Total human water withdrawal in the Huangshui River Basin.

This pattern suggests that while human activity in the basin has been generally consistent since the 1980s, there have been notable fluctuations, particularly in the late 1990s and early 2000s. These changes likely reflect shifts in water management practices, agricultural demands, and possibly climate-related factors affecting water availability and demand.

4.2. Analysis of Ecological Flow Assurance Rate

The Tennant method is used to determine the ecological flow at the Xiangtang and Minhe sections by designating 30% of the natural average flow from 1956 to 2016 as the ecological flow for each section. As per the Implementation Rules of the Yellow River Water Dispatching Regulations, the minimum flow requirement for a 95% assurance rate at the Xiangtang Station of the Datong River is set at 10 m^3/s. Based on this method, the annual average ecological flow at the Xiangtang Station is calculated to be 28.42 m^3/s, with an annual ecological water demand of 896 million m^3.

Similarly, for the Minhe section in the main stream of the Huangshui River, 30% of the average natural flow from 1956 to 2016 is used to calculate the ecological flow. With the warning flow rate of 8 m^3/s as stipulated in the same regulations, the annual average ecological flow for the Minhe section is determined to be 19.98 m^3/s, with an annual ecological water demand of 631 million m^3.

The monthly ecological flow calculation results of the two sections are shown in Table 6. These calculations ensure that the ecological needs of the river systems are met while also considering the regulatory requirements for water flow management.

The ecological flow assurance rate for the Xiangtang section of the Datong River is 98% in January, classifying the ecological flow satisfaction as 'good' for that month. For all other months, the rate is 100%, indicating an 'excellent' ecological flow satisfaction level. Similarly, the Minhe section of the Huangshui River has ecological flow assurance rates of 98% in March, August, and September, with a 'good' satisfaction degree for those months, and 100% in other months, reflecting 'excellent' satisfaction. The specific results are shown in Table 7.

These data indicate that the Datong River and the main stream of the Huangshui River have sufficient water volumes to meet ecological water demands throughout the year, with only a slight decrease in assurance rates during specific months. The ecological health of both rivers is generally well-maintained, demonstrating a strong capacity to support their respective ecosystems.

Table 6. Ecological flow rate of cross-section Unit: m^3/s.

Month	Hydrology Stations	
	Xiangtang	Minhe
January	10	8.32
February	10	8.65
March	10	10.13
April	17.8	22.16
May	28.87	18.57
June	37.80	20.72
July	60.87	28.43
August	58.54	31.28
September	52.19	31.14
October	29.95	31.13
November	15	18.43
December	10	10.21

Table 7. Evaluation of ecological flow assurance rate.

Month	Ecological Flow Assurance Rate (%)		Ecological Flow Satisfaction	
	Datong River	Huangshui River	Datong River	Huangshui River
January	98	100	Good	Excellent
February	100	100	Excellent	Excellent
March	100	98	Excellent	Good
April	100	100	Excellent	Excellent
May	100	100	Excellent	Excellent
June	100	100	Excellent	Excellent
July	100	100	Excellent	Excellent
August	100	98	Excellent	Good
September	100	98	Excellent	Good
October	100	100	Excellent	Excellent
November	100	100	Excellent	Excellent
December	100	100	Excellent	Excellent

5. Conclusions and Suggestions

Since 1994, there has been an increase in human activity levels in the Datong River, resulting in a measured runoff that has shown a decreasing trend when compared to the natural runoff. Although human activities were present in the main stream of the Huangshui River before 1972, after 1972, these activities intensified, leading to a more pronounced decrease in measured runoff compared to the natural runoff.

Ecological flow analysis indicates that the ecological flow assurance rates for both the main stream of the Huangshui River and the Datong River are 100% for all months except for a few, where the rates are 98%. This suggests that the water volume is relatively abundant and capable of meeting the ecological water demand.

The Datong-to-Huangshui River diversion Project marked significant milestones with successful water supply in the main canal by the end of 2015 and in the North and West main canals in September 2023. These achievements underscore the importance of monitoring the hydrological changes and ecological flow of the Datong and Huangshui rivers, influenced by the project, to ensure the stability and health of the river ecosystems.

Ensuring stable water supply and ecological flow assurance rates involves a combination of management strategies, technological solutions, and regulatory measures. Here are some specific measures that can be taken to maintain these aspects:

Water Allocation Management: Implementing a comprehensive water allocation plan that prioritizes ecological needs while balancing human consumption and agricultural demands.

Flow Regulation: Adjusting the operation of reservoirs and dams to maintain minimum ecological flows, especially during dry seasons. This can involve releasing water from reservoirs to supplement natural flows.

River Connectivity: Ensuring that rivers and their tributaries remain connected to facilitate the movement of aquatic species and maintain ecological integrity.

Monitoring Systems: Establishing real-time monitoring systems to track water levels, flow rates, and water quality. These data can be used to make informed decisions about water management.

Climate Resilience Planning: Developing strategies to adapt to climate change impacts, such as increased variability in precipitation and higher temperatures, which can affect the flow of rivers and water availability and quality.

Stakeholder Engagement: Involving local communities, industries, and other stakeholders in water management decisions to ensure that all perspectives are considered and that solutions are sustainable and equitable.

Regulatory Frameworks: Enforcing regulations that limit water extraction and protect water resources, ensuring that ecological flow requirements are met.

Restoration Projects: Undertake river restoration projects to improve habitat conditions, reconnect fragmented habitats to ensure the continuity of flow, and enhance the overall health of river ecosystems.

Education and Awareness: Raising public awareness about the importance of maintaining ecological flows and the role of water in supporting biodiversity and ecosystem services.

Research and Innovation: Supporting research into new technologies and methods for water management, such as artificial intelligence for predicting water needs and optimizing water allocation.

By integrating these measures, it is possible to ensure that water supply and ecological flow assurance rates remain stable, supporting both human needs and the health of river ecosystems.

Author Contributions: L.L.: Conceptualization, Methodology, Writing–original draft. L.F.: Investigation, Validation. J.H.: Writing–review and editing. Y.Y. Revision and Funding acquisition. C.L.: Resources, Methodology, Writing–review and editing. All authors have read and agreed to the published version of the manuscript.

Funding: This work was supported by the Joint Funds of the National Natural Science Foundation of China (U2243236), the National Science Fund for Distinguished Young Scholars (52025092), and the Qinghai Haidong Urban-rural Eco-development projection (L3443-PRC-HD-CB-CS4).

Data Availability Statement: The data used in this study is mainly sourced from the National Hydrological Statistical Yearbook.

Conflicts of Interest: The authors declare that the research was conducted in the absence of any commercial or financial relationships that could be construed as a potential conflict of interest.

References

1. Kimmany, B.; Visessri, S.; Pech, P.; Ekkawatpanit, C. The Impact of Climate Change on Hydro-Meteorological Droughts in the Chao Phraya River Basin, Thailand. *Water* **2024**, *16*, 1023. [CrossRef]
2. Sadio, C.A.A.S.; Faye, C.; Pande, C.B.; Tolche, A.D.; Ali, M.S.; Cabral-Pinto, M.M.S.; Elsahabi, M. Hydrological response of tropical rivers basins to climate change using the GR2M model: The case of the Casamance and Kayanga-Géva rivers basins. *Environ. Sci. Eur.* **2023**, *35*, 113. [CrossRef]
3. Lv, X.; Liu, S.; Li, S.; Ni, Y.; Qin, T.; Zhang, Q. Quantitative Estimation on Contribution of Climate Changes and Watershed Characteristic Changes to Decreasing Streamflow in a Typical Basin of Yellow River. *Front. Earth Sci.* **2021**, *9*, 752425. [CrossRef]
4. Wang, S.; Yan, M.; Yan, Y.; Shi, C.; He, L. Contributions of climate change and human activities to the changes in runoff increment in different sections of the Yellow River. *Quat. Int.* **2012**, *282*, 66–77. [CrossRef]
5. Li, W.; Wang, W.; Wu, Y.; Quan, Q.; Zhao, S.; Zhang, W. Impact of Human Activities on Hydrological Drought Evolution in the Xilin River Basin. *Atmosphere* **2022**, *13*, 2079. [CrossRef]
6. Tennant, D.L. Instream flow regimens for fish, wildlife, recreation and related environmental resources. *Fisheries* **1976**, *1*, 6–10. [CrossRef]
7. Boner, M.C.; Furland, L.P. Seasonal treatment and variable effluent quality based on assimilative capacity. *J. Water Pollut. Control Field* **1982**, *54*, 1408–1416.
8. Foufoula-Georgiou, E.; Kumar, P. *Wavelets in Geophysics*; Academic Press: Cambridge, MA, USA, 1994.
9. Kumar, P.; Foufoula-Georgiou, E. Wavelet analysis for geophysical applications. *Rev. Geophys.* **1997**, *35*, 385–412. [CrossRef]

10. Richter, B.; Baumgartner, J.; Powell, J.; Braunilz, D.P. A method for assessing hydrologic alteration within ecosystems. *Conserv. Biol.* **1996**, *10*, 1163–1174. [CrossRef]
11. Richter, B.; Baumgartner, J.; Wigington, R.; Braun, D. How much water does a river need. *Freshw. Biol.* **1997**, *37*, 231–249. [CrossRef]
12. Beven, K.J. *Distributed Hydrological Modelling: Applications of the TOPMODEL Concept*; John Wiley and Sons Ltd.: Hoboken, NJ, USA, 1997.
13. Martz, L.W.; Garbrecht, J. Numerical definition of drainage network and subcatchment areas from Digital Elevation Models. *Comput. Geosci.* **1992**, *18*, 747–761. [CrossRef]
14. Fodini, E. The ARNO rainfall-runoff model. *J. Hydrol.* **1996**, *175*, 339–382.
15. Beven, K.J.; Calver, A.; Morris, E.M. *The Institute of Hydrology Distributed Model*; Water Resources Publications: Littleton, CO, USA, 1995.
16. Mohammed, H.; Hansen, T.A. Spatial heterogeneity of low flow hydrological alterations in response to climate and land use within the Upper Mississippi River basin. *J. Hydrol.* **2024**, *632*, 130872. [CrossRef]
17. Wang, X.; Ma, W.; Lv, J.; Li, H.; Liu, H.; Mu, G.; Bian, D. Analysis of changes in the hydrological regime in Lalin River basin and its impact on the ecological environment. *Front. Earth Sci.* **2022**, *10*, 987296. [CrossRef]
18. Gao, B.; Li, J.; Wang, X. Analyzing Changes in the Flow Regime of the Yangtze River Using the Eco-Flow Metrics and IHA Metrics. *Water* **2018**, *10*, 1552. [CrossRef]
19. Yan, M.; Fang, G.H.; Dai, L.H.; Tan, Q.F.; Huang, X.F. Optimizing reservoir operation considering downstream ecological demands of water quantity and fluctuation based on IHA parameters. *J. Hydrol.* **2021**, *600*, 126647. [CrossRef]
20. Kannan, N.; Anandhi, A.; Jeong, J. Estimation of Stream Health Using Flow-Based Indices. *Hydrology* **2018**, *5*, 20. [CrossRef]
21. Weidong, W.; Zheng, Y.; Dou, C. Mann-Kendall Mutation Analysis of Temporal Variation of Apparent Stress in Qinba Mountains and Its Adjacent Areas. *IOP Conf. Ser. Earth Environ. Sci.* **2021**, *660*, 012112.

water

MDPI

Article

Simulation and Optimal Scheduling of Water Quality in Urban and Rural Water Supply Systems: A Case Study in the Northwest Arid Region of China

Youjia Zhang [1], Tao Hu [1], Hongqin Xue [2] and Xiaodong Liu [1,*]

[1] Ministry of Education Key Laboratory of Integrated Regulation and Resource Development on Shallow Lakes, College of Environment, Hohai Universities, Nanjing 210098, China; zhangyoujia0923@163.com (Y.Z.); huuuutao@163.com (T.H.)

[2] School of Civil Engineering, Nanjing Forestry University, Nanjing 210037, China; xhq819@163.com

* Correspondence: xdliu@hhu.edu.cn

Abstract: The Northwest Arid Region faces the most serious resource-based water shortage in China, with challenges from engineering-, structural- and management-based water shortages. This water scarcity critically limits the socio-economic development of the region. Rational allocation of scarce water resources to achieve sustainable development of the ecological environment and economy has become a key issue in water resources research in the Northwest Arid Region. South-Central Ningxia, part of the Northwest Arid Region, exemplifies these challenges. This paper examines the urban and rural water supply projects in South-Central Ningxia. The current scheduling scheme focuses primarily on the distribution of water demand, with inadequate attention paid to water-quality requirements. Localized exceedances of water-quality standards indicate the existing scheduling scheme has failed to effectively control water-quality issues while ensuring water quantity. This study is the first to systematically evaluate the impact of the South-Central Ningxia Water Supply Project on water quality alongside field surveys and data analysis and propose an optimized scheduling scheme that addresses both water quantity and quality needs. The main findings are as follows: 1. Overall water quality is good, except for consistently high total nitrogen levels. 2. The optimized scheme significantly reduced total nitrogen levels, achieving a maximum reduction rate of 78.81%, and met all Class III standards.

Keywords: water quantity and quality; scheduling; water transfer projects; environmental impact assessment; optimal allocation of water resources

1. Introduction

Effective and equitable water distribution is an important means for ensuring the sustainable use of water resources [1], and for achieving overall regulation of a basin's water cycle [2]. As water resource shortages and environmental pollution intensify globally, a water resource model focused solely on water quantity allocation no longer meets the needs of society. The integrated allocation of water quantity and quality has become a major concern for the sustainable use of water resources in various countries [3]. However, compared to single-objective dispatching, this combined allocation approach is more complex and requires a comprehensive consideration of water distribution, the assimilation capacity of water functional areas, and the improvement of water efficiency, thereby unifying the mechanisms and joint optimization of water consumption and pollutant capacity [4].

Regarding water quantity and quality evaluation prediction and optimal scheduling, two primary methods are currently used: field monitoring and numerical simulation. Masse first proposed the issue of reservoir optimization scheduling in the 1940s, aiming to achieve a rational allocation of water resources [5]. With advances in computer technology, water resource system analysis, optimization, and simulation technologies have rapidly developed [6]. By

1971, Marks had introduced linear programming for water resource systems, promoting the widespread application of mathematical programming and simulation techniques in water resource optimization scheduling [7]. In 1978, Shafer and Labadie proposed a watershed management model [8], marking a significant breakthrough in the technical application of water resource management. The water resources management model developed by the Canadian Inland Waters Centre, one based on linear programming and network flow algorithms, was successfully applied to the Ottawa River Basin and the Great Lakes region in 1982 [9]. In the late 1980s, research on water resource allocation began incorporating hierarchical theory and gradually shifted towards multi-objective optimization [10]. In 1987, Willis et al. [11] used linear programming to accurately simulate the joint scheduling of surface water, groundwater, and reservoirs and solved the integrated management problem of a single reservoir and its underlying aquifer using the SUMT algorithm. Percia et al. [12] developed a multi-source integrated scheduling model that includes the utilization of groundwater, surface water, and recycled wastewater, aiming to maximize economic benefits. In the 1990s, the application of visualization technology and decision support systems promoted the development of multi-user, multi-objective water resource allocation models [3]. To achieve visualization of water resources decision analysis, Camara et al. [13] developed a multidimensional simulation decision model based on logical relationships and vector calculations. Hamalalnen et al. [14] investigated multi-criteria water resource management and Multi-Stakeholder Decision Support systems. At the end of the 20th century, further development was made in water resources allocation. In 2000, Rosegrant et al. [15] combined hydrological models with economic models to assess the benefits of optimizing water resource allocation and applied this approach to the Maipo River Basin in Chile. In 2002, McKinney et al. [16] proposed a framework for simulating watershed water resource allocation based on Geographic Information System (GIS) technology. The application of GIS technology enabled managers to better understand and analyze the spatial distribution and trends of water resources, leading to decisions which were more scientific. These studies made meaningful attempts at optimizing water resource allocation, applying emerging optimization techniques to water resource system models and providing additional tools and methods for achieving rational water resource allocation and management.

In the early stages of water resource optimization, excessive emphasis was placed on rapid economic and social development, neglecting the crucial attribute of water quality, which led to a failure to address the diverse water-quality requirements of different users and the impact of wastewater discharge on the water environment within the social water cycle [17]. With the advancement of theoretical theories in water resource management, studies on water resource optimization have evolved from solely analyzing water quantity allocation to models integrating both water quantity and quality and have shifted from pursuing economic optimization to seeking overall benefit optimization while paying more attention to the coordinated development of ecology and economy [18]. By 1990, Pingryd et al. [19] had developed a joint water quantity and quality scheduling Decision-Support system to address issues related to water resource allocation and water pollution treatment balance. In 1992, Mehrez [20] employed a nonlinear programming model to establish a multisource water supply system, incorporating various regional reservoirs and groundwater wells. That same year, Afzalet al. [21] used linear programming models to determine irrigation strategies for each crop based on differentiated water quality. In 1997, Avogadro et al. [22] created a water resource planning decision procedure that, considering water-quality constraints, simulated both water quantity and quality. This procedure analyzed the extent to which different allocation schemes met the temporal and spatial water-quality targets and pollutant reduction progress goals within a watershed, thereby determining the optimal water resource allocation scheme. To manage the combined use of surface water and groundwater, Wong [23] incorporated measures to prevent groundwater degradation within a two-step nonlinear optimization model. By 2002, Campbell et al. [24] had coupled the HEC-5C water-quality model with the MODSIM water quantity model to study the impacts of surface water and groundwater diversion mixing and dilution on water quality, and examined the results under different scenarios.

The Northwest Region of China, despite possessing the majority of land and mineral resources, faces severe water scarcity due to poor water resource combination, which constrains its economic development [25]. The increasing population, coupled with rising water demands from industrial and agricultural sectors, exacerbates water pollution and results in the irrational exploitation and utilization of water resources. Consequently, there is a substantial shortfall in socio-economic water use, and the ecological environment in the arid northwest has become increasingly fragile [26]. Ningxia, part of this Northwest Arid Region, exemplifies these challenges. The current water supply projects in Ningxia have significant deficiencies, leading to water-quality exceedances that hinder the region's development. To address these issues, this paper conducts a study on the optimization of the water allocation scheme based on water quantity and quality demand for the South-Central Ningxia Urban and Rural Water Supply Project to ensure regional water security and provide a reference for developing and implementing optimal water resource allocation schemes in other regions.

2. Materials and Methods

2.1. Research Area and Project Overview

The South-Central Ningxia Urban and Rural Water Supply Project (Figure 1) is a water resource optimization and allocation project that transports abundant surface water from the Jing River basin on the eastern slopes of the Liupan Mountains in the south to the arid and water-scarce regions in the central and northern parts of Guyuan. The first intake point of the project is the Longtan Reservoir, located in the source area of the Jing River, a first-order tributary of the Wei River. The project's average annual water intake is 39.8 M m^3, with seven intake points distributed along the route. The main regulating reservoir, Zhongzhuang Reservoir, is located 10 km south of Guyuan in a primary tributary of the Qingshui River. The auxiliary regulating reservoir, Nuanshui River Reservoir, is situated at the outlet of the Qinjia Valley.

2.2. Methodologies

This study evaluates the water quality of the South-Central Ningxia Urban and Rural Water Supply Project and simulates an optimized scheduling scheme based on water quality and water-quantity standards.

2.2.1. Water-Quality Analysis Methods

From 2019 to 2022, intake points were monitored every six months for fluoride, pH, sulfate, dissolved solids, and total hardness, with focused monthly monitoring from March 2020 to February 2021 for 29 factors including temperature, pH, permanganate index, dissolved oxygen, ammonia nitrogen, chloride, and nitrate. Zhongzhuang Reservoir was monitored monthly during the same period for the same 29 factors, providing a comprehensive four-year water-quality assessment. Monitoring sections were set up at each intake point (Table 1), with water quality monitored according to China's "Environmental Quality Standards for Surface Water" (GB3838-2002) [27]. Table 2 outlines the specific testing methods used to monitor various water-quality parameters. All of the data are provided by the Liupanshui Water Authority in Ningxia.

Table 1. Surface water quality monitoring sections at intake points.

Section Name	River	Section Property
Longtan Reservoir	Jing River mainstream	Reservoir Center
Shi Ju Zi	Cedi River mainstream	Intake Point
Hongjia Canyon	Jing River branch	Intake Point
Qingjia Gully	Nuanshui River mainstream	Reservoir Front
Baijia Gully	Nuanshui River branch	Intake Point

Table 1. *Cont.*

Section Name	River	Section Property
Qingshui Gully	Jie River branch	Intake Point
Woyang Valley	Jie River mainstream	Intake Point
Longtan Reservoir	Jing River mainstream	Reservoir Center

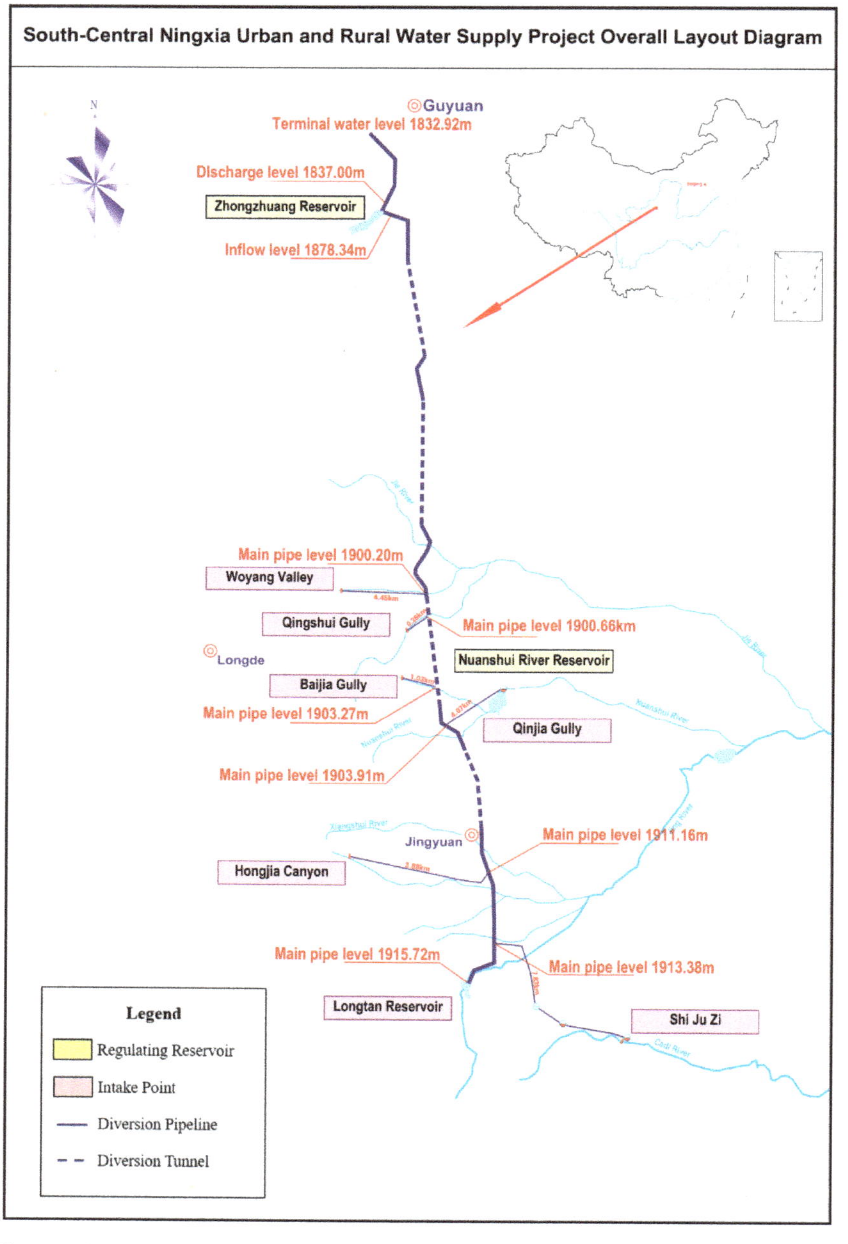

Figure 1. Engineering site location map.

Table 2. Water quality parameters and corresponding testing methods.

Data Type	Testing Method
Permanganate Index (mg/L)	Acid Process
Bio-chemical Oxygen Demand (BOD5, mg/L)	Dilution and Inoculation Test
pH	Glass Electrode Method
Fluoride (mg/L)	Fluoride Reagent Spectrophotometry
Ammonia Nitrogen (mg/L)	Nessler's Reagent Spectrophotometry
Total Phosphorus (mg/L)	Molybdate Spectrophotometry
Nitrate (mg/L)	Phenol Disulfonic Acid Spectrophotometer
Sulfate (mg/L)	Ion Chromatograph
Chloride (mg/L)	Silver Nitrate Titration
Chemical Oxygen Demand (COD)	Dichromate Titration
Total Hardness	EDTA Titration

The equipment used included an Atomic Absorption Spectrophotometer (A3, Shimadzu, Kyoto, Japan); Ion Chromatograph (ICS-90, Dionex, Sunnyvale, CA, USA); Gas Chromatograph (Agilent 7890B GC, Agilent Technologies, Santa Clara, CA, USA); Electronic Balance (BSA224S-CW, Sartorius, Göttingen, Germany); Electro-Optical Analytical Balance (TG328A, Shimadzu, Kyoto, Japan); Spectrophotometer (UV751-GD model, Shanghai Metash Instruments Co., Ltd., Shanghai, China); Spectrophotometer (721 model, Shanghai Analytical Instrument Factory, Shanghai, China); Spectrophotometer (722S model, Shanghai Precision & Scientific Instrument Co., Ltd., Shanghai, China); Infrared Oil Meter (IR-200A, Infralyt, Berlin, Germany); pH Meter (PB-10 model, Sartorius, Göttingen, Germany); Conductivity Meter (DDS-307A, Shanghai INESA Scientific Instrument Co., Ltd., Shanghai, China); Atomic Absorption Spectrophotometer (4530F, Beijing Ruili Analytical Instrument Co., Ltd., Beijing, China).

2.2.2. Simulation of Water Supply System Optimization Model

The optimization of the water supply system was simulated using the Storm Water Management Model (SWMM Version 5.1), which is specifically designed to simulate urban hydrological and hydraulic conditions. SWMM effectively handles various flow scenarios such as surface runoff, infiltration, pipe network flow, and river flow [28]. In this model, the drainage system is represented by a network of nodes and connecting pipes, in which each node can serve as an inflow point, confluence point, or outflow point. The model accurately tracks and simulates the quantity and quality of runoff generated by each sub-basin across different time steps, as well as the flow and water-quality changes within pipes and tunnels.

Pipe Network Water Quantity Model

SWMM simulates water flow movement based on the Saint-Venant equations for unsteady free-surface flow, which conserves mass and momentum [29]:

$$\frac{\partial A}{\partial t} + \frac{\partial Q}{\partial x} = 0 \tag{1}$$

$$\frac{\partial Q}{\partial t} + \frac{\partial\left(\frac{Q^2}{A}\right)}{\partial x} + gA\frac{\partial H}{\partial x} + gAS_f = 0 \tag{2}$$

A is the cross-sectional area of flow; t is the time coordinate; Q is the flow rate; x is the spatial coordinate; H is the hydraulic head; S_f is the friction angle; g is the gravitational acceleration.

Nodes in the system are categorized into non-storage types (such as junction nodes) and storage types (such as ponds and tanks). The model ensures flow conservation at system nodes using the node continuity equation, where the total area of a node is composed of its storage surface area and the surface area contributions of connected pipes [30]. The change in the hydraulic head at a node is approximated by the following equation [31]:

$$\frac{dH}{dt} = \frac{\sum Q_{in} - \sum Q_{out}}{A} \tag{3}$$

where Q_{in} and Q_{out} are inflow and outflow rates, and A is the total area of the node.

The model uses the surface areas of the node and connected pipes in a finite difference computation with an iterative method, solving implicit solutions at a set time step. Outflow boundary conditions are user-defined and can be constants, time-series extractions, or based on critical or normal flow depths [32].

Pipe Network Water-Quality Model

In the SWMM model, the concentration change of dissolved components along the pipe is simulated through the mass conservation equation:

$$\frac{\partial c}{\partial t} = -\frac{\partial (uc)}{\partial x} + \frac{\partial}{\partial x}\left(D\frac{\partial c}{\partial x}\right) + r(c) \tag{4}$$

c is the constituent concentration (ML^{-3}); u is the longitudinal velocity (LT^{-1}); D is the longitudinal dispersion coefficient; and $r(c)$ is the reaction rate term.

Boundary conditions in SWMM are defined by the water-quality concentration at the end nodes of the transport network [31]. For non-storage nodes, the concentration is the flow-weighted average of inflow and outflow masses. For storage nodes, complete mixing is assumed, and concentrations are updated using a simplified mixing equation based on the mass conservation equation:

$$c(t+\Delta t) = \frac{c(t)V(t)e^{-K_1\Delta t} + C_{in}Q_{in}\Delta t}{V(t) + Q_{in}\Delta t}. \tag{5}$$

$V(t)$ is the water volume in the reactor and $e^{-K_1\Delta t}$ is the decay factor, with K_1 being the first-order reaction rate constant.

Model Calibration Validation

The model was calibrated using data collected from the water source intake points on 8 April 2020. Calculations were performed to determine the water volumes and pollutant distributions at various depths and concentrations along different pipelines over the corresponding period. The simulation results show an average relative error of only 3.8% compared to the measured data.

Further validation, using monitoring results from 9–17 April 2020, showed an average error of just 1% (Figure 2). These results indicate that the model has high precision and reliability and can be used for joint optimization and scheduling simulation of water quantity and water quality.

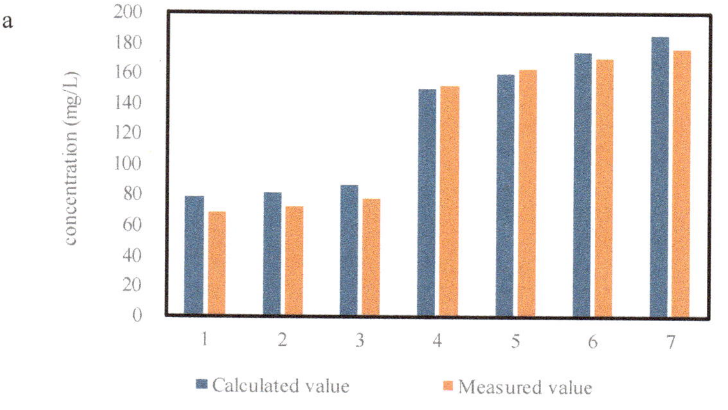

Figure 2. *Cont.*

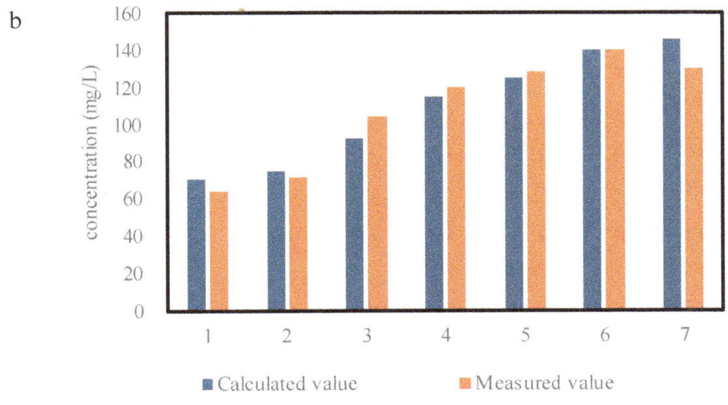

Figure 2. Model calibration and verification at intake points: (**a**) calibration results and (**b**) validation results.

3. Results and Discussion

3.1. Overall Water Quality

Figure 3 illustrates the basic water-quality parameters along the project route. Monthly monitoring data were collected from each intake point of the water supply project and the Zhongzhuang reservoir from March 2020 to February 2021. The water temperature ranged from 1 °C to 24.2 °C, with a median of 12.0 °C. Dissolved oxygen levels were between 6.15 mg/L and 11.01 mg/L (median: 8.4 mg/L). The permanganate index varied from 0.73 mg/L to 3.54 mg/L (median: 1.25 mg/L), while biochemical oxygen demand (BOD) was from 0.03 mg/L to 2.44 mg/L (median: 0.82 mg/L). The pH values ranged from 6.99 to 9.37, with a median of 8.33. For total phosphorus and fluoride, some sections and periods showed values below detection limits, with medians of 0.03 mg/L and 0.23 mg/L, respectively. Ammonia nitrogen concentrations ranged from 0.01 mg/L to 0.68 mg/L (median: 0.1 mg/L). Total nitrogen levels were between 0.33 mg/L and 6.08 mg/L (median: 1.30 mg/L). Nitrate concentrations varied from 0.2 mg/L to 5.92 mg/L (median: 1.135 mg/L). Chloride levels ranged from 6 mg/L to 157 mg/L, with a median of 15 mg/L, and sulfate levels ranged from 8 mg/L to 741 mg/L, with a median of 124.5 mg/L.

When compared with the concentration limits of the Surface Water Environmental Quality Standards, the levels of dissolved oxygen, permanganate index, biochemical oxygen demand (BOD), pH, fluoride, nitrate, and chloride were mostly at Class I levels. Approximately 80% of sulfate measurements were at Class I levels. As for total phosphorus, about 60% of measurements were at Class II levels and 40% at Class I. Ammonia nitrogen was at Class I levels for around 70% of the measurements, with 30% at Class II. Total nitrogen was generally at Class IV levels, but this is typically not used as an assessment indicator. Overall, the water quality over the monitoring period of one year was good, with most key indicators meeting Class II standards or higher.

3.2. Water-Quality Investigation and Evaluation of Zhongzhuang Reservoir

To understand the current water quality of the water source in the project area, a surface-water-quality survey of Zhongzhuang Reservoir was conducted. Monitoring was carried out monthly from 2019 to 2022. The results (Figure 4) indicated that the pH value (8.0), permanganate index concentration (1.20–1.80 mg/L), five-day biochemical oxygen demand (BOD_5) concentration (0.71–1.77 mg/L), ammonia nitrogen (NH_3-N) concentration (0.04–0.21 mg/L), chemical oxygen demand (COD) concentration (3.75–11.03 mg/L), dissolved oxygen concentration (8.00 mg/L), total phosphorus (as P) concentration (0–0.04 mg/L), and sulfate (SO_4^{2-}) concentration (89–125 mg/L) all met the Class III water-quality standards set by the "Environmental Quality Standards for Surface Water". However, the total nitrogen (as N) concentration (0.84–1.40 mg/L) consistently exceeded the standard but showed a fluctuating downward trend, indicating the potential to meet the Class III water-quality standards in the future. Dissolved oxygen concentration exhibited clear seasonal variation, being lower in summer and higher in winter, while the permanganate index and ammonia nitrogen concentrations were higher in summer and lower in winter. Other indicators did not show significant seasonal variations throughout the year.

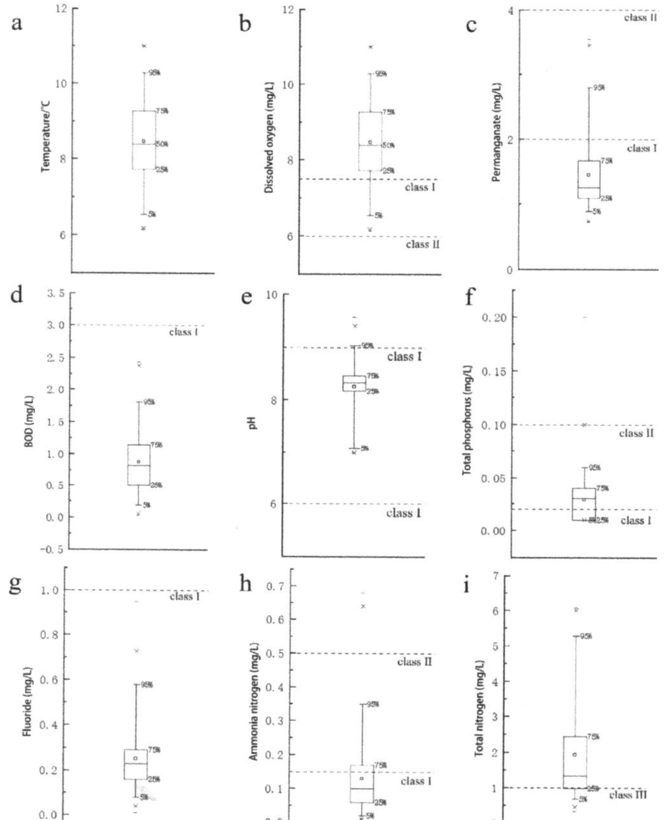

Figure 3. Water-quality parameter comparison: (**a**) Temperature (°C); (**b**) Dissolved Oxygen (mg/L); (**c**) Permanganate Index (mg/L); (**d**) Biochemical Oxygen Demand (BOD5, mg/L); (**e**) pH; (**f**) Fluoride (mg/L); (**g**) Ammonia Nitrogen (NH_3-N, mg/L); (**h**) Total Phosphorus (mg/L); (**i**) Nitrate (mg/L); (**j**) Total Nitrogen (mg/L); (**k**) Sulfate (SO_4^{2-}, mg/L); (**l**) Chloride (mg/L), as compared to standards in the urban and rural areas of southern Ningxia.

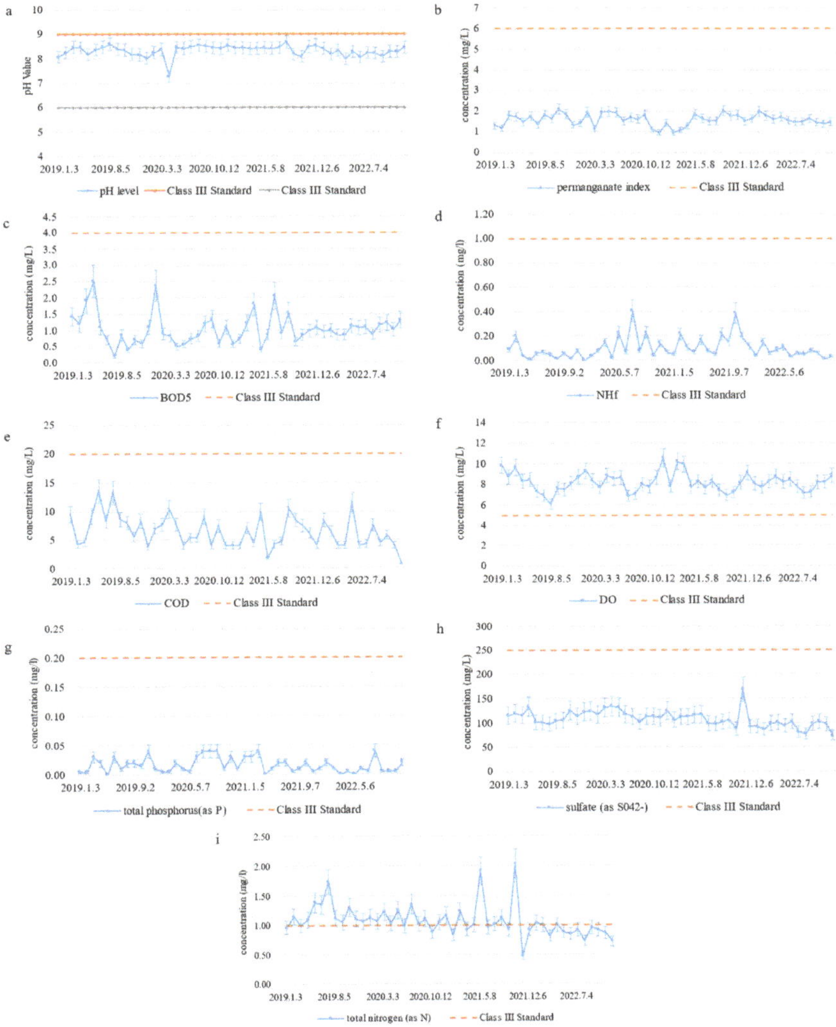

Figure 4. Water-quality trends for (**a**) pH Values; (**b**) Permanganate Index; (**c**) Biochemical Oxygen Demand (BOD5); (**d**) Ammonia Nitrogen (NH_3-N); (**e**) Chemical Oxygen Demand (COD); (**f**) Dissolved Oxygen (DO); (**g**) Total Phosphorus; (**h**) Sulfate (SO_4^{2-}); and (**i**) Total Nitrogen at Zhongzhuang Reservoir.

3.3. Water-Quality Investigation and Evaluation of Intake Points

Based on China's "Environmental Quality Standards for Surface Water" (GB3838-2002) [27], water-quality monitoring was conducted at various intake points over four years from 2019 to 2022, with monitoring conducted every 6 months. Additionally, from March 2019 to March 2021, intensive monthly monitoring focused on Total Nitrogen (TN) as a key factor.

3.3.1. Water-Quality Changes from 2019 to 2022

The pH monitoring results (Figure 5a) show that all monitoring points exhibited similar variation patterns, with most results meeting the Class III water-quality standards. Only Shi Ju Zi and Longtan Reservoir sections exceeded the standards, and only for a few months. The pH values peaked and hit their lowest point in April 2020, with Longtan Reservoir reaching its peak and other points, including Baijia Gully, hitting their lowest. Baijia Gully recorded the lowest pH value, at 7.04. Fluoride concentration monitoring results (Figure 5b) indicate that most levels were far below Class III standards, except for the Baijia Gully section in August 2019, where the fluoride levels were close to the upper limit of Class III standards. Dissolved total-solids pollution was relatively severe across the sections (Figure 5c), and these concentrations were generally higher in the autumn across various sections. Baijia Gully section was the most prominent, showing concentrations between 947 and 1586 mg/L, except for the August 2019 measurement, while all other measurements failed to meet Class III standards. Baijia Gully section also exhibited significant total hardness pollution (214–625 mg/L) (Figure 5d) and sulfate concentration pollution (300–621 mg/L) (Figure 5e), while these types of pollution were relatively less severe at other sections.

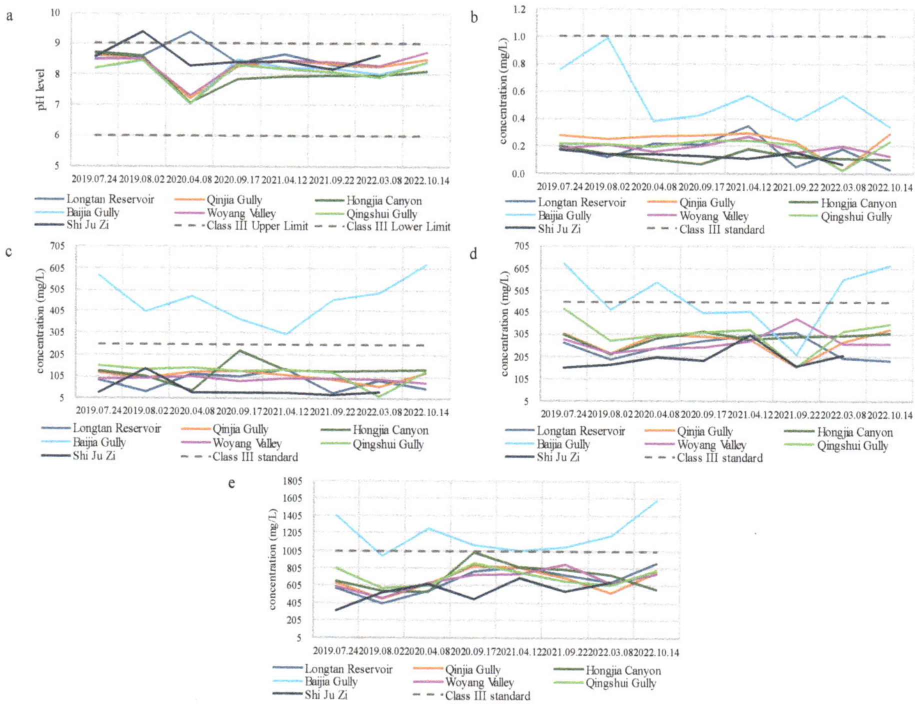

Figure 5. Water-quality trends of (**a**) pH Values; (**b**) Fluoride Concentration; (**c**) Total Dissolved Solid Concentration; (**d**) Total Hardness; and (**e**) Sulfate Concentration at different intake points.

3.3.2. Water-Quality Trend of Key Factor (TN) from 2020 to 2021

Based on the overall water-quality changes, further monitoring and analysis of Total Nitrogen (TN) concentrations were conducted from 2020 to 2021 (Figure 6). The results indicate that, except for the Hongjia Canyon section, most sections exhibited serious TN pollution, failing to meet the Class III water-quality standards set by the "Environmental Quality Standards for Surface Water" (GB3838-2002) [27]. The concentration of total nitrogen showed obvious differences between the different points, and the fluctuation was significant. The Qingshui Gully and Woyang Valley sections were particularly affected, with all measurements falling short of the Class III standards. In contrast, the Hongjia Canyon section showed relatively low TN pollution, meeting the Class III standards in all monitoring results except for June 2020.

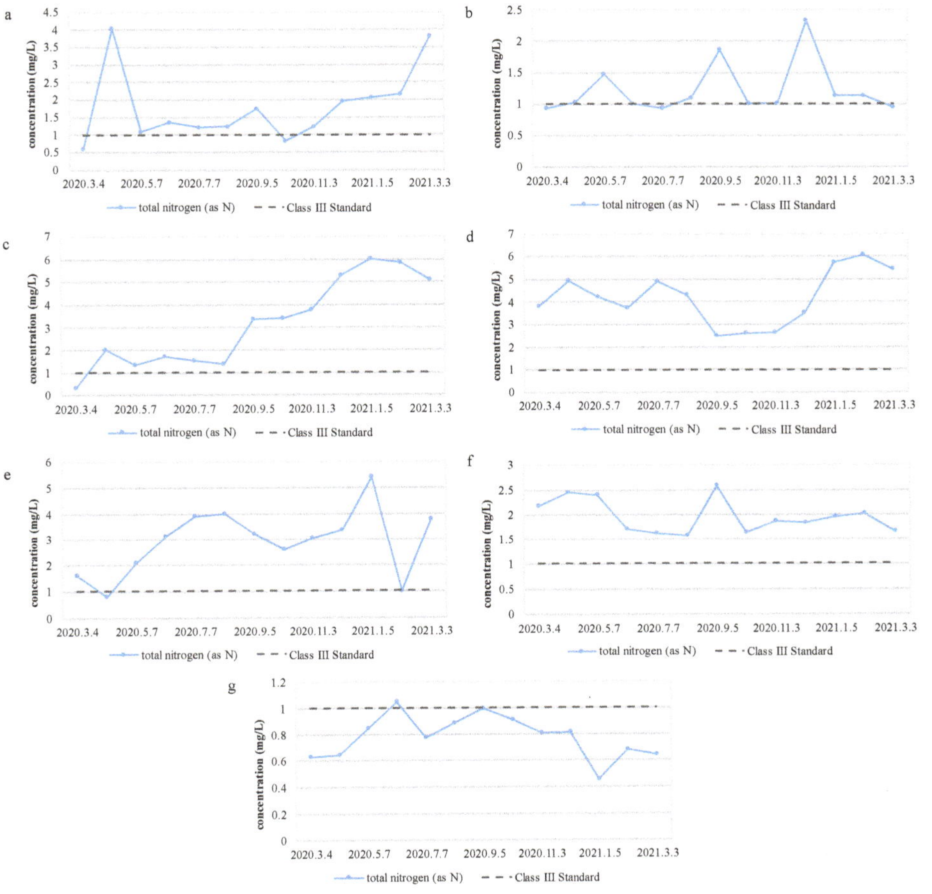

Figure 6. Water-quality trends of total nitrogen (as N) concentration at (**a**) Baijia Gully; (**b**) Longtan Reservoir; (**c**) Qinijia Gully; (**d**) Qingshui Gully; (**e**) Shi Ju Zi; (**f**) Woyang Valley; and (**g**) Hongjia Canyon.

3.4. Optimization Scheduling of the Water Supply System

3.4.1. Design Schemes

The water intake volume determined in the "Preliminary Design of Urban and Rural Drinking Water Safety Source Project 2012" was used as the baseline scheduling scheme. To address issues such as the widespread exceedance of Total Nitrogen (TN) concentrations at some intake points, an optimized scheme was designed by adjusting the water intake volume. The optimization schemes follow the principle of "more water diversion in wet season, less water diversion in normal season, and no water diversion, as far as possible, in dry season", and all intake points and reservoirs give priority to ensuring 10% ecological water quantity in the river and water demand outside the river. According to water-quality monitoring data, when the TN concentration at an intake point exceeds 3 mg/L, water intake at that point will cease, and its flow will be redistributed among other intake points to ensure the total diverted water volume remains unchanged (Optimization Scheme 1). If the TN concentration exceeds 2 mg/L, water intake at that point will cease (Optimization Scheme 2). Although the Class III standard for TN concentration is 1 mg/L, the threshold values of 2 mg/L and 3 mg/L are thought to be responsible for degradation of TN concentrations along the way.

The water intake volumes for the three design schemes are shown in Table 3.

Table 3. Optimization of water intake scheme (m^3/s).

Scheme	Month	Shi Ju Zi	Longtan Reservoir	Hongjia Canyon	Qinjia Gully	Baijia Gully	Qingshui Gully	Woyang Valley	Total
Baseline Scheme	1	28	74	8	42	7	11	5	175
	2	23	54	6	40	7	10	4	144
	3	26	65	7	48	8	13	5	171
	4	47	139	16	42	7	11	4	266
	5	56	195	18	40	6	11	20	345
	6	47	168	17	42	6	13	21	314
	7	59	266	29	66	8	28	35	492
	8	51	288	32	78	9	40	45	544
	9	51	278	33	90	10	52	55	568
	10	33	183	20	78	9	40	43	405
	11	36	142	20	72	10	28	20	327
	12	24	101	13	55	8	17	11	229
	total	481	1950	218	692	96	275	268	3980
Optimization Scheme 1	1	0	74	8	0	7	0	5	94
	2	23	54	6	0	7	0	4	94
	3	0	65	7	0	0	0	5	77
	4	68	139	16	53	0	0	4	280
	5	61	195	18	129	6	0	20	429
	6	0	168	17	119	6	0	21	571
	7	0	313	29	94	8	0	35	479
	8	0	376	32	109	9	0	45	571
	9	0	478	33	0	10	52	55	628
	10	33	271	20	0	9	40	43	416
	11	0	217	26	0	12	44	36	335
	12	0	214	13	0	8	0	11	246
	total	185	2564	225	504	82	136	284	3980
Optimization Scheme 2	1	0	74	8	0	0	0	5	87
	2	23	54	6	0	0	0	4	87
	3	0	65	7	0	0	0	17	89
	4	79	193	22	0	0	0	0	294
	5	68	195	18	131	13	0	0	425
	6	0	192	17	91	14	0	48	362
	7	0	353	34	131	13	0	53	584
	8	0	375	37	144	14	0	66	636
	9	0	539	58	0	13	0	0	610
	10	0	279	29	0	12	0	45	365
	11	0	219	26	0	13	0	36	294
	12	0	0	26	0	14	0	0	40
	total	170	2538	288	497	106	0	274	3980

3.4.2. Simulation Analysis of Scheduling Scheme

Zhongzhuang Reservoir Simulation Results Analysis

The annual Total Nitrogen (TN) concentration simulation results of different scheduling schemes for Zhongzhuang Reservoir are shown in Figure 7, and the monthly average TN concentration simulation results are presented in Table 4. Using the baseline scheduling scheme, the predicted TN concentration in Zhongzhuang Reservoir exceeded the Class III standard for 192 days, with a non-compliance rate of 52.89%. The baseline scheme simulation results display significant TN concentration peaks, particularly in January, September, and December.

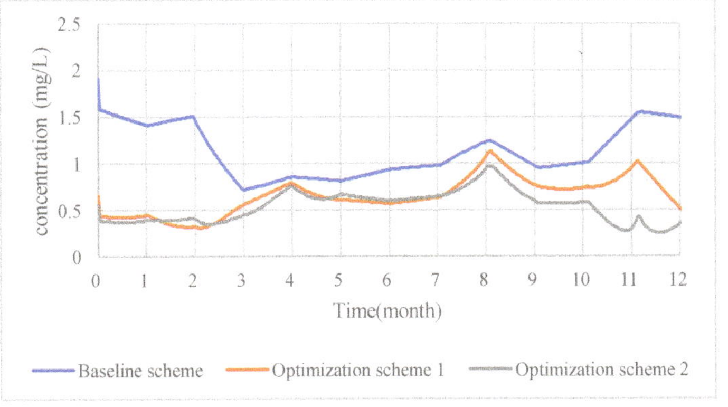

Figure 7. Annual total nitrogen (TN) concentration (mg/L): simulation results for Zhongzhuang Reservoir under different schemes.

Table 4. Monthly average total nitrogen (TN) concentration simulation result (mg/L) for Zhongzhuang Reservoir under different schemes.

Month	Baseline Scheme	Optimization Scheme 1	Optimization Scheme 2	Class Standard III
1	1.51	0.44	0.38	1
2	1.46	0.36	0.39	1
3	1.04	0.41	0.38	1
4	0.79	0.68	0.59	1
5	0.84	0.67	0.65	1
6	0.88	0.59	0.63	1
7	0.96	0.60	0.62	1
8	1.11	0.84	0.76	1
9	1.11	0.99	0.78	1
10	0.98	0.73	0.57	1
11	1.22	0.80	0.39	1
12	1.52	0.76	0.31	1

The simulation results of Optimization Schemes 1 and 2 indicate that by reasonably adjusting the intake volumes and water diversion month at the intake points, the TN concentration in Zhongzhuang Reservoir can be effectively reduced, leading to a more stable annual water quality. Particularly during the autumn and winter seasons, the annual predicted TN concentration shows a more noticeable decrease. Specifically, in Optimization Scheme 1, the annual TN concentration exhibited a noticeable downward trend, with February showing the most significant change, achieving a maximum reduction rate of 78.81%. The predicted maximum reduction rate reached 78.81%, and the monthly average concentration decreased from 1.46 mg/L to 0.36 mg/L, a difference of 1.1 mg/L. However, the reduction during the spring and summer seasons was minimal, with 21 days of TN concentration exceeding the standard in August and September, while the maximum exceedance concentration reached 1.19 mg/L. The average monthly TN concentration in September was 0.99 mg/L, slightly below the standard. In contrast, the average monthly TN concentration under Optimization Scheme 2 was only 0.76 mg/L in September. The predicted maximum annual reduction rate reached 83.66%, and all simulation results met the Class III standard. In critical months like October to December, Optimization Scheme 2 avoided significant TN concentration peaks. In December, the average monthly TN concentration was 0.31 mg/L under Scheme 2, compared to 1.52 mg/L under the baseline scheme. These lower concentrations in the autumn months help reduce the risk of water-quality deterioration during the dry season (January to March), ensuring better management of potential pollution peaks. Therefore, Optimization Scheme 2 is recommended for Zhongzhuang Reservoir to ensure compliance with water-quality standards and sustainable use of the water source.

Simulation Analysis of Water Diversion Tunnels

The simulated section location of the water diversion tunnel is shown in Figure 8. To understand the water-quality distribution along the water diversion system, a simulation analysis was conducted for the water transfer tunnels (Table 5). The simulation results for Optimization Scheme 1 indicate significant water-quality improvements in Tunnels 1, 4, 6, and 8. Compared to the baseline scheme, the number of days exceeding the standard in Tunnel 1 decreased from 317 to 135 days. In Tunnel 4, the exceedance days were reduced to 104, dropping the exceedance rate to 28.57%. For Tunnel 6, the exceedance days decreased to 131, with the rate falling to 35.99%. Tunnel 8, the most downstream, showed a substantial reduction to 46 exceedance days, lowering the rate to 12.64%. Further comparison of the simulation results for Optimization Scheme 2 reveals even more obvious improvements in Tunnels 1, 4, 6, and 8. In Tunnel 1, exceedance days were reduced to 83, with an exceedance rate of 22.80%. For Tunnel 4, the exceedance days dropped to 57, lowering the exceedance rate to 15.66%. In Tunnel 6, the exceedance days were reduced to 34, and the exceedance rate was 9.43%. Finally, in Tunnel 8, the exceedance days significantly decreased to 11, with an exceedance rate of only 3.02%. In summary, Optimization Scheme 1 and Optimization Scheme 2 demonstrated obvious advantages over the baseline scheme in improving water quality, with Optimization Scheme 2 maintaining much lower exceedance rates and demonstrating improvements which were remarkable and more stable.

Table 5. Simulation results for water intake tunnel.

Tunnel Number	Baseline Scheme		Optimization Scheme 1		Optimization Scheme 2	
	Exceedance Days	Exceedance Rate	Exceedance Days	Exceedance Rate	Exceedance Rate	Exceedance Rate
1	317	87.09%	135	37.09%	83	22.80%
4	347	95.33%	104	28.57%	57	15.66%
6	364	100%	131	35.99%	34	9.43%
8	282	77.47%	46	12.64%	11	3.02%

The annual simulation results of the cross-sections of the water diversion tunnels under different scheduling schemes are shown in Figure 9. Optimization Scheme 2 demonstrates a more stable degradation of total nitrogen (TN) along the route, leading to a consistent improvement in water quality. Both the baseline scheme and optimization Scheme 1 exhibit an increase in TN concentration along Tunnel 4, mainly due to the severely high TN concentrations from the Baijia Valley intake point. During October to December, these two schemes experienced water-quality deterioration and a surge in TN concentration, with consistently high TN peaks in September. Optimization Scheme 2 effectively addresses these issues by adjusting the water intake scheme, resulting in a more stable TN degradation along the route and significant improvement in water quality. Compared to the baseline scheme and Optimization Scheme 1, Optimization Scheme 2 achieved more noticeable reductions in TN concentration from upstream (Tunnel 1) to downstream (Tunnel 8) in October to December, indicating that early interventions effectively prevented high-TN water from entering the system, resulting in a more stable water quality across all tunnels. Starting from Tunnel 6 in September, the TN concentration peaks were significantly lower in Optimization Scheme 2. The average monthly TN concentration in September was reduced by 0.47 mg/L compared to the baseline scheme, whereas Optimization Scheme 1 only achieved a reduction of 0.21 mg/L. This indicates that Optimization Scheme 2's overall reduction strategy is robust, with cumulative effects becoming significant downstream. Even if some TN enters the system, it can be effectively managed and reduced as the water flows through each tunnel.

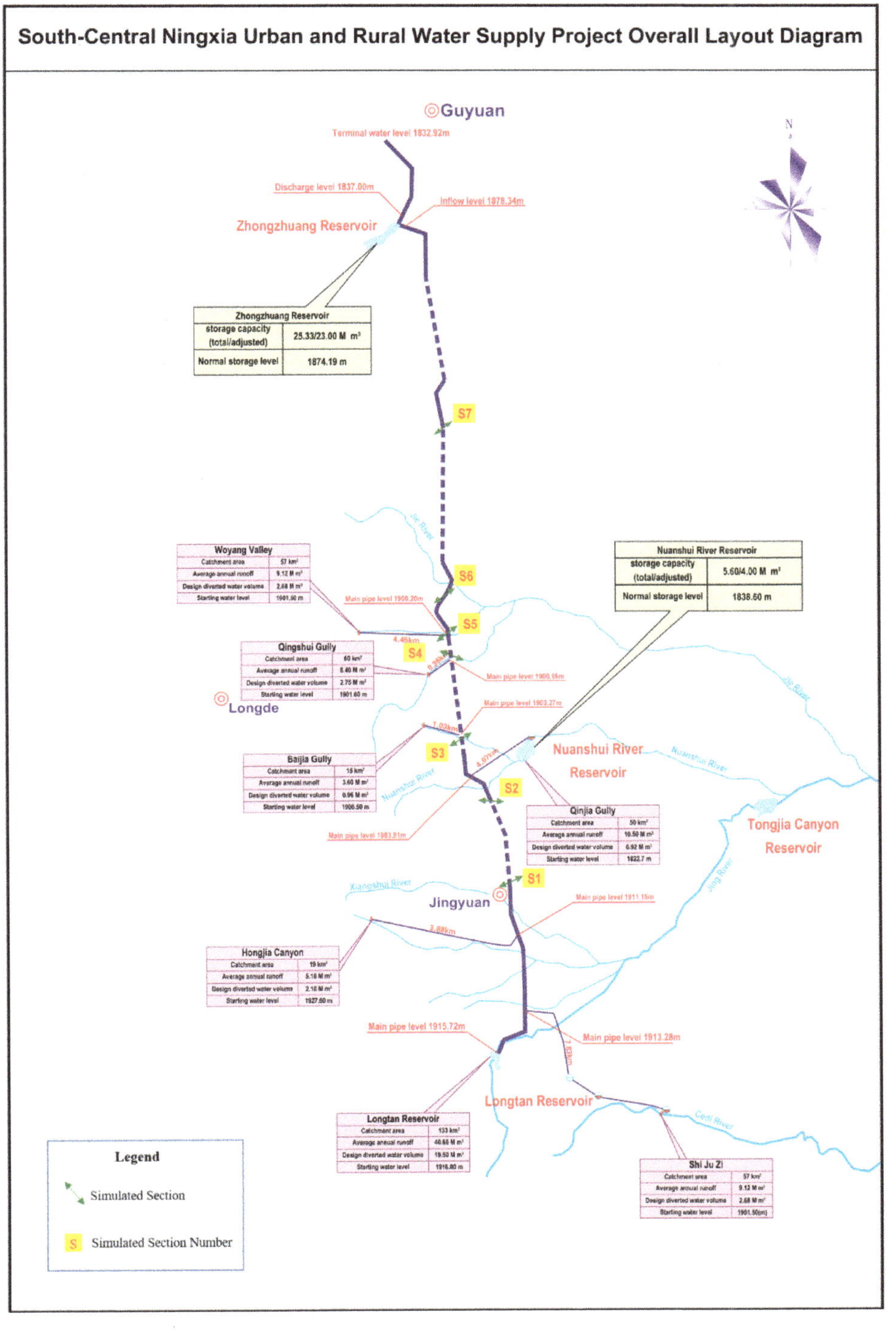

Figure 8. Simulated cross-section location of water intake tunnel.

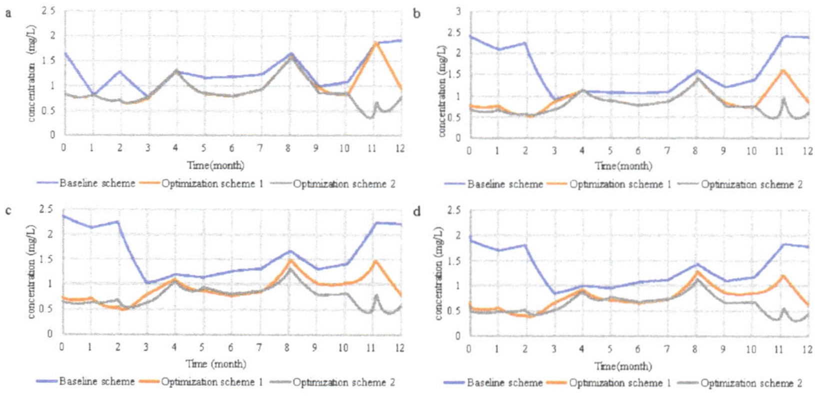

Figure 9. Simulation results for (**a**) No.1; (**b**) No.4; (**c**) No.6; and (**d**) No.8 intake tunnels under different schemes.

The key factors contributing to the superiority of Optimization Scheme 2 include strategic increases in water diversion during critical months and effective redistribution of water in high-TN areas. In August, the TN concentration in Qinjia Gully approached the Class III standard, and Optimization Scheme 2 increased the water diversion from Qinjia Gully (from 109 to 144 m3/s), coupled with a significant increase in diversion in September (from 568 to 628 m3/s). These measures ensured that low-N water diluted the TN concentration, helping to flush out accumulated pollutants and maintain better water quality. These actions ensured that TN levels remained low not only in August and September but also in the following months. This further proves the superiority of Optimization Scheme 2 over Scheme 1, indicating that Optimization Scheme 2 is more suitable as the final optimized scheme to ensure improved water quality and reliable scheduling effects.

In summary, Optimization Scheme 2 provides stable and consistent TN degradation, effectively solving the water-quality decline caused by excessive TN concentrations at the Baijia Valley intake point. However, this study focused solely on the optimization of total nitrogen, without addressing other water-quality indicators. While other indicators generally met quality standards, some monitoring points exhibited short-term exceedances. Future studies should comprehensively consider multiple indicators, including pH, total dissolved solids, and sulfates. Additionally, the improvements in water quality achieved by altering water transfer plans are temporary. Long-term solutions should involve proactive protection of the water source area to prevent water-quality issues from occurring at the source.

4. Conclusions

Using the South-Central Ningxia Urban and Rural Water Supply Project as a case study, this research employs the Storm Water Management Model to develop an optimized scheduling scheme based on water quantity and quality demand, following a thorough evaluation of the water quality at intake points and the water source. The main conclusions are as follows:

1. Water-Quality Evaluation of Intake Points: The results indicate that the overall water quality in the intake area is good, with most indicators meeting Class III water-quality standards. However, there are instances of excessive total nitrogen and sulfate levels, particularly in Baijia Valley, where sulfate and dissolved solids concentrations exceed the standards to a considerable extent. Therefore, further efforts are needed to enhance water environment management and governance.

2. Water-Quality Evaluation of Zhongzhuang Reservoir: The results show that the overall water quality of Zhongzhuang Reservoir is good, except for consistently high total nitrogen levels. Other monitored factors meet Class III water-quality standards. After water from the intake points mixes and degrades along the route, the total nitrogen concentration upon reaching Zhongzhuang Reservoir is close to the Class III standard.
3. Water-Quality Simulation Results: The simulation results reveal that using the design water intake volume specified in the "Preliminary Design of Urban and Rural Drinking Water Safety Source Project 2012," the predicted annual total nitrogen concentration in Zhongzhuang Reservoir exceeds the standards throughout the year, with an over-standard rate of up to 52.89%. After the optimization scheme was adopted, the annual predicted total nitrogen concentration in Zhongzhuang Reservoir significantly decreased, with the maximum reduction rate reaching 78.81% and all simulation results meeting the Class III standards of the "Environmental Quality Standards for Surface Water".

In conclusion, according to the "Technical guideline for delineating source water protection areas" (HJ 338-2018) [33], issued by theMinistry of Ecology and Environment of the People's Republic of China, it is essential to scientifically delineate each intake point and regulate reservoirs as water source protection areas and give them priority protection. Additionally, there is a need to strengthen environmental risk assessments for water sources, which includes screening potential risk sources and identifying potential risk types, as well as assessing their risk levels. Finally, a comprehensive water-quality monitoring and early warning system needs to be established, including regular patrols of water bodies within the protection areas and continuous water-quality monitoring to prevent pollution. Implementing these measures will likely enhance scientific management and scheduling and provide valuable insights for other regions, promoting the sustainable utilization of regional water resources.

Author Contributions: Conceptualization, X.L. and Y.Z.; methodology, Y.Z.; software, T.H.; validation, T.H. and H.X.; investigation, H.X.; data curation, Y.Z.; writing—original draft preparation, Y.Z.; writing—review and editing, X.L.; visualization, Y.Z.; supervision, X.L. All authors have read and agreed to the published version of the manuscript.

Funding: This research received no external funding.

Data Availability Statement: Data are contained within the article.

Conflicts of Interest: The authors declare no conflicts of interest.

References

1. Ali, M.A.; Kamraju, M. Water Resources Allocation and Governance. In *Natural Resources and Society: Understanding the Complex Relationship Between Humans and the Environment*; Springer Nature: Cham, Switzerland, 2023; pp. 99–113, ISBN 978-3-031-46720-2.
2. Ray, R.L.; Abeysingha, N.S.; Ray, R.L.; Abeysingha, N.S. Introductory Chapter: Water Resources Planning, Monitoring, Conservation, and Management. In *River Basin Management-Under a Changing Climate*; Ram, L.R., Dionysia, P., Nimal, A., Eds.; IntechOpen: London, UK, 2023; ISBN 978-1-80355-559-1.
3. Zhang, S.; Hou, L.; Wei, C.; Zhou, X.; Wei, N. Study on Water Quantity and Quality-Integrated Evaluation Based on the Natural-Social Dualistic Water Cycle. *Pol. J. Environ. Stud.* **2015**, *24*, 829–840. [CrossRef] [PubMed]
4. Guo, X.; Wu, Z.; Wang, X.; Lv, C.; Gu, C.; Li, Y.; Gao, M. The Joint Optimal Allocation Study of Regional Total Water Consumption and Pollutant Carrying Capacity of Water Function Areas Based on Emergy Theory. *Water* **2020**, *12*, 1101. [CrossRef]
5. Huang, X.; Fang, G.; Gao, Y.; Dong, Q. Chaotic Optimal Operation of Hydropower Station with Ecology Consideration. *Energy Power Eng.* **2010**, *2*, 182–189. [CrossRef]
6. Loucks, D.P.; van Beek, E. Water Resource Systems Modeling: Its Role in Planning and Management. In *Water Resource Systems Planning and Management: An Introduction to Methods, Models, and Applications*; Loucks, D.P., van Beek, E., Eds.; Springer International Publishing: Cham, Switzerland, 2017; pp. 51–72, ISBN 978-3-319-44234-1.
7. Cohon, J.L.; Marks, D.H. A Review and Evaluation of Multiobjective Programing Techniques. *Water Resour. Res.* **1975**, *11*, 208–220. [CrossRef]
8. Shafer, J.; Labadie, J. Synthesis and Calibration of a River Basin Water Management Model. 3 January 2007. Available online: https://api.semanticscholar.org/CorpusID:134890146 (accessed on 16 April 2023).

9. Li, X.; Liu, P.; Cheng, L.; Cheng, Q.; Zhang, W.; Xu, S.; Zheng, Y. Strategic Bidding for a Hydro-Wind-Photovoltaic Hybrid System Considering the Profit beyond Forecast Time. *Renew. Energy* **2023**, *204*, 277–289. [CrossRef]
10. Yao, Z.; Wang, Z.; Cui, X.; Zhao, H. Research on Multi-Objective Optimal Allocation of Regional Water Resources Based on Improved Sparrow Search Algorithm. *J. Hydroinform.* **2023**, *25*, 1413–1437. [CrossRef]
11. Willis, R.; Yeh, W. *Groundwater Systems Planning and Management*; Pearson College Div: Upper Saddle River, NJ, USA, 1987.
12. Percia, C.; Oron, G.; Mehrez, A. Optimal Operation of Regional System with Diverse Water Quality Sources. *J. Water Resour. Plann. Manag.* **1997**, *123*, 105–115. [CrossRef]
13. Camara, A.S.; Ferreira, F.C.; Loucks, D.P.; Seixas, M.J. Multidimensional Simulation Applied to Water Resources Management. *Water Resour. Res.* **1990**, *26*, 1877–1886. [CrossRef]
14. Hämäläinen, R.; Kettunen, E.; Marttunen, M.; Ehtamo, H. Evaluating a Framework for Multi-Stakeholder Decision Support in Water Resources Management. *Group Decis. Negot.* **2001**, *10*, 331–353. [CrossRef]
15. Rosegrant, M.W.; Ringler, C.; McKinney, D.C.; Cai, X.; Keller, A.; Donoso, G. Integrated Economic–Hydrologic Water Modeling at the Basin Scale: The Maipo River Basin. *Agric. Econ.* **2000**, *24*, 33–46. [CrossRef]
16. McKinney, D.C.; Cai, X. Linking GIS and Water Resources Management Models: An Object-Oriented Method. *Environ. Model. Softw.* **2002**, *17*, 413–425. [CrossRef]
17. Kang, A.; Li, J.; Lei, X.; Ye, M. Optimal Allocation of Water Resources Considering Water Quality and the Absorbing Pollution Capacity of Water. *Water Resour.* **2020**, *47*, 336–347. [CrossRef]
18. Wang, G.; Mang, S.; Cai, H.; Liu, S.; Zhang, Z.; Wang, L.; Innes, J.L. Integrated Watershed Management: Evolution, Development and Emerging Trends. *J. For. Res.* **2016**, *27*, 967–994. [CrossRef]
19. Pingry, D.E.; Shaftel, T.L.; Boles, K.E. Role for Decision-Support Systems in Water-Delivery Design. *J. Water Resour. Plan. Manag.* **1991**, *117*, 629–644. [CrossRef]
20. Mehrez, A.; Percia, C.; Oron, G. Optimal Operation of a Multisource and Multiquality Regional Water System. *Water Resour. Res.* **1992**, *28*, 1199–1206. [CrossRef]
21. Afzal, J.; Noble, D.H.; Weatherhead, E.K. Optimization Model for Alternative Use of Different Quality Irrigation Waters. *J. Irrig. Drain. Eng.* **1992**, *118*, 218–228. [CrossRef]
22. Avogadro, E.; Minciardi, R.; Paolucci, M. A Decisional Procedure for Water Resources Planning Taking into Account Water Quality Constraints. *Eur. J. Oper. Res.* **1997**, *102*, 320–334. [CrossRef]
23. Wong, H.S.; Sun, N.; Yeh, W. *A Two-Step Nonlinear Programming Approach to the Optimization of Conjunctive Use of Surface Water and Ground Water*; University of California Water Resources Center: Lisbon, Portugal, 1997.
24. Campbell, J.E.; Briggs David, A.; Denton Richard, A. Gartrell Gregory Water Quality Operation with a Blending Reservoir and Variable Sources. *J. Water Resour. Plan. Manag.* **2002**, *128*, 288–302. [CrossRef]
25. Han, X.; Zhao, Y.; Gao, X.; Jiang, S.; Lin, L.; An, T. Virtual Water Output Intensifies the Water Scarcity in Northwest China: Current Situation, Problem Analysis and Countermeasures. *Sci. Total Environ.* **2021**, *765*, 144276. [CrossRef]
26. Yan, H.; Tao, W.; Shao, F.; Su, L.; Wang, Q.; Deng, M.; Zhou, B. Spatiotemporal Patterns and Evolutionary Trends of Eco-Environmental Quality in Arid Regions of Northwest China. *Environ. Monit. Assess.* **2024**, *196*, 176. [CrossRef]
27. *GB 3838-2002*; Environmental Quality Standards for Surface Water. Ministry of Ecology and Environment of the People's Republic of China: Beijing, China, 2002. Available online: https://www.mee.gov.cn/ywgz/fgbz/bz/bzwb/shjbh/shjzlbz/200206/t20020601_66497.shtml (accessed on 6 July 2024).
28. Ahiablame, L.M.; Engel, B.A.; Chaubey, I. Effectiveness of Low Impact Development Practices: Literature Review and Suggestions for Future Research. *Water Air Soil Pollut.* **2012**, *223*, 4253–4273. [CrossRef]
29. Pachaly, R.L.; Vasconcelos, J.G.; Allasia, D.G.; Bocchi, J.P.P. Evaluating SWMM Capabilities to Simulate Closed Pipe Transients. *J. Hydraul. Res.* **2022**, *60*, 74–81. [CrossRef]
30. Rossman, L.A. *Storm Water Management Model User's Manual Version 5.1.*; United States Environmental Protection Agency (USEPA): Cincinnati, OH, USA, 2015.
31. Reyes-Lúa, A.; Backi, C.J.; Skogestad, S. Improved PI Control for a Surge Tank Satisfying Level Constraints. *IFAC-PapersOnLine* **2018**, *51*, 835–840. [CrossRef]
32. Rossman, L.A.; Bernagros, J.T. *National Stormwater Calculator User's Guide—Version 1.2.0.1*; EPA/600/R-13/085e; Office of Research and Development, United States Environmental Protection Agency (USEPA): Cincinnati, OH, USA, 2018.
33. *HJ 338-2018*; Technical Guideline for Delineating Source Water Protection Areas. Ministry of Ecology and Environment of the People's Republic of China: Beijing, China, 2018. Available online: https://www.mee.gov.cn/ywgz/fgbz/bz/bzwb/jcffbz/201803/t20180321_432813.shtml (accessed on 6 July 2024).

water

MDPI

Article

Dynamic Spatiotemporal Evolution and Driving Mechanisms of Vegetation in the Lower Reaches of the Tarim River, China

Qiang Han [1], Lianqing Xue [1,2,*], Tiansong Qi [3], Yuanhong Liu [1], Mingjie Yang [1], Xinyi Chu [1] and Saihua Liu [1]

[1] College of Hydrology and Water Resources, Hohai University, Nanjing 210098, China
[2] School of Hydraulic Engineering, Wanjiang University of Technology, Ma'anshan 243000, China
[3] Department of Civil, Construction and Environmental Engineering (Dept 2470), North Dakota State University, P.O. Box 6050, Fargo, ND 58108-6050, USA
* Correspondence: lqxue@hhu.edu.cn

Abstract: Analyzing the changes in vegetation under different factors is crucial for ecological protection in arid areas. The spatial-temporal variations of vegetation in the lower reaches of the Tarim River (LRTR) from 2000 to 2020, were analyzed using the Theil-Sen estimator and the Mann-Kendall test. The future trends of NDVI are projected to use the Hurst exponent method. The driving mechanisms of vegetation changes were analyzed using the GeoDetector method and multivariate residual analysis. The NDVI values in the LRTR significantly increased during the study period, indicating good vegetation recovery. The overall vegetation level remains poor and was primarily concentrated around the riverine areas. There is still a risk of vegetation degradation in most areas of the future LRTR. Compared to climate change, vegetation was more affected by human activities. Human activities have helped restore the riparian vegetation and prevented the degradation of vegetation far from the river. Therefore, distance from river channels is the strongest explanatory factor (q = 0.078) for vegetation changes, followed by precipitation, and temperature, while changes in slope have minimal impact on vegetation. Statistics have found that when two factors are combined, their impact on vegetation change is stronger. These findings are beneficial for identifying vegetation evolution patterns in LRTR and providing theoretical support for the government to carry out ecological restoration.

Keywords: vegetation dynamics; remote sensing; environmental restoration; driving mechanisms; Geodetector method

1. Introduction

Vegetation plays a crucial role in regulating the regional climate, and maintaining soil and water conservation, making it a vital component of the ecological environment [1]. Because of the intensification of climate change and frequent human activities, ecological problems in arid areas are becoming increasingly severe [2]. Evaluating vegetation dynamics and further analyzing the driving mechanisms behind them have become crucial for the management and restoration of ecosystems in arid areas [3]. In order to achieve sustainable development and ecological conservation, the issue of vegetation restoration has garnered widespread attention worldwide [4–6].

In the study of vegetation dynamics, remote sensing data are widely used due to the advantages of continuous time series, wide coverage, and high spatial resolution [7,8]. By performing statistical analysis on the Normalized Difference Vegetation Index (NDVI), information on vegetation coverage and growth conditions can be obtained [9,10]. The Theil-Sen median estimator [11] coupled with the Mann-Kendall test [12,13] is a commonly used method for analyzing spatiotemporal changes in vegetation. This method is less susceptible to the influence of outliers and can more accurately describe the spatial and temporal variation characteristics of surface objects. Emamian et al. [14] discovered that

NDVI in northern Iran exhibited a declining trend from 2004 to 2015. Sun et al. [15] indicated that in the Haihe River Basin, NDVI increased in 2000–2013. Therefore, exploring the driving mechanisms behind vegetation spatiotemporal changes has become a new research focus.

Residual analysis is utilized to examine the driving factor that influences NDVI, [16,17] and has been widely applied to research the contributions of human activity and climate change to vegetation cover changes. Yang, et al. [18] found that in the Han River Basin, human activity and climate change have generally promoted NDVI increase in non-urban areas. Liu, et al. [19] analyzed that climate change is the primary influence factor of NDVI increase on the Qinghai-Tibet Plateau, accounting for 68.05% of the observed increase. The Geodetector method is commonly used to analyze the spatial changes of vegetation and accurately identify the driving factors behind them [20]. Numerous researchers have employed the Geodetector method to investigate the impact factors driving the spatial and temporal changes of NDVI in different areas, such as Inner Mongolia [21], the Loess Plateau [22], and the Heihe River Basin [23]. This statistical method uncovers the factors influencing the temporal and spatial changes of vegetation cover, providing a comprehensive analysis of the variables [24].

The research area of this article is located in the lower reaches of the Tarim River (LRTR) in northwest China. The local drought and water shortage have led to a very fragile ecosystem [25]. Since 2000, the basin management organization has implemented ecological water conveyance (EWC) to artificially intervene in the vegetation of the LRTR, aiming to restore the degraded ecological environment. This paper investigates the spatiotemporal change process of NDVI in the LRTR using the Theil-Sen Median estimator coupled with the Mann-Kendall test. The impact of human activities and climate change on vegetation change was quantitatively evaluated through residual analysis. The impact of different driving factors on vegetation spatial change was evaluated using the Geodetector method, which helped clarify the mechanisms driving vegetation change and the research results provide theoretical support for the implementation of ecological restoration and management in LRTR.

2. Study Area and Data Processing

2.1. Study Area

The Tarim River, situated in northwest China, is the longest inland river in the country [26,27]. The climate in the basin is arid, with low annual precipitation (116 mm), high average annual temperature (10 °C), and extremely high evaporation (2200 mm) [28]. The geographical scope of LRTR is the area between the Daxihaizi Reservoir and the Taitema Lake. The local vegetation is dominated by species such as Populus euphratica and Hippophae rhamnoides, with a low vegetation coverage rate [29,30]. Vegetation growth heavily relies on groundwater, rendering the ecological environment extremely fragile. The Tarim River divides into two channels at the Daxihaizi Reservoir. The northern channel is called the Qiwenkuoer River, while the southern one is the Old Tarim River. These two channels converge at Alagan. Since the late 20th century, because of the unreasonable exploitation and allocation of water resources in the upstream area [31], the available water volume in the downstream has sharply decreased. This has led to serious ecological problems such as the decline of groundwater level, the reduction of vegetation, and the decrease of the area of the terminal lake [29,32]. In 2000, the Tarim River Basin Authority launched EWC to restore the ecosystem in the LRTR [33], conveying water from the Daxihaizi Reservoir through the old Tarim River and Qiwenkuoer River, to the downstream area. The river water ultimately flows into the terminal Taitema Lake, playing a pivotal role in restoring riparian vegetation [34]. See Figure 1.

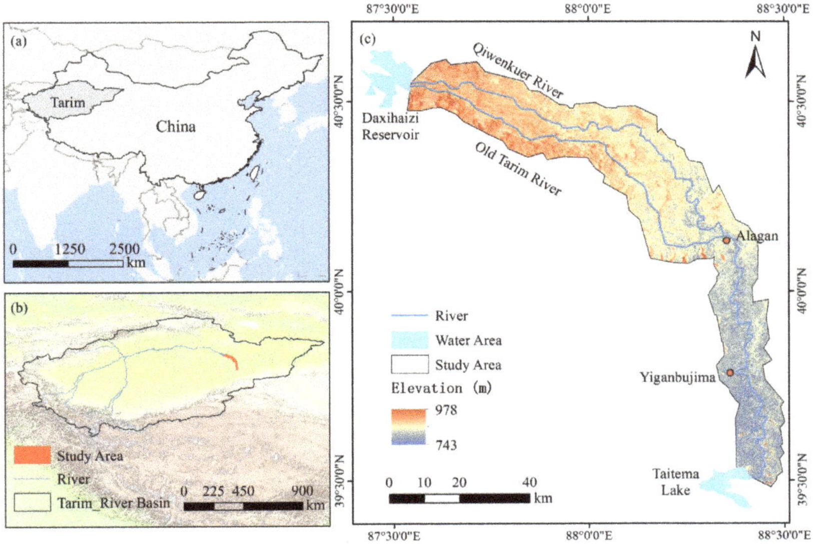

Figure 1. Schematic diagram of the LRTR: (**a**) geographical location of the Tarim River Basin in China; (**b**) geographical location of the LRTR in the Tarim River Basin; (**c**) elevation and river distribution in the LRTR.

2.2. Data Source and Preprocessing

This study utilized high-resolution NDVI data (16-day, 500 m) from the MOD13Q1 product, and analyzed vegetation changes by calculating the annual average NDVI data. The precipitation and temperature data (monthly, 1 km) were sourced from the National Tibetan Plateau/Third Pole Environment Data Center (http://data.tpdc.ac.cn, accessed on 28 July 2024) [35,36]. This study converted the monthly scale data to an annual scale in order to calculate the influence of climatic factors on vegetation.

The DEM data (90 m resolution) was obtained from the Geospatial Data Cloud website (https://www.gscloud.cn, accessed on 28 July 2024). ArcGIS 10.4 software was used to calculate slope data using the DEM data. Further calculations were made to determine the distance between different grids and river channels, in order to analyze the impact of EWC on riparian vegetation. The Tarim River Basin Authority provided the EWC data, including the volume and the duration. The time range of this study spanned from 2000 to 2020, and the grid accuracy is unified at 1 km. See Figure 2.

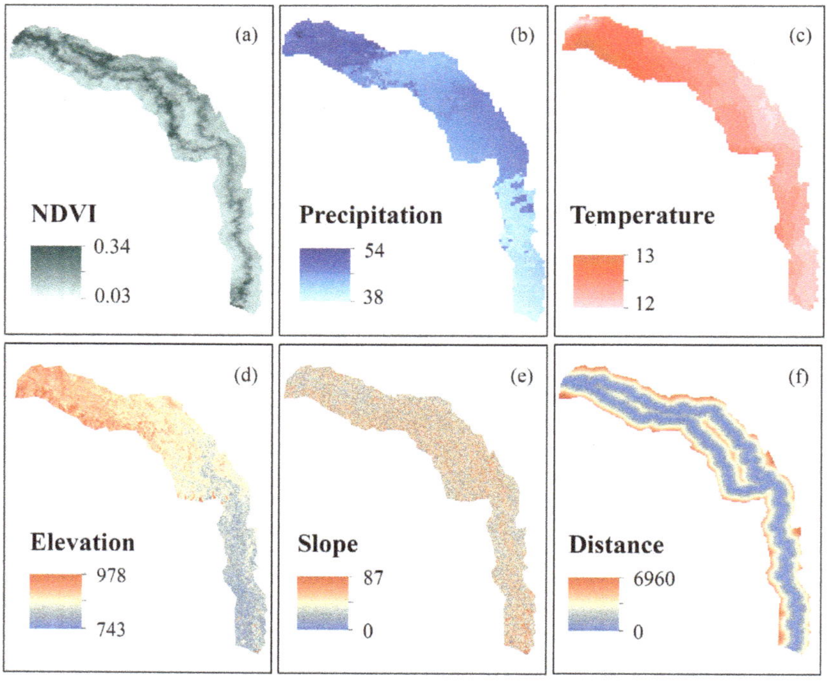

Figure 2. Diagram of the study data: (**a**) average NDVI; (**b**) annual precipitation (mm); (**c**) annual average temperature (°C); (**d**) elevation (m); (**e**) slope (°); (**f**) distance to the river channel (m).

3. Method

3.1. Theil-Sen Estimator

The Theil-Sen estimator is recognized as a reliable non-parametric approach for trend analysis [37], and was proposed by Pranab K. Sen [38]. In contrast to approaches that rely on time series conformity, this method adeptly manages minor outliers and missing data. This study used it to compute the slope between successive data pairs within the NDVI dataset. The overall trend within the NDVI changes is represented by the median slope computed from the following calculations:

$$Slope_{NDVI} = Median\left(\frac{x_j - x_i}{j - i}\right),\ \forall j > i \tag{1}$$

where $Slope_{NDVI}$ represents the median of all slope data. If $Slope_{NDVI} > 0$, reveals a growing tendency in NDVI; if $Slope_{NDVI} < 0$, shows a decline in NDVI over time. x_i and x_j represent the two variables at time i and j, respectively.

3.2. Mann-Kendall (M-K) Significance Test

The Mann-Kendall significance test is combined with Theil-Sen estimator to assess the significance of time series trends [39,40]. In this study, it was employed to evaluate vegetation trends which were calculated as follows:

$$Z = \begin{cases} \dfrac{S}{\sqrt{V(S)}}(S > 0) \\ 0(S = 0) \\ \dfrac{S+1}{\sqrt{V(S)}}(S < 0) \end{cases} \tag{2}$$

$$S = \sum_{i=1}^{n-1} \sum_{j=i+1}^{n} Sign(x_j - x_i) \tag{3}$$

$$V(S) = \frac{n(n-1)(2n+5)}{18} \tag{4}$$

$$sign(NDVI_j - NDVI_i) = \begin{cases} 1, & NDVI_j - NDVI_i > 0 \\ 0, & NDVI_j - NDVI_i = 0 \\ -1, & NDVI_j - NDVI_i < 0 \end{cases} \tag{5}$$

where n refers to the length of the dataset. If the $|Z|$ is greater than 1.65, 1.96, or 2.58 respectively, it indicates that the trend has passed the significance test at a confidence level of 90%, 95%, and 99%, respectively.

The integration of the Theil-Sen estimator and the M-K significance test effectively captured the diverse spatial distribution of vegetation change characteristics. By overlaying the Sen's slope results with those of the M-K test, the changes in NDVI can be classified into nine categories. See Table 1.

Table 1. The classification of NDVI changes.

Sen's Slope	Z Value	Trend Characteristics
$\beta > 0$	$2.58 < Z$	Extremely significant increased
	$1.96 < Z \leq 2.58$	significant increased
	$1.65 < Z \leq 1.96$	slightly significant increased
	$Z \leq 1.65$	Non-significant increased
$\beta = 0$	0	No changes
$\beta < 0$	$Z \leq 1.65$	Non-significant decreased
	$1.65 < Z \leq 1.96$	slightly significant decreased
	$1.96 < Z \leq 2.58$	significant decreased
	$2.58 < Z$	Extremely significant decreased

3.3. Hurst Exponent Method

The Hurst exponent, originally established by Hurst [41] and later improved by Mandelbrot [42], is a measure used to assess the persistence of changes in time series. It is calculated using the rescaled range analysis (R/S) method and assisted in revealing the autocorrelation within the time series, particularly emphasizing long-term trends that are might otherwise remain concealed. Recent research has increasingly incorporated this exponent into the analysis of long-term vegetation dynamics over time [43–45]. The process for conducting R/S analysis encompasses the following steps:

$$X_{t,a} = \sum_{z=1}^{t} x_{z,a} - e_a,\ t = 1,\ 2,\ \cdots,\ n \tag{6}$$

$$R_a = max(X_{t,a}) - min(X_{t,a}),\ 1 \leq t \leq m \tag{7}$$

$$(R/S)_m = \frac{1}{A} \sum_{a=1}^{A} R_a/S_a \tag{8}$$

$$(R/S)_m = D \times m^H \tag{9}$$

$X_{t,a}$ denotes the cumulative deviation; x is the annual NDVI; e_a is the average NDVI; R_a is the extreme deviation; and S_a is the sample standard deviation. H ranges from 0 to 1 and can be categorized into three types. When $H = 0.5$, the future changes of NDVI are random. When $H > 0.5$, the future changes of NDVI remain the same as in the past. Conversely, if $H < 0.5$, the future changes of NDVI are opposite to the past.

By combining the Theil-Sen median slope with the Hurst exponent method, the trend of vegetation changes in the future can be analyzed. The classification indicators are as follows (Table 2):

Table 2. Classification of vegetation change trends.

Hurst	Sen's Slope	Future Trends
$0.5 < H < 1$	$\beta > 0$	Improvement
	$\beta < 0$	Degradation
$H = 0.5$	-	Uncertain
$0 < H < 0.5$	$\beta > 0$	Degradation
	$\beta < 0$	Improvement

3.4. Multivariate Residual Analysis

The impact of human activities and climate change on vegetation changes was evaluated using multivariate residual analysis. This approach is based on the premise that changes in vegetation primarily driven by human activities can be discerned once the influence of climate factors has been eliminated as seen in the following:

$$NDVI_{CC} = a \times P + b \times T + c \tag{10}$$

$$NDVI_{HA} = NDVI_{obs} - NDVI_{CC} \tag{11}$$

where a and b are the regression coefficients, and c is the constant. P represents the precipitation time series and T represents the temperature time series. $NDVI_{CC}$ represents the calculated NDVI data with the regression equation, $NDVI_{obs}$ represents the observed NDVI data, and $NDVI_{HA}$ represents the result of subtracting observed from the calculated NDVI data. After conducting residual analysis, the factors influencing vegetation can be categorized into six classes. The classification criteria used are shown in the following (Table 3):

Table 3. Classification criterion of the driving factors of NDVI change.

Slope ($NDVI_{obs}$)	Driving Factors	Division Criteria		Contribution Rate/%	
		Slope ($NDVI_{CC}$)	Slope ($NDVI_{HA}$)	Climate Change (CC)	Human Activity (HA)
>0	CC&HA	>0	>0	$\frac{Slope(NDVI_{CC})}{Slope(NDVI_{obs})}$	$\frac{Slope(NDVI_{HA})}{Slope(NDVI_{obs})}$
	CC	>0	<0	100	0
	HA	<0	>0	0	100
<0	CC&HA	<0	<0	$\frac{Slope(NDVI_{CC})}{Slope(NDVI_{obs})}$	$\frac{Slope(NDVI_{HA})}{Slope(NDVI_{obs})}$
	CC	<0	>0	100	0
	HA	>0	<0	0	100

3.5. Geodetector Method

The Geodetector model, which can be freely downloaded from https://www.geodetector.cn/ (accessed on 28 July 2024), was utilized in conjunction with Excel. The model comprised of three types of detections: factor, ecological, and interactive. The factor detector identified the spatial differentiation of NDVI and quantified the extent to which a driving factor explained this differentiation. By calculating the q-values of various driving factors, the relative strength of their impacts on NDVI changes can be compared. The specific calculation process is as follows:

$$q = 1 - \frac{\sum_{h=1}^{L} N_h \sigma_h^2 b}{N\sigma^2} \tag{12}$$

where L represents the variable of NDVI, N_h and σ_h^2 represent the number of units and variance of layer h, respectively, N and σ^2 refer to the number of units and overall variance.

The range of the q value is [0,1]. The magnitude of the q value and driving factors on NDVI is positively proportional. At the same time, interaction detection identifies the interplay between various influencing factors and assesses the explanatory power when double driving factors interact.

4. Results

4.1. Temporal Variation Characteristics of the NDVI

From Figure 3a, it is observed that the annual average NDVI during 2000–2020 was about 0.007. NDVI exhibited an overall increasing trend, increasing by an average of 0.001 per year. At the same time, the accumulative ecological water conveyance volume (AEWCV) also increased year by year. In years when the slope of AEWCV increase and decreased, such as from 2006–2009, when ecological water conveyance volume (EWCV) declined, NDVI also decreased, this indicated a high correlation between the two. To evaluate areas with different vegetation cover, NDVI values were divided into three categories: 0–0.1, 0.1–0.2, and 0.2–1. Figure 3b shows that the area of the region with NDVI values between 0.1 and 0.2 significantly increases, while the area with NDVI values between 0.2 and 1 also showed a small increase. From this, it can be observed that EWC played a significant role in the vegetation restoration of LRTR.

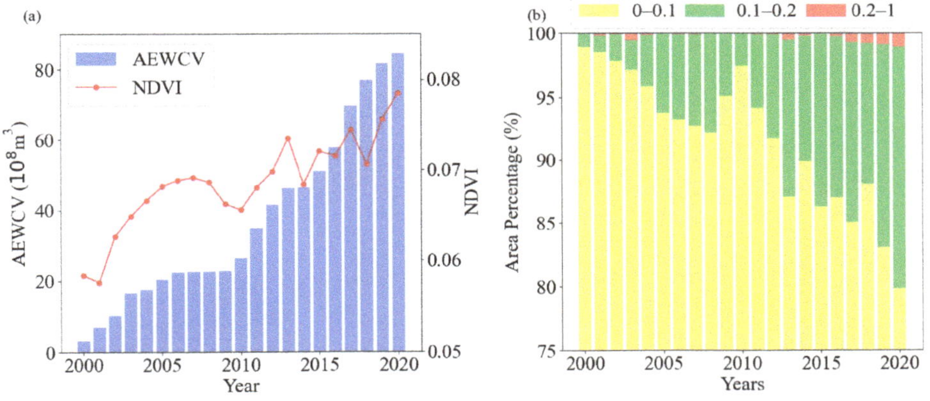

Figure 3. Interannual variation of NDVI: (**a**) overall changes in NDVI and AEWCV; (**b**) changes in the proportion of areas classified by different NDVI.

4.2. Spatial Variation Characteristics of the NDVI

From the vegetation change trends in Figures 4 and 5, it can be observed that vegetation showed significant recovery during the first period after the initiation of EWC (2000–2005), with a recovery area proportion reaching 91%. The vegetation in areas proximate to the river channel has recovered better, indicating that EWC has a positive impact on vegetation growth. However, as it can be seen in Figure 4b,d and Figure 5b,d, during the periods from 2005–2010 and 2015–2020, vegetation degradation reoccurred across the study area, with the proportions of degraded areas reaching 71% and 73.2%, respectively. This was associated with concurrent decreases in EWCV during these periods. Among them, during the period from 2015 to 2020, the NDVI in 8.5% of the area significantly decreased, indicating that not only the reduced EWCV but also other combined factors contributed to vegetation degradation. From Figures 4e and 5e, it can be seen that vegetation has significantly recovered during 2000–2020, with the restored areas accounting for 88.6% of the total, mainly concentrated around river channel and tail lake.

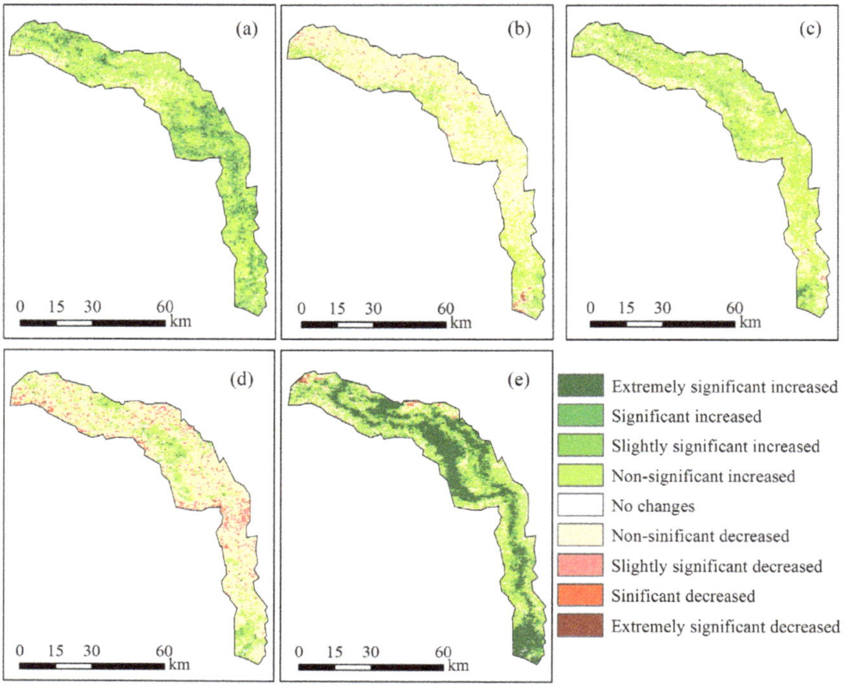

Figure 4. The changing trends of NDVI from: (**a**) 2000 to 2005, (**b**) 2005 to 2010, (**c**) 2010 to 2015, (**d**) 2015 to 2020, and (**e**) 2000 to 2020.

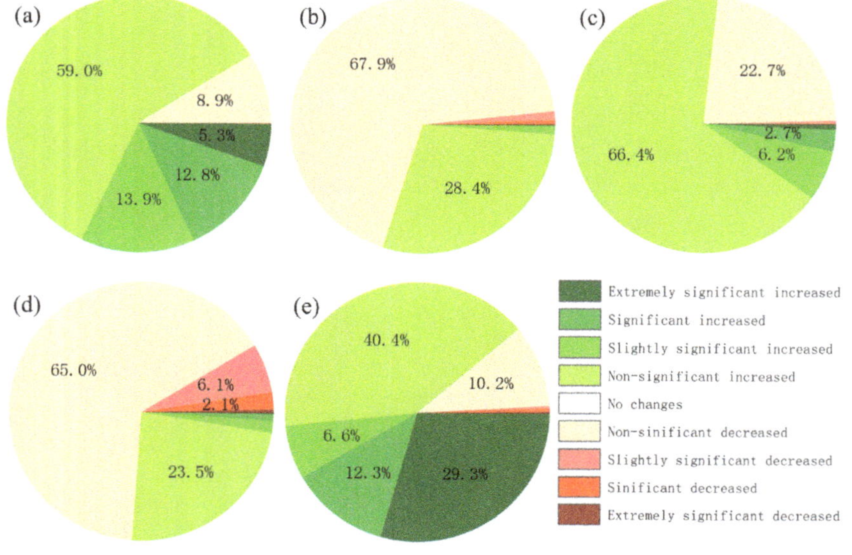

Figure 5. The proportion of areas with different trends in NDVI from: (**a**) 2000 to 2005, (**b**) 2005 to 2010, (**c**) 2010 to 2015, (**d**) 2015 to 2020, and (**e**) 2000 to 2020 (portions less than 2% do not display specific numbers).

Furthermore, Hurst exponent analysis is used to analyze the vegetation change trend of future LRTR. Figure 6 illustrates that the spatial distribution of the trends was rather scattered, with no clear regularity. In total, 33% of the areas had H values between 0 and 0.5, indicating that the future trends would be opposite to the past; 66% of the areas had H values between 0.5 and 1, indicating that the future trends would continue like the past trends. Further, the results of the Theil-Sen estimator coupled with the Hurst exponent method showed three future trends in vegetation change. Among them, 27.8% of the areas would experience vegetation increase, while 71.5% of the areas would face vegetation degradation. In terms of spatial distribution, the upper and terminal regions of the watershed mainly showed an increasing trend in vegetation, while the middle region of the watershed showed a decreasing trend in vegetation. This serves as a warning for future ecological protection efforts.

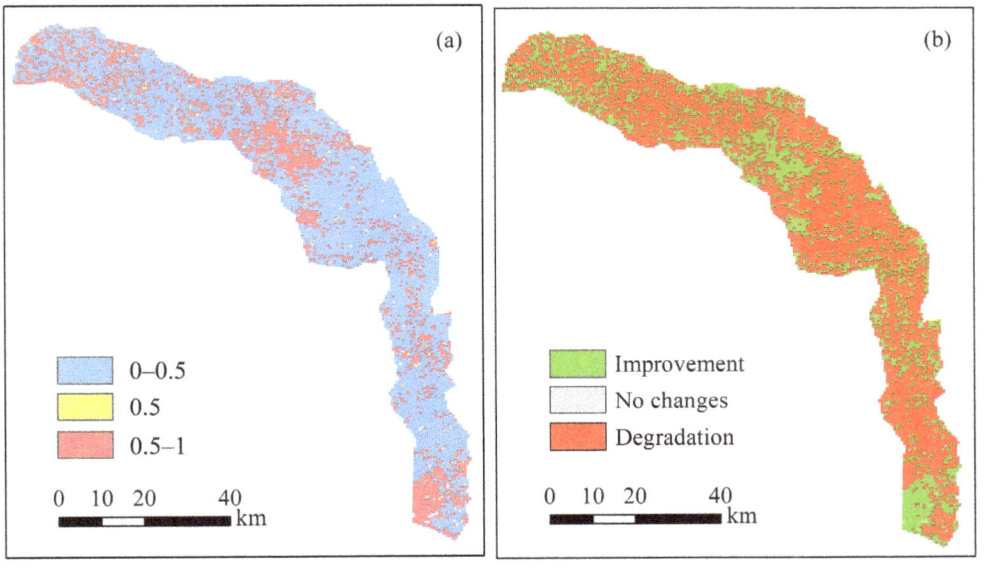

Figure 6. The Hurst exponent of (**a**) NDVI and (**b**) its future trends.

4.3. Contribution of Climate Change and Human Activity to Vegetation Dynamics

Multivariate residual analysis method was used to evaluate the spatial distribution of factors affecting NDVI changes. Figure 7 illustrates substantial spatial variability in how human activity and climate change influence NDVI across the study area. Firstly, it was evident that the area where NDVI has increased was much larger than the area where NDVI has decreased. In the areas surrounding the river channel, the increase in NDVI is due to the combined effects of climate change (CC) and human activity (HA). In contrast, the areas where climate change and human activity independently contribute to the increase of NDVI were scattered around the periphery without any regular pattern. Meanwhile, in areas far from the river channel, NDVI showed a decreasing trend, mostly driven by human activity.

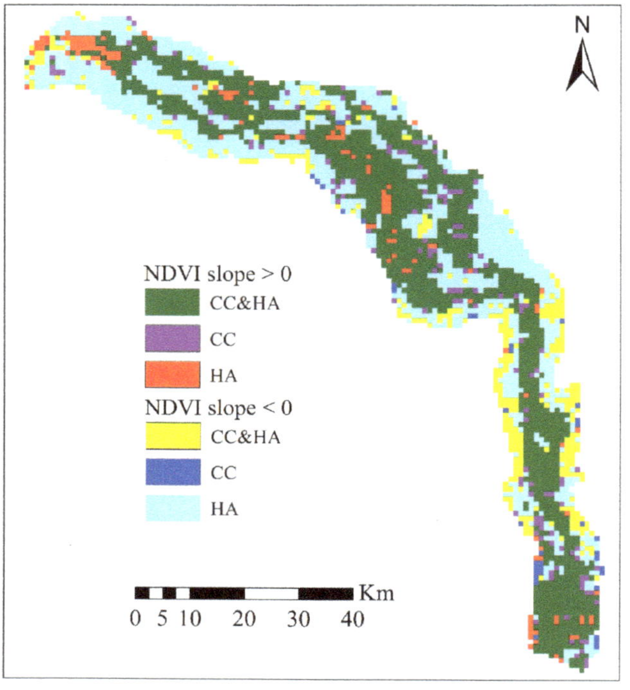

Figure 7. Driving factors of NDVI change in different regions.

Furthermore, the impacts of human activity and climate change on NDVI changes were quantified. The study categorized regions into 11 groups based on their varying contribution levels, specifically: [−100, −80), [−80, −60), . . ., (80, 100]. In Figure 8, the warmer the color tone, the greater the contribution to NDVI changes. Positive contributions indicate a positive effect on NDVI, while negative contributions indicate a negative effect. Figure 8a shows that regions where climate change contributes positively to NDVI changes cover approximately 32.26% of the study area. Notably, areas with contributions falling within the ranges of 0–20% and 80–100% collectively cover a significant portion, approximately 20.2% of the total area. In contrast, climate change had no observable impact on NDVI changes in 62% of the regions. Overall, the effect of climate change on NDVI changes in the LRTR was not significant, with the affected areas being relatively scattered.

As shown in Figure 8b, the effect of human activity on NDVI changes is more pronounced and the affected areas are relatively concentrated. However, the degree of impact is polarized. The regions where human activity contributed 80–100% were mainly concentrated near the river channel. This indicated that in these areas, the NDVI showed a significant increasing trend, and human activity positively contributed to vegetation restoration. The areas where the contribution of human activity ranges from −100% to 80% were mainly concentrated in regions far from the river channel, accounting for 38.15% of the whole study area. Considering that NDVI showed a decreasing trend, it can be understood that human activity has had a significant inhibitory effect on vegetation degradation.

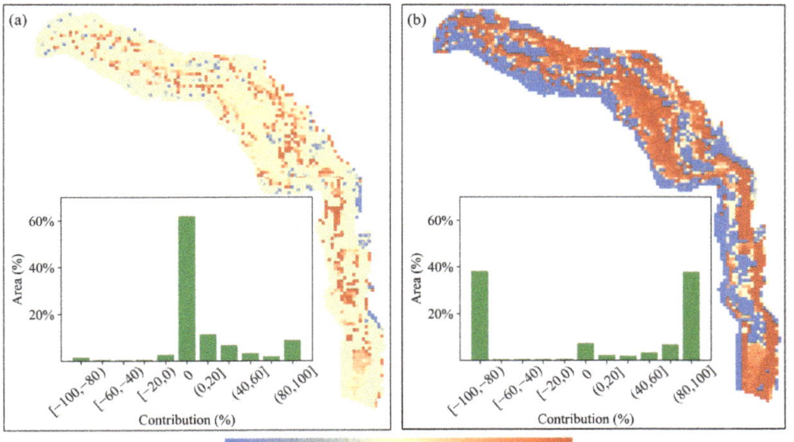

Figure 8. Spatial distribution of the contributions of (**a**) climatic change and (**b**) human activity to NDVI change.

4.4. Detection of Factors Influencing Vegetation Distribution

This study considered five factors to analyze the driving mechanisms of vegetation spatial variation, as shown in Table 4. These factors were annual precipitation, annual average temperature, elevation, slope, and distance to the river channel. According to the factor detection results (Figure 9), the order of the impact of driving factors on NDVI was: X5 > X3 > X2 > X1 > X4. The distance to the river channel factor has the strongest explanatory power for NDVI, with a Q value of 0.078, much higher than other driving factors. The impact of slope on the spatiotemporal variation of NDVI is minimal, with a Q value of only 0.001.

Table 4. Symbols for each factor.

X1	X2	X3	X4	X5
annual precipitation	annual average temperature	elevation	slope	distance to the river channel

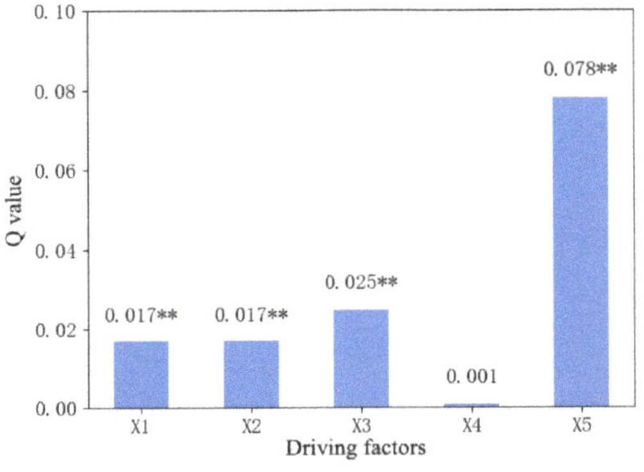

Figure 9. The result of factor detection (** indicates $p < 0.01$, significant by testing).

Ecological detection results (Table 5) show that only the distance to the river channel, when combined with other factors, showed a significant difference ($\alpha = 0.05$) in the impact on the change of NDVI. The combinations of additional driving factors did not exhibit an insignificant variation in their influence on the spatial distribution of NDVI.

Table 5. The result of ecological detection.

	X1	X2	X3	X4	X5
X1	\				
X2	N	\			
X3	N	N	\		
X4	N	N	N	\	
X5	Y	Y	Y	Y	\

Note: Y indicates a significant difference exists between the two driving factors at the $\alpha = 0.05$ level, while N indicates there is none.

The comprehensive impact of driving factors on NDVI is evaluated using the interaction detector in the Geodetector method. The interactive detection results (Figure 10) demonstrated a notable synergistic enhancement effect between different factors, with the combination of two factors significantly amplifying their influence on NDVI's spatial distribution. Among them, the interaction between the distance to the river channel factor and the precipitation factor, temperature factor, and elevation factor showed better explanatory power for NDVI spatial distribution, with Q values of 0.1101, 0.1099, and 0.1058, respectively. This result showed that the distance to the river channel was the primary factor influencing the NDVI changes. The study demonstrated that the driving factors influencing NDVI were interdependent and exhibited greater explanatory power when they interacted.

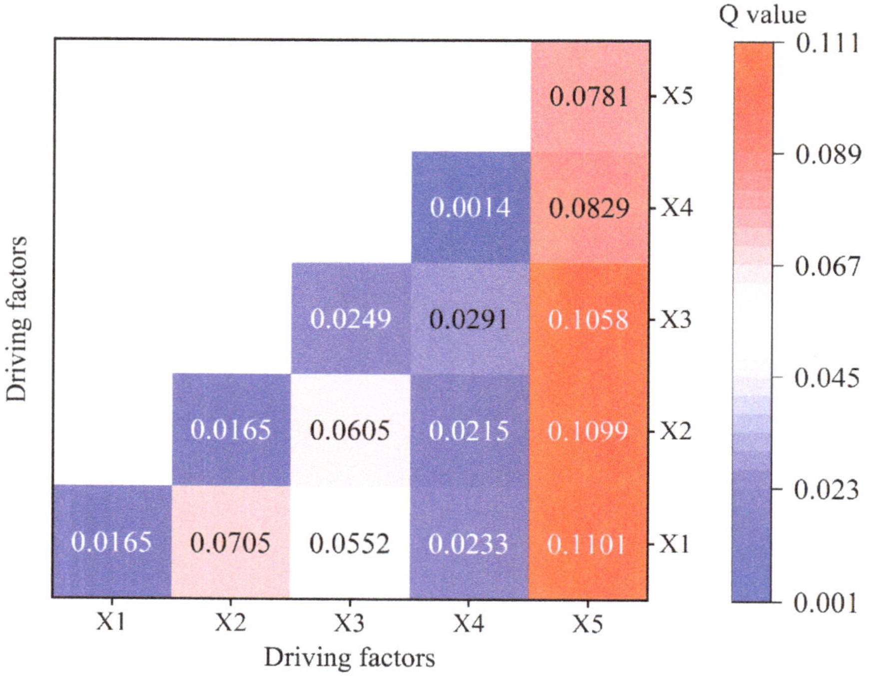

Figure 10. Interaction between factors.

5. Discussion

5.1. Vegetation Variation Characteristics in the LRTR

This study revealed a continuous increase in NDVI in the LRTR in 2000–2020. Hu et al. [46] reported a significant vegetation increase in 65.9% of the Tarim River Basin, while Yu et al. [47] found that the proportion of increasing NDVI in China's endorheic basins exceeded the proportion of decreasing NDVI. These findings were consistent with our results. The restoration in vegetation conditions in the LRTR can be largely attributed to the EWC [48]. Using the Theil-Sen estimator and the M-K significance test, we analyzed vegetation changes of the LRTR from 2000 to 2020 and found that the vegetation conditions in most areas (88.6%) have undoubtedly improved. From this perspective, the Chinese government's ecological water conveyance has been very successful in effectively preventing vegetation degradation.

5.2. Effect of Driving Factors on LRTR Vegetation Change

The vegetation cover in the LRTR was affected by the EWC, resulting in a human activity contribution to NDVI changes significantly outweighing that of climate change. Due to the scarcity of precipitation in the LRTR, vegetation growth mainly depends on groundwater. At the same time, the main growth of drought-resistant tree species in the area has lower leaf evaporation and is less sensitive to high temperatures. Therefore, climate change has a small impact on vegetation. Meanwhile, human beings have made continuous efforts in ecological restoration, so under the comprehensive impact of climate change and human activities, vegetation is showing a trend of growth. Besides meteorological factors, the distance to the river channel was a key driver of vegetation distribution in the LRTR. Rivers mainly impact vegetation by recharging groundwater, which significantly influences NDVI in arid regions [49,50]. The closer the distance to the river channel, the higher the groundwater level, which better supplements water for plants. Additionally, elevation is a major factor influencing vegetation distribution. It indirectly affects vegetation growth by altering precipitation and temperature. However, in arid regions, due to the scarcity of precipitation, the differences in vegetation growth at varying elevations are not as pronounced as in humid areas [51]. Additionally, the slope has a weaker explanatory power for vegetation growth. The study results indicate that slope had little impact on vegetation distribution in the LRTR. Generally, slope affects vegetation growth through solar radiation. However, in arid regions, due to intense sunlight and high temperatures, vegetation growth may be inhibited due to water scarcity instead.

We found that combining low q statistics driving factors with other driving factors greatly enhances explanatory power. Moreover, interactions between two driving factors were not simply linear additions; instead, they exhibited nonlinear amplification. Specifically, factors like distance to the river channel, precipitation, and temperature, which are significantly correlated with vegetation, exhibited higher explanatory power after interaction. This suggests that the factors are not independent of each other and typically act together to collectively influence vegetation distribution.

6. Conclusions

This study employed multi-source geospatial data and used the Theil-Sen estimator and the Mann-Kendall significance test to assess historical changes in NDVI in the LRTR. Despite a gradual increase in vegetation cover from 2000–2020, the NDVI values in the LRTR remain relatively low. The future trend of NDVI change was analyzed using the Hurst exponent method. The findings showed that at the upper and lower parts of LRTR, most areas would further increase in vegetation, while in the middle reaches, vegetation was more likely to degrade. Subsequently, residual analysis was used for quantitative evaluation of the contributions of human activity and climate change to NDVI changes. The results showed that human activities, such as EWC, were the primary contributing factors to vegetation growth in the LRTR. Meanwhile, the contribution of climate change is minimal. Using the Geodetector method, we analyzed the driving mechanism of NDVI change in the LRTR. The top three driving factors affecting vegetation coverage were the distance to the river channel, precipitation, and temperature. Slope has the least effect on vegetation. The interaction effects between different driving factors showed significant nonlinear enhancement, surpassing the explanatory power of individual factors for vegetation changes. The study area is a typical arid region, and human intervention in local vegetation growth is very obvious. The issue of vegetation protection is a common problem faced by various arid regions around the world. Different human activities affect whether vegetation is restored or degraded. Research by scholars in northern Oman [52], the Loess Plateau in China [53], and Central Asia [54] has also confirmed that human activities are the main factors affecting vegetation growth in arid regions. Due to the positive impact of river leakage, riparian vegetation that relies on groundwater has emerged in arid areas, which is significantly denser than vegetation in other regions. Studying the mechanisms behind vegetation restoration and degradation is of great significance for protecting fragile ecosystems. These findings offer theoretical support for ecological conservation and sustainable development in arid areas.

Author Contributions: Methodology, Q.H., X.C. and S.L.; Software, T.Q. and X.C.; Validation, T.Q.; Formal analysis, Q.H.; Investigation, Y.L.; Resources, L.X. and S.L.; Data curation, Y.L.; Writing—original draft, Q.H.; Writing—review & editing, L.X.; Visualization, M.Y.; Supervision, M.Y.; Funding acquisition, L.X. All authors have read and agreed to the published version of the manuscript.

Funding: This study was supported by the National Key Research and Development Program of China (2023YFC3206804), Xinjiang Production and Construction Corps (No. 2022BC001), National Scientific Foundation of China (No. 51779074).

Data Availability Statement: The raw data supporting the conclusions of this article will be made available by the authors on request.

Conflicts of Interest: The authors declare that they have no known competing financial interests or personal relationships that could have appeared to influence the work reported in this paper.

References

1. Karnieli, A.; Qin, Z.; Wu, B.; Panov, N.; Yan, F. Spatio-Temporal Dynamics of Land-Use and Land-Cover in the Mu Us Sandy Land, China, Using the Change Vector Analysis Technique. *Remote Sens.* **2014**, *6*, 9316–9339. [CrossRef]
2. Han, Q.; Xue, L.; Liu, Y.; Yang, M.; Chu, X.; Liu, S. Developing a multi-objective simulation-optimization model for ecological water conveyance in arid inland river basins. *J. Hydrol. Reg. Stud.* **2023**, *50*, 101551. [CrossRef]
3. Xu, Y.; Dai, Q.-Y.; Lu, Y.-G.; Zhao, C.; Huang, W.-T.; Xu, M.; Feng, Y.-X. Identification of ecologically sensitive zones affected by climate change and anthropogenic activities in Southwest China through a NDVI-based spatial-temporal model. *Ecol. Indic.* **2024**, *158*, 111482. [CrossRef]
4. Xiao, X.; Guan, Q.; Zhang, Z.; Liu, H.; Du, Q.; Yuan, T. Investigating the underlying drivers of vegetation dynamics in cold-arid mountainous. *Catena* **2024**, *237*, 107831. [CrossRef]
5. Zhan, Y.; Liu, X.; Li, Y.; Zhang, H.; Wang, D.; Fan, J.; Yang, J. Trends and contribution of different grassland types in restoring the Three River Headwater Region, China, 1988–2012. *Sci. Total Environ.* **2024**, *908*, 168161. [CrossRef]
6. Khanaum, M.M.; Qi, T.; Boutin, K.D.; Otte, M.L.; Lin, Z.; Chu, X. Assessing the Impacts of Wetlands on Discharge and Nutrient Loading: Insights from Restoring Past Wetlands with GIS-Based Analysis and Modeling. *Wetlands* **2023**, *43*, 103. [CrossRef]
7. Xiong, Y.; Zhang, Z.; Fu, M.; Wang, L.; Li, S.; Wei, C.; Wang, L. Analysis of Vegetation Cover Change in the Geomorphic Zoning of the Han River Basin Based on Sustainable Development. *Remote Sens.* **2023**, *15*, 4916. [CrossRef]
8. Gao, C.; Ren, X.; Fan, L.; He, H.; Zhang, L.; Zhang, X.; Li, Y.; Zeng, N.; Chen, X. Assessing the Vegetation Dynamics and Its Influencing Factors in Central Asia from 2001 to 2020. *Remote Sens.* **2023**, *15*, 4670. [CrossRef]
9. Ding, J.; Zhao, W.; Daryanto, S.; Wang, L.; Fan, H.; Feng, Q.; Wang, Y. The spatial distribution and temporal variation of desert riparian forests and their influencing factors in the downstream Heihe River basin, China. *Hydrol. Earth Syst. Sci.* **2017**, *21*, 2405–2419. [CrossRef]
10. Shijie, P.; Lei, W.; Yongkun, L.; Ruowen, W.; Tianming, G.; Zongjun, G. A study on ecohydrological mutual feedback relationship of the Shangdong River basin based on hydrological connectivity. *Sci. Total Environ.* **2024**, *927*, 171957. [CrossRef]
11. Wang, J.; Wang, K.; Zhang, M.; Zhang, C. Impacts of climate change and human activities on vegetation cover in hilly southern China. *Ecol. Eng.* **2015**, *81*, 451–461. [CrossRef]
12. Mehmood, K.; Anees, S.A.; Rehman, A.; Pan, S.; Tariq, A.; Zubair, M.; Liu, Q.; Rabbi, F.; Khan, K.A.; Luo, M. Exploring spatiotemporal dynamics of NDVI and climate-driven responses in ecosystems: Insights for sustainable management and climate resilience. *Ecol. Inform.* **2024**, *80*, 102532. [CrossRef]

13. Faheem, Z.; Kazmi, J.H.; Shaikh, S.; Arshad, S.; Mohammed, S. Random forest-based analysis of land cover/land use LCLU dynamics associated with meteorological droughts in the desert ecosystem of Pakistan. *Ecol. Indic.* **2024**, *159*, 111670. [CrossRef]
14. Emamian, A.; Rashki, A.; Kaskaoutis, D.G.; Gholami, A.; Opp, C.; Middleton, N. Assessing vegetation restoration potential under different land uses and climatic classes in northeast Iran. *Ecol. Indic.* **2021**, *122*, 107325. [CrossRef]
15. Sun, Y.-L.; Shan, M.; Pei, X.-R.; Zhang, X.-K.; Yang, Y.-L. Assessment of the impacts of climate change and human activities on vegetation cover change in the Haihe River basin, China. *Phys. Chem. Earth* **2020**, *115*, 102834. [CrossRef]
16. Zhang, H.; Li, L.; Zhao, X.; Chen, F.; Wei, J.; Feng, Z.; Hou, T.; Chen, Y.; Yue, W.; Shang, H.; et al. Changes in Vegetation NDVI and Its Response to Climate Change and Human Activities in the Ferghana Basin from 1982 to 2015. *Remote Sens.* **2024**, *16*, 1296. [CrossRef]
17. Yan, W.; Wang, H.; Jiang, C.; Jin, S.; Ai, J.; Sun, O.J. Satellite view of vegetation dynamics and drivers over southwestern China. *Ecol. Indic.* **2021**, *130*, 108074. [CrossRef]
18. Yang, S.; Liu, J.; Wang, C.; Zhang, T.; Dong, X.; Liu, Y. Vegetation dynamics influenced by climate change and human activities in the Hanjiang River Basin, central China. *Ecol. Indic.* **2022**, *145*, 109586. [CrossRef]
19. Liu, Y.; Zhang, X.; Du, X.; Du, Z.; Sun, M. Alpine grassland greening on the Northern Tibetan Plateau driven by climate change and human activities considering extreme temperature and soil moisture. *Sci. Total Environ.* **2024**, *916*, 169995. [CrossRef]
20. Wang, J.; Xu, C. Geodetector: Principle and prospective. *Acta Geogr. Sin.* **2017**, *72*, 116–134.
21. Kang, Y.; Guo, E.; Wang, Y.; Bao, Y.; Bao, Y.; Mandula, N. Monitoring Vegetation Change and Its Potential Drivers in Inner Mongolia from 2000 to 2019. *Remote Sens.* **2021**, *13*, 3357. [CrossRef]
22. He, P.; Bi, R.-T.; Xu, L.-S.; Wang, J.-S.; Cao, C.-B. Using geographical detection to analyze responses of vegetation growth to climate change in the Loess Pla-teau, China. *Ying Yong Sheng Tai Xue Bao = J. Appl. Ecol.* **2022**, *33*, 448–456.
23. Zhu, L.; Meng, J.; Zhu, L. Applying Geodetector to disentangle the contributions of natural and anthropogenic factors to NDVI variations in the middle reaches of the Heihe River Basin. *Ecol. Indic.* **2020**, *117*, 106545. [CrossRef]
24. Whetton, R.; Zhao, Y.; Shaddad, S.; Mouazen, A.M. Nonlinear parametric modelling to study how soil properties affect crop yields and NDVI. *Comput. Electron. Agric.* **2017**, *138*, 127–136. [CrossRef]
25. Han, Q.; Xue, L.; Qi, T.; Liu, Y.; Yang, M.; Chu, X.; Liu, S. Assessing the Impacts of Future Climate and Land-Use Changes on Streamflow under Multiple Scenarios: A Case Study of the Upper Reaches of the Tarim River in Northwest China. *Water* **2024**, *16*, 100. [CrossRef]
26. Liao, S.; Xue, L.; Dong, Z.; Zhu, B.; Zhang, K.; Wei, Q.; Fu, F.; Wei, G. Cumulative ecohydrological response to hydrological processes in arid basins. *Ecol. Indic.* **2020**, *111*, 106005. [CrossRef]
27. Xue, L.; Wang, J.; Zhang, L.; Wei, G.; Zhu, B. Spatiotemporal analysis of ecological vulnerability and management in the Tarim River Basin, China. *Sci. Total Environ.* **2019**, *649*, 876–888. [CrossRef] [PubMed]
28. Huang, T.; Pang, Z. Changes in groundwater induced by water diversion in the Lower Tarim River, Xinjiang Uygur, NW China: Evidence from environmental isotopes and water chemistry. *J. Hydrol.* **2010**, *387*, 188–201. [CrossRef]
29. Chen, Y.; Chen, Y.; Xu, C.; Ye, Z.; Li, Z.; Zhu, C.; Ma, X. Effects of ecological water conveyance on groundwater dynamics and riparian vegetation in the lower reaches of Tarim River, China. *Hydrol. Process.* **2010**, *24*, 170–177. [CrossRef]
30. Hao, X.; Li, W.; Huang, X.; Zhu, C.; Ma, J. Assessment of the groundwater threshold of desert riparian forest vegetation along the middle and lower reaches of the Tarim River, China. *Hydrol. Process.* **2010**, *24*, 178–186. [CrossRef]
31. Zhu, C.; Shen, Q.; Zhang, K.; Zhang, X.; Li, J. Multiscale Detection and Assessment of Vegetation Eco-Environmental Restoration following Ecological Water Compensation in the Lower Reaches of the Tarim River, China. *Remote Sens.* **2022**, *14*, 5855. [CrossRef]
32. Wang, S.; Zhou, K.; Zuo, Q.; Wang, J.; Wang, W. Land use/land cover change responses to ecological water conveyance in the lower reaches of Tarim River, China. *J. Arid. Land* **2021**, *13*, 1274–1286. [CrossRef]
33. Yan, W.; Ma, X.; Liu, Y.; Qian, K.; Yang, X.; Li, J.; Wang, Y. Ecological Assessment of Terminal Lake Basins in Central Asia under Changing Landscape Patterns. *Remote Sens.* **2022**, *14*, 4842. [CrossRef]
34. Zhao, X.; Xu, H.; Zhang, Q.; Liu, K. Whether the ecological benefits will continue to increase as usual and improve under the background of continuous ecological water delivery?—Taking the Lower Tarim River in China as an example. *Ecol. Indic.* **2024**, *159*, 111733. [CrossRef]
35. He, J.; Yang, K.; Tang, W.; Lu, H.; Qin, J.; Chen, Y.; Li, X. The first high-resolution meteorological forcing dataset for land process studies over China. *Sci. Data* **2020**, *7*, 25. [CrossRef] [PubMed]
36. Yang, K.; He, J.; Tang, W.; Qin, J.; Cheng, C.C.K. On downward shortwave and longwave radiations over high altitude regions: Observation and modeling in the Tibetan Plateau. *Agric. For. Meteorol.* **2010**, *150*, 38–46. [CrossRef]
37. Ren, Z.; Tian, Z.; Wei, H.; Liu, Y.; Yu, Y. Spatiotemporal evolution and driving mechanisms of vegetation in the Yellow River Basin, China during 2000–2020. *Ecol. Indic.* **2022**, *138*, 108832. [CrossRef]
38. Sen, P.K. Estimates of the Regression Coefficient Based on Kendall's Tau. *J. Am. Stat. Assoc.* **1968**, *63*, 1379–1389. [CrossRef]
39. Kendall, M.G. Rank Correlation Methods. *Br. J. Psychol.* **1990**, *25*, 86–91. [CrossRef]
40. Mann, H.B. Non-parametric tests against trend. *Econometrica* **1945**, *13*, 245. [CrossRef]
41. Hurst, H.E. Long-Term Storage Capacity of Reservoirs. *Trans. Am. Soc. Civ. Eng.* **1951**, *116*, 770–799. [CrossRef]
42. Mandelbrot, B.B.; Wallis, J.R. Robustness of the rescaled range R/S in the measurement of noncyclic long run statistical dependence. *Water Resour. Res.* **1969**, *5*, 967–988. [CrossRef]

43. Bai, Y. Analysis of vegetation dynamics in the Qinling-Daba Mountains region from MODIS time series data. *Ecol. Indic.* **2021**, *129*, 108029. [CrossRef]
44. Gu, Z.; Duan, X.; Shi, Y.; Li, Y.; Pan, X. Spatiotemporal variation in vegetation coverage and its response to climatic factors in the Red River Basin, China. *Ecol. Indic.* **2018**, *93*, 54–64. [CrossRef]
45. Wang, B.; Xu, G.; Li, P.; Li, Z.; Zhang, Y.; Cheng, Y.; Jia, L.; Zhang, J. Vegetation dynamics and their relationships with climatic factors in the Qinling Mountains of China. *Ecol. Indic.* **2020**, *108*, 105719. [CrossRef]
46. Hu, R.; Wang, Y.; Chang, J.; Istanbulluoglu, E.; Guo, A.; Meng, X.; Li, Z.; He, B.; Zhao, Y. Coupling water cycle processes with water demand routes of vegetation using a cascade causal modeling approach in arid inland basins. *Sci. Total Environ.* **2022**, *840*, 156492. [CrossRef] [PubMed]
47. Yu, Z.; Zhang, Y.; Wang, P.; Yu, J.; Wang, T.; Shi, S. Detection of the nonlinear response of vegetation to terrestrial water storage changes in central Asian endorheic basins. *Ecol. Indic.* **2023**, *154*, 110901. [CrossRef]
48. Song, J.; Betz, F.; Aishan, T.; Halik, U.; Abliz, A. Impact of water supply on the restoration of the severely damaged riparian plants along the Tarim River in Xinjiang, Northwest China. *Ecol. Indic.* **2024**, *158*, 111570. [CrossRef]
49. Liu, Q.; Dai, H.; Gui, D.; Hu, B.X.; Ye, M.; Wei, G.; Qin, J.; Zhang, J. Evaluation and optimization of the water diversion system of ecohydrological restoration megaproject of Tarim River, China, through wavelet analysis and a neural network. *J. Hydrol.* **2022**, *608*, 127586. [CrossRef]
50. Huang, F.; Ochoa, C.G.; Chen, X.; Zhang, D. Modeling oasis dynamics driven by ecological water diversion and implications for oasis restoration in arid endorheic basins. *J. Hydrol.* **2021**, *593*, 125774. [CrossRef]
51. Liu, L.; Wang, Y.; Wang, Z.; Li, D.; Zhang, Y.; Qin, D.; Li, S. Elevation-dependent decline in vegetation greening rate driven by increasing dryness based on three satellite NDVI datasets on the Tibetan Plateau. *Ecol. Indic.* **2019**, *107*, 105569. [CrossRef]
52. Brinkmann, K.; Dickhoefer, U.; Schlecht, E.; Buerkert, A. Quantification of Aboveground Rangeland Productivity and Anthropogenic Degradation on the Arabian Peninsula Using Landsat Imagery and Field Inventory Data. *Remote Sens. Environ.* **2011**, *115*, 465–474. [CrossRef]
53. Shi, S.; Yu, J.; Wang, F.; Wang, P.; Jin, K. Quantitative contributions of climate change and human activities to vegetation changes over multiple time scales on the loess plateau. *Sci. Total Environ.* **2021**, *755 Pt 2*, 142419. [CrossRef]
54. Jiang, L.; Jiapaer, G.; Bao, A.; Guo, H.; Ndayisaba, F. Vegetation dynamics and responses to climate change and human activities in Central Asia. *Sci. Total Environ.* **2017**, *599*, 967–980. [CrossRef]

water

MDPI

Article

Evolution Characteristics of Meteorological and Hydrological Drought in an Arid Oasis of Northwest China

Yier Dan [1,2,†], Hao Tian [1,2,†], Muhammad Arsalan Farid [1,2], Guang Yang [1,2,*], Xiaolong Li [1,2], Pengfei Li [1,2], Yongli Gao [3], Xinlin He [1,2], Fadong Li [4], Bing Liu [1,2] and Yi Li [5]

1 College of Water Conservancy & Architectural Engineering, Shihezi University, Shihezi 832000, China; 18990618877@139.com (Y.D.); haotian1996_china@163.com (H.T.); engr.arsalan.406@gmail.com (M.A.F.); lixiaolong409@shzu.edu.cn (X.L.); pengfei@shzu.edu.cn (P.L.); hexinlin2002@163.com (X.H.); liubing198306@163.com (B.L.)

2 Key Laboratory of Cold and Arid Regions Eco-Hydraulic Engineering of Xinjiang Production & Construction Corps, Shihezi 832000, China

3 Department of Earth and Planetary Sciences, University of Texas at San Antonio, San Antonio, TX 78249, USA; yongli.gao@utsa.edu

4 Institute of Geographic Sciences and Natural Resources Research, Beijing 100101, China; lifadong@igsnrr.ac.cn

5 College of Water Resources and Architectural Engineering, Northwest Agriculture and Forestry University, Xianyang 712100, China; liyi@nwafu.edu.cn

* Correspondence: mikeyork@163.com; Tel.: +86-150-0993-0733

† These authors contributed equally to this work.

Abstract: In the context of global warming, the acceleration of the water cycle increases the risk of meteorological drought (MD) and hydrological drought (HD) in the arid region of Northwest China. The Manas River Basin is a typical agricultural oasis and the largest oasis farming area in Xinjiang, Northwest China. Droughts in this basin have significant implications for both agricultural production and the livelihoods of inhabitants. To evaluate the MD and HD and provide information for drought relief in the MRB, the standardized precipitation evapotranspiration index (SPEI) and standardized runoff index (SRI) were calculated using long-term rainfall and runoff data. Subsequently, combined with ArcGIS 10.3 software and the trend analysis method, the SPEI and SRI characteristics were evaluated at different time scales (1-, 3-, 6-, and 12-month). There were three main findings. First, both MD and HD were alleviated, with significantly more HD alleviation. MDs in spring and autumn exhibited a trend of aggravation. The SRIs in summer, autumn, and winter increased significantly at a confidence level of $p < 0.01$, with an insignificant decline in spring. In the 2010s, the frequency of light drought of MD was stable at 10% to 20%, while severe and extreme droughts increased. The frequency of HDs has decreased since the 1990s. Second, on annual and seasonal scales, MDs occurred mainly as light and moderate droughts. The highest frequency of MD was 24% of moderate droughts in winter. Spatially, the northern region of the MRB was characterized by more frequent light and extreme droughts. Third, runoff in the Manas River Basin increased significantly during the 1990s, which may have been related to the acceleration of glacial retreat in the Tianshan Mountains. This study can effectively reveal the changes in meteorological and hydrological drought in NWC and provide the basis for risk decision-making and management for watershed managers.

Keywords: meteorological and hydrological droughts; spatiotemporal distribution characteristics; Manas River Basin

1. Introduction

Drought is a disaster with far-reaching effects and prolonged durations worldwide. Accurately monitoring a drought is challenging because of its repetitive nature [1,2]. This calamity damages the natural environment, including vegetation loss, land desertification, and oasis shrinkage, and affects the development of human societies due to a lack of water resources, crop reduction, and resulting sandstorms [3,4]. A highlight of the sixth report of the Intergovernmental Panel on Climate Change [5] was that global warming has led to an increased atmospheric evaporation demand and intensity of drought events. As the temperature on land increases more than that of the ocean, the relative humidity near the surface decreases, resulting in regional drought. Water resources are indispensable for sustainable economic development [6]. Recent studies have emphasized that extreme drought events caused by climate change threaten the water balance in various regions, resulting in the drying of rivers, declines in lake water levels, restriction of water use, and destruction of ecosystems [7,8]. In California, the worst drought in a millennium [9] has resulted in the death of large areas of coniferous forest. Water shortages in the arid and semi-arid areas of Northwest China have resulted in vegetation degradation and severe desertification [10]. Tropical plateau glaciers in East Africa have receded [11], and hydropower generation in West African rivers has decreased. Droughts have unpredictable consequences for both humans and nature. Therefore, to mitigate the consequences of drought, it is necessary to investigate the temporal and spatial variations in drought conditions. This knowledge can provide guidance for drought relief efforts and significantly impact agricultural production, social life, and the environment [12,13].

Drought is a serious problem that affects many global regions. Three types of droughts have been studied: meteorological, agricultural, and hydrological [14]. Meteorological drought is the cause of almost all droughts. This drought occurs when precipitation is below the normal threshold for an extended period. Agricultural drought occurs when soil water and rainfall are insufficient to maintain crop growth during the growing season. The occurrence of hydrological drought is related to surface water and groundwater and is caused by a lack of effluent, which is closely linked to the water cycle [15,16]. Generally, meteorological droughts have a significant impact on large regional scales [17]. As rainfall deficits and evaporation rates increase, river runoff and groundwater recharge decrease, leading to hydrological droughts. Several scholars have conducted extensive research on meteorological and hydrological droughts. These studies include the use of Pearson's linear or copula function-based conditional probability nonlinear methods [18] to calculate the drought propagation time, trigger the threshold of meteorological drought to hydrological drought propagation in water-limited areas [19], and determine the relationship between meteorological and hydrological droughts based on wavelet analysis [20].

Data obtained from meteorological stations are characterized by long time series and high accuracy. Numerous researchers have utilized meteorological station data to calculate meteorological drought indices such as the standardized precipitation index (SPI), Palmer drought severity index (PDSI), and standardized precipitation evapotranspiration index (SPEI). Among these indices, the SPEI is widely used in drought assessment because it incorporates the advantages of the multiple time scales of the SPI and the sensitivity to evapotranspiration of the PDSI [21]. The SPEI is particularly useful in arid and semi-arid regions. The standardized runoff index (SRI), which is based on runoff volume, can be used to monitor and evaluate hydrological droughts and determine the drought status in a basin. To ensure efficient use and management of water resources, it is essential to study the characteristics of meteorological and hydrological droughts. Muhammad et al. (2020) used the SPEI and SRI to predict the impact of climate change on the propagation of meteorological to hydrological drought [22]. The authors found that the propagation probability of meteorological to hydrological drought in Korea has increased significantly under the influence of climate change. Using the SPEI and SRI, Li et al. (2023) analyzed the spatiotemporal variations of meteorological and hydrological droughts in various regions, including the upper reaches of the Yangtze River, over the past 120 years [23]. The authors reported that the two indices are effective in identifying relatively continuous dry and wet periods. The collective findings indicate the reliability of SPEI and SRI for assessing meteorological and hydrological drought conditions.

The Manas River Basin (MRB) is the largest oasis farming area in Xinjiang [24], the fourth largest irrigated agricultural region in China, and an essential part of the local economic belt. However, in recent years, the basin has been affected by the increased frequency of drought events due to global warming. This has restricted water resource usage, created ecological security barriers, and hindered socioeconomic development in the region. Water resources are the foundation of production and life in the MRB. In this context, it is essential to understand how drought conditions in the MRB are changing. Meteorological and hydrological droughts in the MRB have not been investigated. This study aimed to analyze the drought situation in the MRB using the SPEI and SRI, with the ultimate goal of guiding production and life in the river basin. The study objectives were to clarify the temporal and spatial evolution characteristics of meteorological and hydrological droughts, periodic characteristics of the meteorological and hydrological droughts, and runoff trends in the MRB. By gaining a better understanding of the drought situation in the MRB, it will be possible to enhance the region's ability to prevent and manage drought disasters.

2. Materials and Methods

2.1. Study Area

The MRB is located in the Xinjiang territory of China between 43°27′~45°21′ N and 85°01′~86°32′ E (Figure 1). This is an arid region of Northwest China on the southern edge of the Junggar Basin. The landscape of the basin inclines from southeast to northwest, with the highest elevation at 5145 m and the lowest at 241 m. Its mountainous oasis plains and desert areas constitute a "mountain-oasis-desert" topographical structure. Rainfall in the basin is minimal, occurs mainly during summer, and is unevenly distributed in both time and space. Evaporation is high, and evapotranspiration is much greater than rainfall. The study area has a typical continental temperate climate.

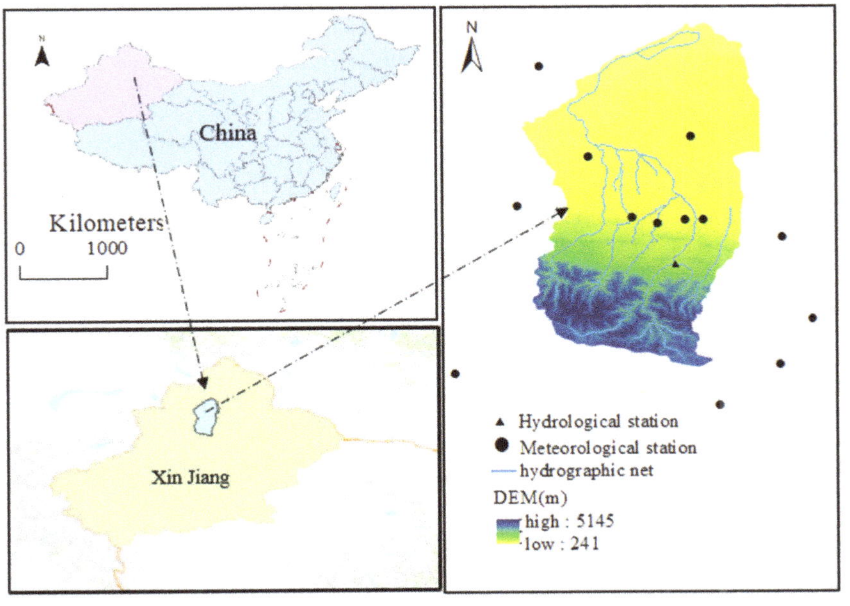

Figure 1. The Manas River Basin study area.

2.2. Data Sources and Processing

Meteorological data from 1970 to 2019, including the monthly average temperature and rainfall, were from 13 meteorological stations in and around the MRB from 1970 to 2019. The data were obtained from the Resources and Environment Science and Data

Center (https://www.resdc.cn/ accessed on 29 November 2022). Hydrological data were obtained from the Kenswat Station, a control hydrological station for the drainage pass. Because the construction of the reservoir began in 2013, the discharge volume was not the natural discharge volume [25] but rather an artificially controlled discharge volume, which was unsuitable for analysis. Therefore, the monthly runoff volume from 1970 to 2012 was selected for the analyses. Meteorological and hydrological data were processed using the SPEI package in R to obtain the SPEI and SRI, which represent the meteorological and hydrological droughts, respectively. In this study, Origin2021 and ArcMap10.3 were used to visualize all the processed data. This study used the data processed in this manner to examine the meteorological and hydrological drought conditions in the region.

2.3. Drought Indices

The Thornthwaite equation [26] was used to calculate the SPEI with multi-time scale characteristics after the standardization of precipitation and evapotranspiration. First, the TH equation calculates the potential evapotranspiration (PET) of the basin using three general methods: the Penman–Monteith (PM), Thornthwaite (TW), and Hargreaves–Samani (HS) equations. However, the PM equation requires many parameters, such as temperature, wind speed, relative humidity, and net radiation, which are not readily available because of the limited number of meteorological stations in Xinjiang and their uneven spatial distribution. The HS equation only requires the highest and lowest temperatures. However, it has limitations under certain weather conditions, such as low humidity or strong winds ($u > 3$ m/s), which can result in inaccurate evapotranspiration (ET0) estimates. In contrast, under conditions of high relative humidity or a low evapotranspiration rate, the ET0 estimate can be relatively high, which can cause errors in the calculation results. Previous researchers have calibrated HS models. However, these calibrations were site-specific and could not be extrapolated to locations with completely different weather conditions [27,28]. Therefore, the TH equation was chosen to calculate the PET in this study. Precipitation data were standardized to represent the water deficit or surplus and were then accumulated into precipitation states at different time scales. Finally, a logarithmic logistic probability distribution was used to fit the sequence data, and the SPEI values at the different time scales were calculated. A negative SPEI value represents a water deficit, whereas a positive value represents a water surplus. The calculation method for the SRI was similar to that for the SPI [29]. Precipitation data replaced the runoff data, and the runoff probability distribution model was selected for normal standardized calculations, which showed short- and long-term drought characteristics. A negative SRI value indicates a lower runoff level, whereas a positive value indicates higher runoff levels. The drought classifications of both the SPEI and SRI are listed in Table 1.

Table 1. Classification schemes of drought intensity.

Drought Classes	SPEI	SRI
No drought/wet	>0.5	>0.5
Near normal	(−0.5, 0.5]	(−0.5, 0.5]
Light drought	(−1.0, −0.5]	(−1.0, −0.5]
Moderate drought	(−1.5, −1.0]	(−1.5, −1.0]
Severe drought	(−2.0, −1.5]	(−2.0, −1.5]
Extreme drought	≤−2.0	≤−2.0

2.4. Trend and Mutation Analyses

Domestic and foreign scholars have extensively researched the trend changes in hydrometeorological time series data. The research methods are mainly divided into parametric and non-parametric test methods. Although parametric methods are typically more efficient than non-parametric techniques, they are usually strict with data that should be distributed, continuous, and independent. Conversely, if the hydrological or meteorological time series do not conform to the normality, independence, and linearity assumptions of the parametric test, non-parametric techniques are employed.

This study utilized the Mann–Kendall non-parametric trend analysis method [30,31] to investigate the meteorological and hydrological drought conditions in the MRB. This method, recommended by the World Meteorological Organization for meteorological research, is widely used to statistically analyze and assess hydrological series trends without requiring adherence to a specific distribution. The Mann–Kendall test features a high level of quantization and is highly effective in determining the direction of sequence changes. Under the observation background of long-term meteorological data, the Mann–Kendall test can also determine whether there is an abrupt climate change in the climate series. If such a change is apparent, the mutation occurrence time is determined. When the UF and UB crossing points appear, and the crossing points are between the critical lines, the corresponding moment of the crossing point is the mutation start time. If the value of the UF or UB statistic is >0, the series exhibits an upward trend. If the value is <0, it indicates a downward trend.

2.5. ArcGIS Spatial Interpolation

The inverse distance weight interpolation (IDW) is based on the basic assumption of "close similarity" in the first law of geography [32]. IDW estimates the pixel value by calculating the average value of sample data points near the pixels to be processed. The closer the data point is to the center of the pixel being estimated, the greater its influence or weight in the average calculation, resulting in a higher degree of similarity as the distance between the points decreases. As a reliable method, IDW combines the advantages of the Tyson polygon adjacent-point method and the trend analysis gradient method while simultaneously accounting for the spatial distribution of regional connections between various factors [33]. This straightforward approach has proven useful in data processing and is well-suited for addressing spatial distribution challenges. In this study, we employed IDW to interpolate the frequency of meteorological drought and investigate the spatial distribution characteristics of drought.

2.6. Run Theory

To identify the meteorological and hydrological droughts, we used the run-course theory [34]. The identification method is shown in Figure 2. An appropriate drought index was selected as the threshold values, R_0, R_1, and R_2, based on the type of drought. A drought event was initially identified when the drought index was below R_1 for a specified period. Drought events lasting only one month and with a drought index above R_2 were excluded from the identified drought events. When the interval between two drought events was one month and the drought index for that month was below R_0, the two drought events were combined. The duration of the drought event identified by the run-course theory is the length of the drought. The intensity of the drought event was represented by the absolute value of the sum of the drought index values below the threshold R_1 for each month of the drought event. In this study, the values at which the drought index fell below 0, indicating light drought and moderate drought, were selected as the threshold values. Therefore, R_0, R_1, and R_2 were equivalent to 0, -0.5, and -1, respectively.

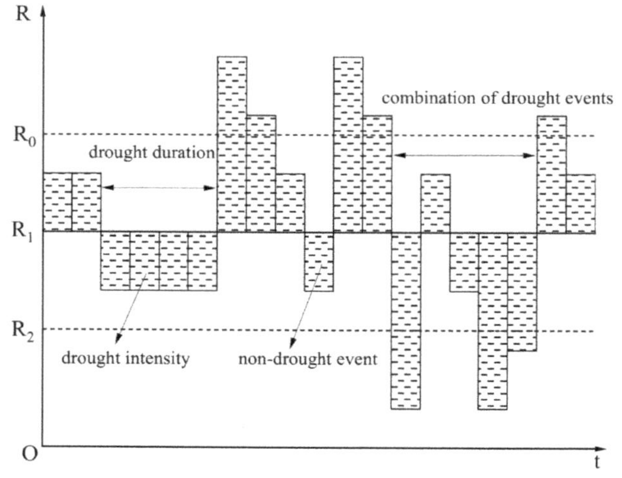

Figure 2. Schematic diagram of drought events identification, pooling, and exclusion.

3. Results

3.1. Temporal Evolution Characteristics of Meteorological and Hydrological Elements

To analyze the interannual variation trends of precipitation, as well as the average temperature in the MRB and discharge of the Kenswart Hydrological Station, meteorological and hydrological elements were selected for linear fitting (Figure 3). The figure illustrates the multiyear values of the three factors. According to the results, both temperature and rainfall passed the significance test of 0.05, whereas rainfall and runoff passed the significance test of 0.01. In the past 50 years, the rainfall in the MRB has increased significantly (Zs = 3.16, $p < 0.01$), with an average of 243.35 mm—increasing by approximately 13.3 mm every 10a—while the average temperature also exhibited a significant rise (Zs = 2.19, $p = 0.02$), with an average temperature of 8.75 °C and growing by about 0.2 °C every 10a. The consistent rainfall and temperature trends are indicative of warmer and wetter climates. The overall fluctuation of the flow of 43a Kenswart Hydrological Station increased significantly (Zs = 3.62, $p < 0.01$), reaching a maximum value of 1946.07 m^3/s in 1999. This increase in flow reduced the occurrence and intensity of hydrological drought events. Although the runoff decreased in the 21st century, the overall growth trend remained the same. These results indicate that the climatic and hydrological conditions in the study area are changing.

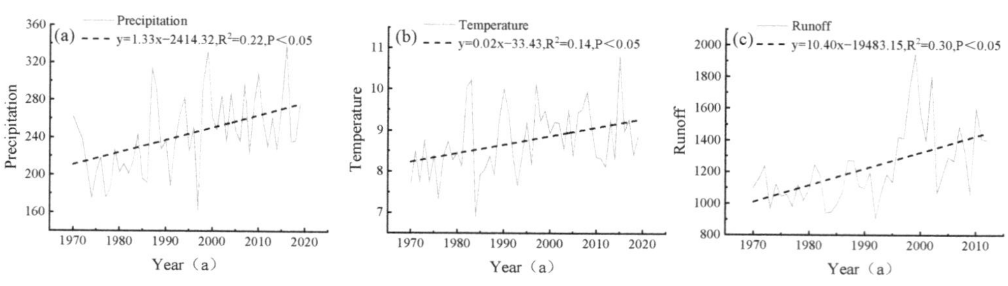

Figure 3. Interannual variation characteristics of precipitation (**a**), mean temperature (**b**), and runoff (**c**) in the MRB.

3.2. Temporal and Spatial Characteristics of Meteorological Drought

3.2.1. Interannual Evolution Trend of Meteorological Drought

Monthly SPEI values on the 1, 3, 6, and 12 scales in the MRB from 1970 to 2019 were calculated to characterize meteorological drought (Figure 4). The figure shows that with the increase in the SPEI time scale, the fluctuation gradually decreases and the continuity becomes stronger. This is mainly due to the high sensitivity of the SPEI to the precipitation of the current month, and the change in meteorological drought is greatly affected by short-term precipitation, resulting in obvious SPEI fluctuations between months. However, with the accumulation of precipitation, the sensitivity of the SPEI to precipitation decreases with the increase in the time scale and demonstrates less randomness and less volatility. The SPEI has a positive linear slope on both short- and long-term scales, and the annual SPEI scale increases at a rate of 0.0037 a^{-1}, demonstrating that meteorological droughts in the MRB have been alleviated in the last 50 years. Notably, in 1997, the SPEI1, SPEI3, SPEI6, and SPEI12 values were all <-1.4, with the minimum value reaching -2.24, indicating severe and extreme droughts. As shown in Figure 4b, the drought events occurred primarily in the 1970s, 2000s, and 2010s, with fewer events occurring in the 1990s. However, drought severity was high during this period, and the SPEI value was the lowest, indicating that relatively serious drought events occurred in the MRB during this period. The interannual variations in drought events are evident in Figure 4d. From 1970 to 2019, there were ten prominent low peaks in the SPEI value, and the average drought duration was 12 months. The longest drought duration lasted 19 months, from November 1977to March 1979, with an average SPEI value of -0.95, which corresponds to a moderate drought. Conversely, the shortest drought duration was two months, from August 1983 to September 1983. The average SPEI value from April 1997 to April 1998 was -1.39, indicating severe drought.

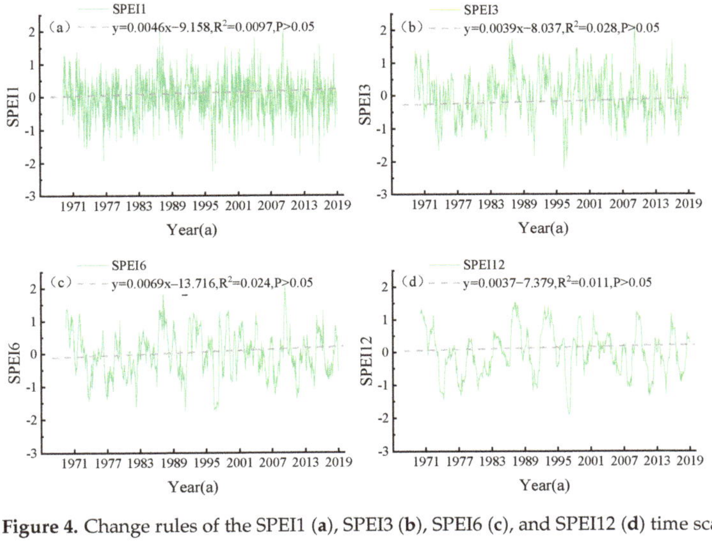

Figure 4. Change rules of the SPEI1 (**a**), SPEI3 (**b**), SPEI6 (**c**), and SPEI12 (**d**) time scales.

3.2.2. Seasonal Variation Trends of Meteorological Drought

To investigate the seasonal variations in meteorological drought in the basin, the seasonal scale SPEI indices of May, August, November, and February of the following year were chosen to represent spring (March–May), summer (June–August), autumn (September–November), and winter (December–February), respectively. The SPEI indices for spring, summer, and autumn decreased at a rate of $-0.063 \times 10\ a^{-1}$, $-0.003 \times 10\ a^{-1}$, and $-0.086 \times 10\ a^{-1}$, respectively, indicating that drought in these three seasons is insignificant (Figure 5a–c). In contrast, the SPEI in winter increased at a rate of $0.154 \times 10\ a^{-1}$. The

Z-values for spring, summer, autumn, and winter were −0.652, −0.05, −1.104, and 2.576, respectively. The results indicate that the drought conditions in spring and autumn were characterized by frequent dry and wet alterations and significantly negative SPEI values after 1997. The summer SPEI values were negative from 1973 to 1983 and were generally >0 after 1984. Severe and extreme drought events in winter were infrequent, with a low degree of drought. The UF statistics in spring and autumn did not exceed the critical line with a significance level of 0.05 (−1.96 < UF < 1.96), indicating that there is no abrupt point in the variation curve of the drought index. The UF and UB curves intersected in approximately 1983 in the summer, indicating that 1983 was a sudden change year. Since 1977, summer droughts in the MRB have significantly intensified. The UF and UB curves intersected in approximately 1993 in winter. Thus, 1993 was a sudden change year, and after 2003, the winter-wetting state was significant.

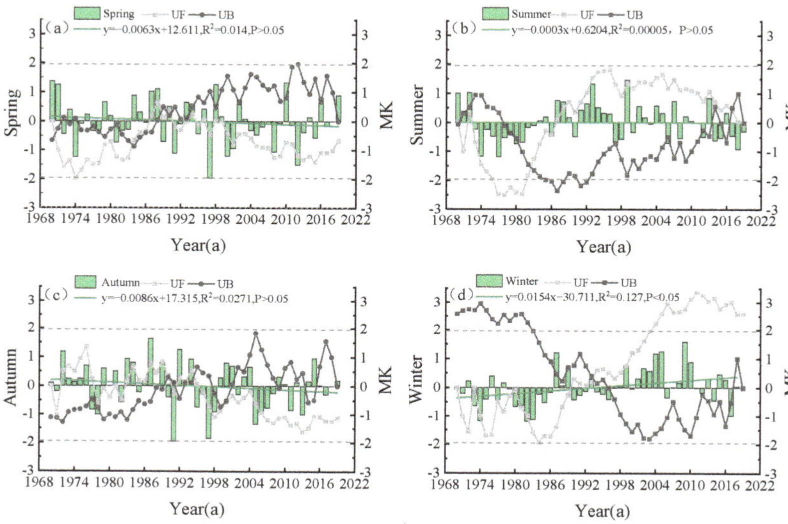

Figure 5. Meteorological drought change characteristics in spring (**a**), summer (**b**), autumn (**c**), and winter (**d**).

3.2.3. Analysis of Annual Scale Meteorological Drought Frequency

Table 2 shows the annual drought frequency in the MRB. TD, LD, MD, SD, and ED represent the total annual scales of drought, light drought, moderate drought, severe drought, and extreme drought, respectively. The total drought frequency ranges from 28% to 32%. Further analyses of the drought conditions of different grades showed that the annual frequency of LD was between 11% and 16%, and the frequency of LD was higher at Shihezi station. The frequency of MDs was between 8% and 12%, and drought was more evident at the Shawan and Wulan Wusu stations. The frequency of SDs was between 4% and 6%, mainly in the Manas. The frequency of EDs was between 1% and 3% and was concentrated in Ulan Usu. In general, LD and MD occurred easily in the MRB. However, the frequencies of SD and ED were low, showing the characteristics of a high frequency of LD and MD and a low frequency of SD and ED.

3.2.4. Analysis of Annual Scale Meteorological Drought Frequency

The IDW method in ArcGIS software was used to analyze the spatial distribution characteristics of drought frequency at different seasonal scales. Figure 6 shows that spring droughts ranged from 26% to 38%, with an average of approximately 31.69%. Among the three stations, Manas experienced the highest frequency of drought (38%), whereas

Shihezi and Ulan Wusu experienced 32%. In spring, the frequency of LDs ranged from 10% to 20%, and all three stations were prone to this phenomenon. The frequency of MDs was between 4% and 14% and was mainly concentrated in the eastern part of the river basin. The frequency of summer droughts varied between 28% and 36%, with a maximum value of 36% in Shihezi and a minimum value of 28% in Shawan, Ulan Wusu, and Manas. Summer droughts were more frequent in the basin, and the LD frequency ranged from 10% to 24%. The frequency of MDs ranged from 4% to 16%, revealing a declining trend from north to south. The frequency of autumn droughts ranged between 26% and 34%, with a higher frequency in the east and a lower frequency in the west. The LD and SD frequencies gradually decreased from north to south. In winter, LDs were concentrated in Paotai, whereas MDs were mainly concentrated in Mossos Bay. SDs and EDs occurred less frequently during this season. The frequency of drought in spring and summer was higher than that in autumn and winter (Figure 6(a1–a4)); The frequency of LDs in autumn was lower than that in other seasons, and the frequency of LDs in winter was more severe (Figure 6(b1,b2)). EDs in the summer season were uncommon but occasionally occurred in the spring and winter months (Figure 6(e1–e4)). In summary, the study area was particularly susceptible to LDs and MDs throughout the year. It is crucial to pay close attention to the Shawan and Mosso Bay regions because they are vulnerable to severe and extreme droughts.

Table 2. Year-scale meteorological drought frequency characteristics.

Station	TD	LD	MD	SD	ED
Paotai	29%	14%	8%	5%	2%
Mosowan	30%	14%	8%	5%	2%
Shihezi	32%	16%	9%	5%	2%
Shawan	30%	13%	12%	4%	2%
Wulanwusu	28%	11%	11%	4%	3%
Manas	31%	13%	10%	6%	1%

3.2.5. Annual Variation Characteristics of Groundwater Table

The study area showed similar drought frequency evolution characteristics on an annual scale. From highest to lowest, it was LD > MD > SD > ED. On a seasonal scale, the frequency of LDs was more prominent in spring and did not change significantly throughout the year. As can be seen from (Figure 7a–c), the MD frequency in spring was lower than that in other seasons from 1970 to 1999 and increased from 2000 to 2019 as shown in (Figure 7d). In summer, MD events occurred easily in the 1970s, but the frequency of SDs decreased from 1970 to 1999 and increased from 2000 to 2019. The probability of SDs in autumn was low from 2000 to 2019. In (Figure 7e) the number of winter MD events decreased between 1970 and 2019. Overall, the frequencies of LDs and MDs did not change significantly, but SDs and EDs increased in this decade compared with the 2000s.

3.3. *Time Variation Characteristics of Hydrological Drought*

3.3.1. Interannual Evolution Trend of Hydrological Drought

Hydrological drought was assessed using SRI1, SRI3, SRI6, and SRI12. (Figure 8a–d) shows that at a significance level of $p < 0.01$, all of the time scales demonstrated a significant increase at rates of 0.0582, 0.0485, 0.0487, and 0.498 a^{-1}, respectively. The values indicate that hydrological droughts have been markedly alleviated over the last 43 years. SD and ED events mainly occurred from 1970 to 2000, with the lowest SRI values of the four scales being −2.05, −2.02, −1.89, and −1.56, respectively. After 2000, the basin mostly experienced LD events. In the short time scales of one and three months, the frequency of drought was higher in the 1970s, 1980s, and 1990s, among which SRI1 (Figure 8a) lasted the longest from September 1974 to July 1975, totaling 11 months with an average of −1.34, indicating a moderate drought level. In SRI3 (Figure 8b), the longest drought lasted from November 1983 to March 1985, reaching 17 months with an average value of −1.11.

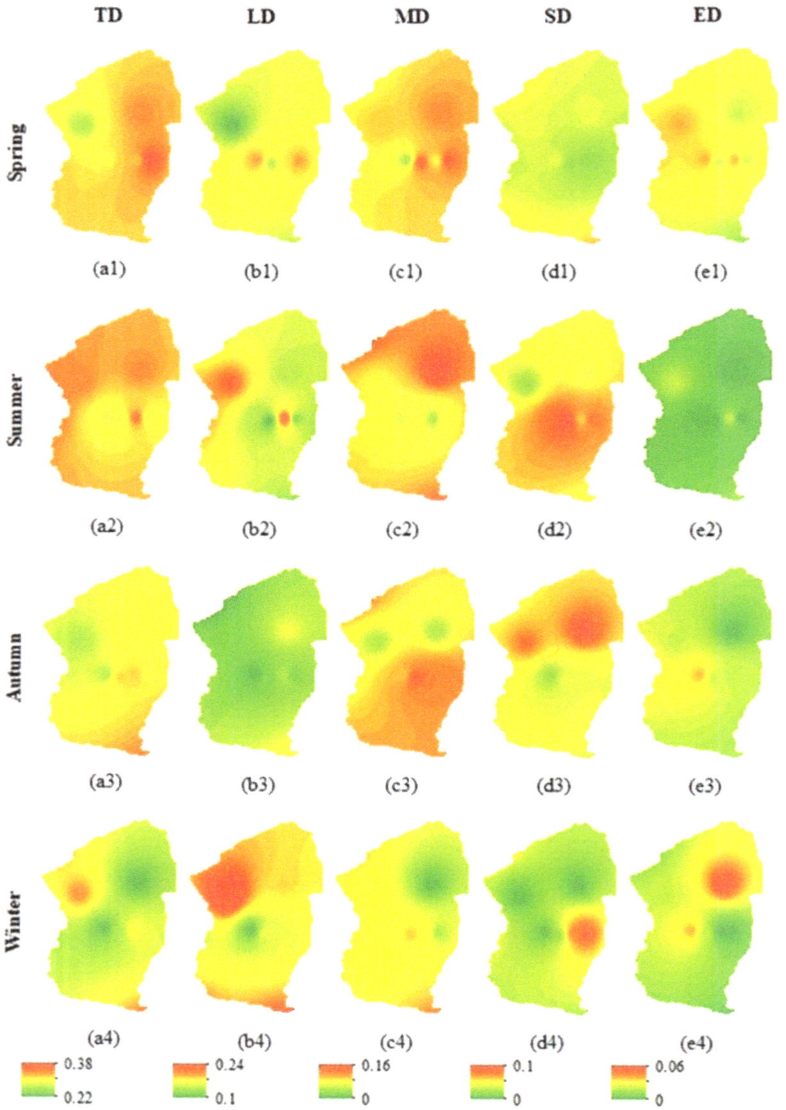

Figure 6. Spatial distribution characteristics of drought frequencies on a seasonal scale. (**a1–a4**) represents the frequency of drought in different seasons of the whole basin, (**b1–b4,c1–c4,d1–d4,e1–e4**) represent the frequency of light drought, medium drought, severe drought and extreme drought in spring, summer, autumn and winter of the basin, respectively.

3.3.2. Seasonal Variation Trends of Hydrological Drought

The seasonal variation trends of hydrological drought were analyzed. A non-significant downward trend was evident in spring SRI (Figure 9a) at the rate of $-0.013a^{-1}$. In contrast, summer (Figure 9b), autumn (Figure 9c), and winter (Figure 9d) showed significant upward trends at confidence levels of 0.04, 0.036, and $0.05a^{-1}$, respectively. The Mann–Kendall trend analysis showed that the Z-values of the four seasons were 1.02, 3.47, 2.97, and 4.5, and all seasons except spring passed the significance test. The hydrological drought-weakening trend was evident in summer, autumn, and winter, whereas the spring SRI values alternated between the dry and wet conditions. The summer SRI was almost negative from 1970 to 1995, with light and moderate droughts. After 1995, the SRI was generally greater than zero, indicating a wet state. MDs easily occur in autumn and winter, and dry and wet states change significantly throughout the year. The UF statistics in spring did not exceed the critical threshold of 0.05 ($-1.96 < UF < 1.96$), indicating no abrupt point in the drought index change curve. In summer, the UF and UB curves intersect around 1993, and abrupt changes occur. Since 1998, summer HDs have decreased significantly. The UF and UB curves in autumn intersected around 1996, indicating that 1996 was a sudden change year and that the wetting state of the fall season was significant after 2001.

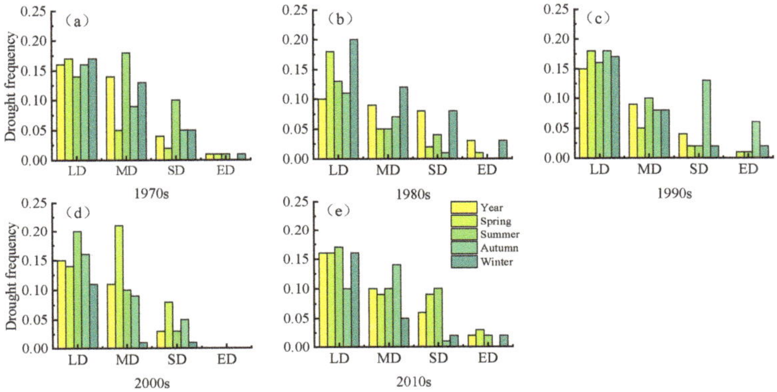

Figure 7. Temporal variation characteristics of meteorological drought frequency. (**a**–**e**) represents the frequency of meteorological drought of different seasons and degrees in five decades.

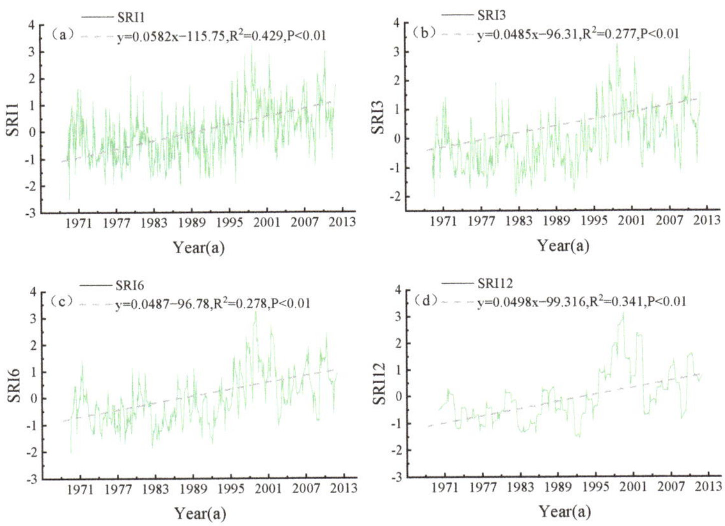

Figure 8. The change rules of the four time scales, SRI1 (**a**), SRI3 (**b**), SRI6 (**c**), and SR12 (**d**).

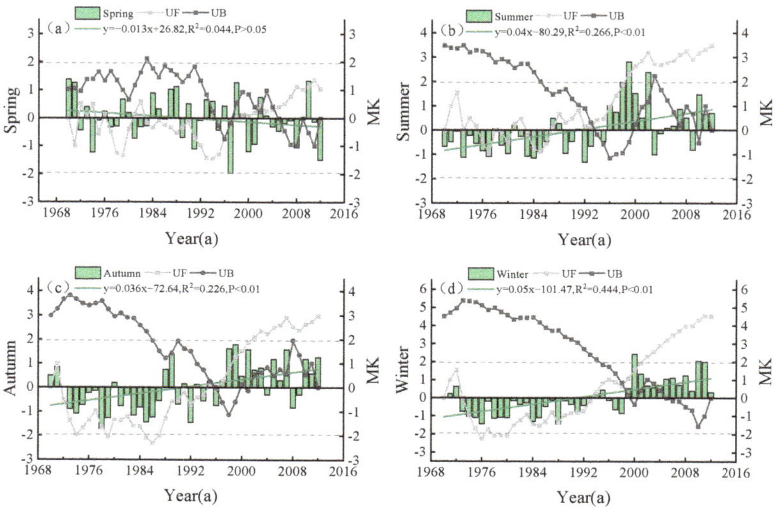

Figure 9. Hydrological drought change characteristics in spring (**a**), summer (**b**), autumn (**c**), and winter (**d**).

3.3.3. Temporal Variation Characteristics of Hydrological Drought Frequency

(Figure 10a–d) shows that from the annual scale, SDs rarely occurred, EDs never occurred, and the frequencies of LDs and MDs decreased each passing year. The change in LDs during spring was negligible, whereas the frequency of LDs decreased during summer. In winter, MDs occurred frequently from 1970 to 1990 (Figure 10a,b) but not after 1991. Notably, the drought frequency of all grades decreased significantly, indicating that hydrological drought was alleviated.

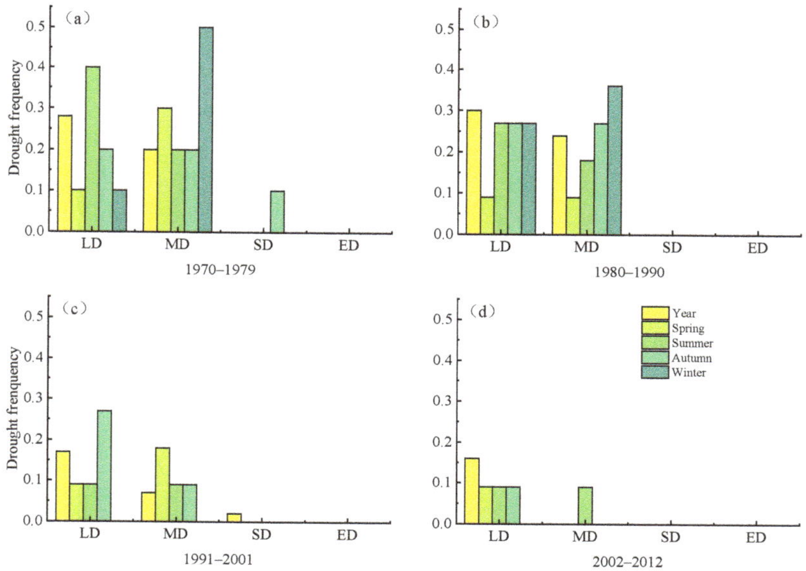

Figure 10. Temporal variation characteristics of hydrological drought frequency. (**a**–**d**) represents the frequency of hydrological drought of different seasons and degrees in recent 43 years.

3.3.4. Drought Characteristic Analysis

The box plot in Figure 11 shows the annual characteristics of the meteorological and hydrological droughts. The duration of the meteorological drought events was relatively short, concentrated within approximately 11 months. The duration of the hydrological drought events was longer, concentrated within approximately 18 months. The average intensity of the meteorological drought events was 5, whereas that of the hydrological drought events was 7. The values indicate that meteorological droughts are characterized by shorter and more concentrated durations and intensities, whereas hydrological droughts have longer and more scattered durations and intensities with greater severity.

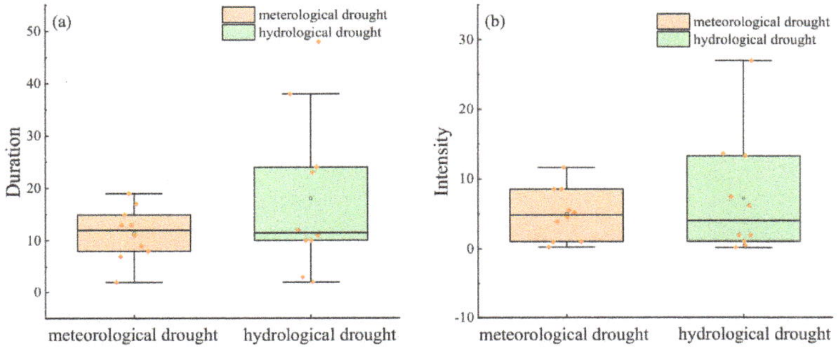

Figure 11. Box chart of annual drought characteristics in the MRB. Figure (**a**) shows the duration months of meteorological and hydrological drought events. Figure (**b**) shows the cumulative index values of meteorological and hydrological drought events from their onset to their end. The orange-yellow dots represent the characteristic values of drought events.

3.3.5. Evolution Trend of Multi-Scale Drought at Different Stations

To comprehensively analyze the multi-timescale evolution of drought in the MRB, data from six meteorological stations and one hydrological station in the basin were selected to calculate the SPEI and SRI values. Over the past 50 years, the frequency of meteorological droughts across the entire river basin (Figure 12a) has been high, with drought events of varying severity occurring in more than half of the years. Yellow and red bands appeared in 1974, 1977, 1992, and 1997, indicating that relatively severe droughts occurred during this period. In these four years, SD, MD, MD, and ED occurred. The Paotai station (Figure 12b) generally exhibits a characteristic drought–humid–drought pattern, with MD events occurring in 1996 and 1997 and ED events occurring in 1974 and 1997, which is consistent with the historical data. Mosso Bay (Figure 12c) experienced five relatively severe droughts, characterized by frequent alternating dry and wet years after 1984. The Shihezi, Shawan, Ulan Wusu, and Manas stations (Figure 12d–g) experienced long drought durations before 1984 and relatively short drought durations after 1984. Among these years, 1987, 1988, and 1993 were wet years. In general, the drought evolution characteristics over multiple time scales at the six meteorological stations were consistent, with droughts occurring relatively easily. From 1970 to 1996, hydrological drought occurred more frequently, except for 1972, 1982, and 1994, when the watershed was wet and the remaining years were dry. After 1996, the watershed changed from dry to wet, with a noticeable increase in moisture content. Overall, the hydrological dry–wet state from 1970 to 2012 showed a trend characteristic of drought–wet–drought.

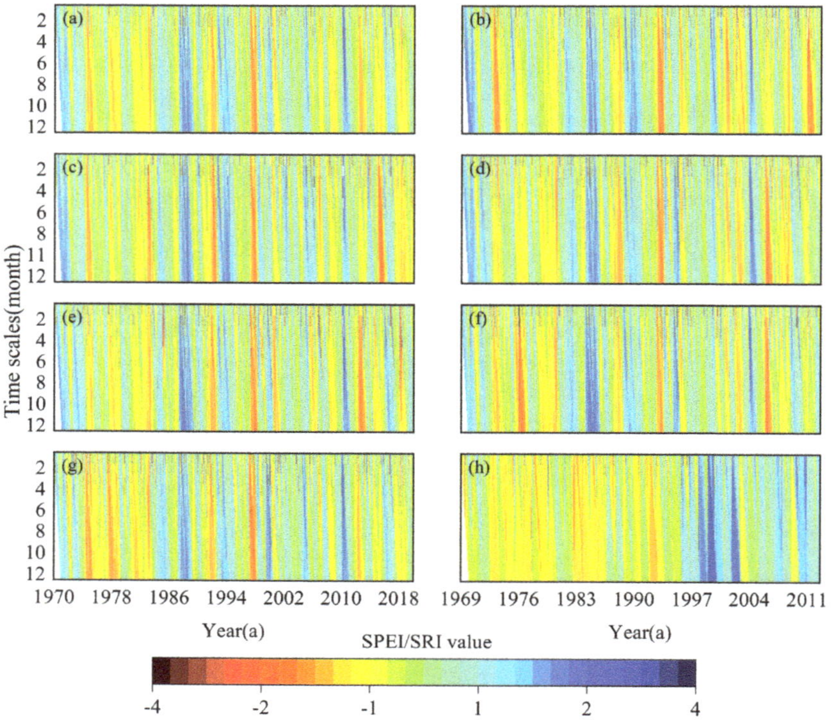

Figure 12. Hovmoller-type maps of the meteorological drought and hydrological drought at different time scales (1–12 months) in the MRB. Subfigures (**a**–**h**) represent the 50–year evolution characteristics of meteorological drought in the whole basin, the 50–year evolution characteristics of meteorological drought at all stations in the basin, and the 50–year evolution characteristics of hydrological drought, respectively.

4. Discussion

4.1. Characteristics of Drought and Runoff Changes

This study found that the precipitation and temperature in the study area showed increasing trends, which aligns with the conclusion of [35] that the northwestern climate would transition to warm and wet conditions. The meteorological drought in the MRB has a weakening trend, but the trend is not obvious, and it maintains a relatively stable state. Seasonally, winter droughts have decreased significantly in recent years, which is related to the increase in rainfall. The weakening trend of hydrological drought is very significant and is closely related to runoff. The MRB, located in the middle of the northern foot of the Tianshan Mountains in Xinjiang, is a typical arid inland area with limited rainfall and intense evaporation. Its river runoff is mainly supplied by mountainous and glacial snowmelt. Some studies have found that with global warming, rising temperatures are causing snowline shifts and mountain snowmelt. Consequently, the overall snow cover area in the MRB showed a downward trend, and the melting of snow and ice significantly increased the recharge of the water inflow of the river, which weakened hydrological droughts and transitioned from drought to wetting [36,37]. Figure 13 shows that the UF and UB runoff curves in the MRB intersected in 1993 and experienced abrupt changes, resulting in a significant increase in runoff after 1993. This finding is consistent with the conclusion of Zhang Zhengyong [38] that the runoff cycle experienced a sudden change in the early 1990s. In Figure 14, the slope of the fitting trend line for runoff from 1993 to

2000 increased significantly, and the slope of the fitting trend line changed from positive to negative from 2000 to 2012; it showed that runoff experienced characteristics of first an increase and then a decrease. Additionally, the drought index SRI in Figure 12h also experienced a period of change from significant wetness to a decreasing wetting degree. The reasons for the significant increase in runoff during the 1990s and the subsequent slight decrease in the 21st century were further analyzed. This increase was primarily attributed to the acceleration of glacier retreat in the Tianshan Mountains during the 1990s. However, glacier retreat slowed or stabilized after the 21st century [39]. Runoff from the Manas River is closely related to the reduction in glacier area and thinning of glacier thickness. Because meltwater is the primary source of runoff in the MRB, its hydrological processes and characteristics are primarily determined by the water stored in ice and snow during winter [40].

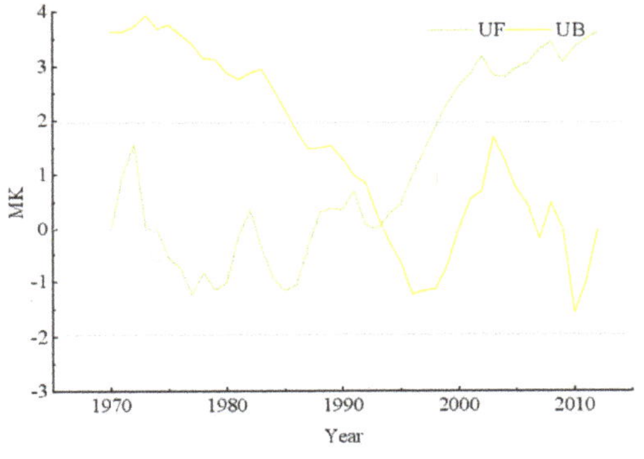

Figure 13. Mann–Kendall mutation detection of runoff sequence in MRB.

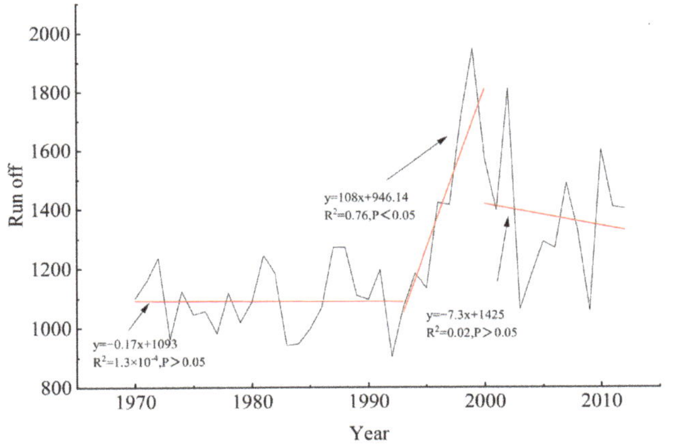

Figure 14. Fractional fitting trend of runoff series in MRB.

4.2. Relationship between Meteorological and Hydrological Droughts

Generally, hydrological drought occurs later than meteorological drought, with a noticeable delay between the two phenomena. According to [41], hydrological drought in the Pearl River Basin lags behind meteorological drought by 2–6 months. The authors of [42] found that hydrological drought in the source region of the Yellow River lags meteorological drought by 3–6 months, and other studies [43] have also demonstrated a lag effect in the propagation of meteorological and hydrological droughts. To examine drought propagation in the MRB, the SPEI and SRI at the 1-, 3-, 6, and 12-month scales of typical drought events from March 1974 to August 1975 were selected for analyses. The yellow- and green-shaded areas represent hydrological drought and meteorological drought, respectively. Drought occurrence is typically defined as a period when the drought index falls below −0.5 for two consecutive months. As shown in Figure 15a, the annual meteorological drought events began in June 1974 and ended in July 1975. Hydrological drought events began in September 1974 and ended in August 1975. Notably, the hydrological drought event occurred three months after the onset of the meteorological drought event and ended one month later. On a seasonal scale, the hydrological drought started five months after the meteorological drought and ended three months after the meteorological drought. In general, the duration of a drought event increases over time. However, except on the annual scale, the peak hydrological drought lagged behind the peak meteorological drought. These findings indicate a time lag between meteorological and hydrological droughts, which are consistent with the previous research results.

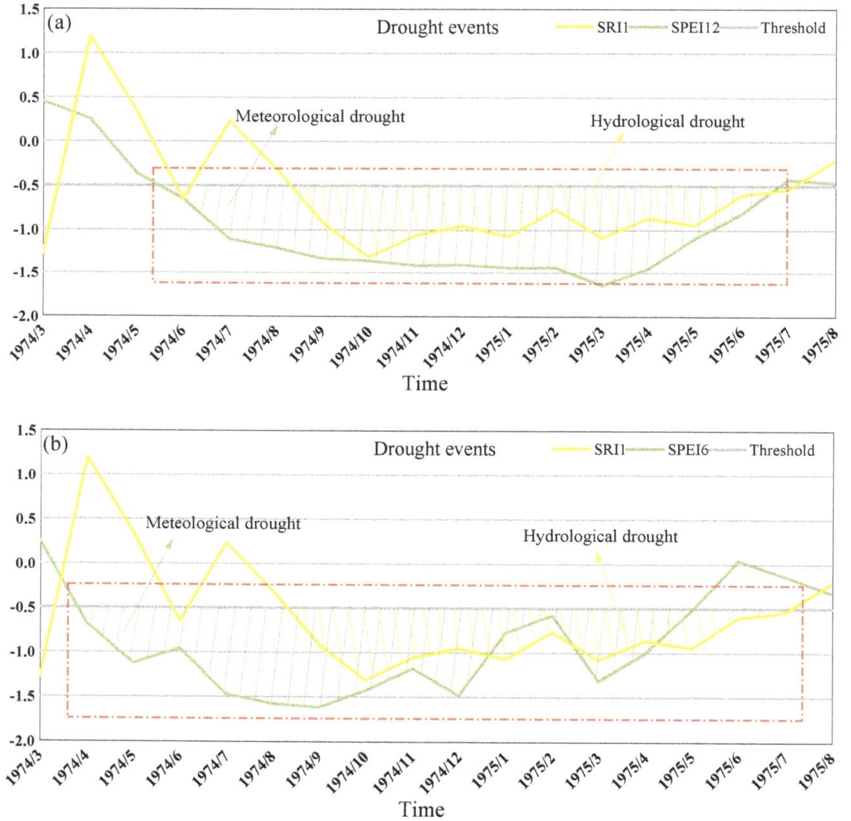

Figure 15. *Cont.*

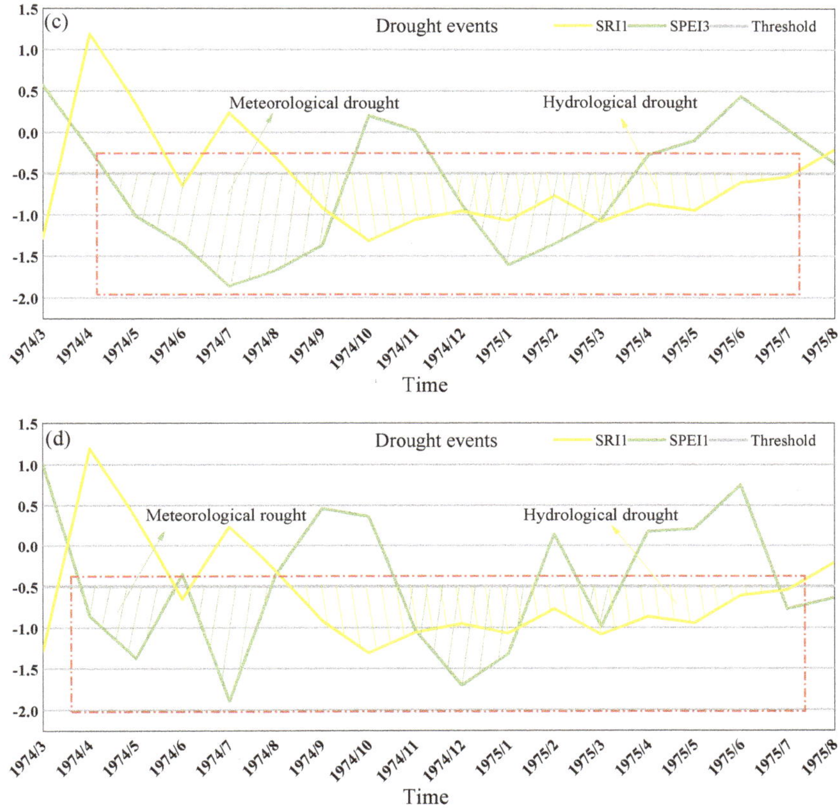

Figure 15. Spatial distribution of the variance of the groundwater table contrast coefficient. (**a–d**) represents the lag relationship between meteorological drought at four time scales and monthly scale hydrological drought in typical drought events. The red dotted frame represent the beginning to the end of the drought.

5. Conclusions

(1) Both meteorological and hydrological droughts decreased, and hydrological droughts decreased significantly. Meteorological droughts tended to increase in spring and autumn, whereas the basin became wetter during winter. The dry and wet conditions in spring and autumn remained relatively stable, with no abrupt changes. Abrupt changes in summer and winter occurred around 1983 and 1993, respectively, from drought to humidity. Except in spring, the SRI did not decrease significantly, and the SRIs in summer, autumn, and winter increased significantly ($p < 0.01$), indicating a shift from drought to humidity and a weakening of seasonal drought.

(2) Meteorological droughts at different time scales and intensities exhibited similar characteristics. Light and moderate droughts are commonly observed at the annual and seasonal scales, whereas severe and extreme droughts are relatively infrequent. In terms of geographic distribution, the northern part of the study region has experienced frequent light and extreme droughts, whereas the southern part has tended to experience more severe droughts during the summer months. Conversely, severe droughts in autumn were more likely to occur in the north. Basin managers can make appropriate water resource allocation plans based on this to effectively reduce the probability of drought. In addition, in this study, Shawan and Mosso Bay are prone to

severe and extreme droughts, so it is important for watershed managers to pay more attention to prevention. From the perspective of meteorological drought frequency, there were no significant changes in the occurrence of light or moderate droughts during the 2010s. However, the frequency of severe and extreme droughts increased during the same period. Hydrological droughts have become less common since the 1990s. Overall, these findings suggest that the frequency of severe and extreme droughts has increased, whereas the frequency of light and moderate droughts has remained relatively stable.

(3) The significant increase in runoff in the MRB during the 1990s could be related to the accelerated glacier retreat that occurred at the same time in the central and western Tianshan Mountains. Hydrological drought, which follows meteorological drought, is inevitably delayed and occurs after meteorological drought.

Author Contributions: Y.D.: Conceptualization, Methodology, Investigation, and Writing the original draft. H.T.: Validation, Data curation, Visualization, and Writing—review and editing. M.A.F.: Validation, Data curation, Visualization, and Writing—review and editing. G.Y.: Visualization and Formal analysis. Y.L. and X.L.: Resources, Supervision and Writing—review and editing. P.L.: Visualization and Formal analysis. Y.G.: wrote, reviewed, and edited the manuscript. X.H.: writing, reviewing, and editing. F.L.: Resources, Supervision, and Writing—reviewing and editing. B.L.: Writing—review and editing. All authors have read and agreed to the published version of the manuscript.

Funding: This research was supported by the National Natural Science Foundation of China (52269006), Projects of Xinjiang Production and Construction Corps (2022BC001, 2023TSYCCX0114, 2022DB023, 2023AB059), Project of Shihezi (2023NY01), and The Third Xinjiang Scientific Expedition Program (2021xjkk0804).

Data Availability Statement: The original contributions presented in this study are included in the article. Further inquiries can be directed to the corresponding authors.

Acknowledgments: We acknowledge the Key Laboratory of Cold and Arid Regions Eco-Hydraulic Engineering of Xinjiang Production and Construction Corps.

Conflicts of Interest: The authors declare no conflicts of interest.

References

1. Wang, R.; Zhang, X.; Guo, E.; Cong, L.; Wang, Y. Characteristics of the Spatial and Temporal Distribution of Drought in Northeast China, 1961–2020. *Water* **2024**, *16*, 234. [CrossRef]
2. Chen, W.H.; Xu, J.; Li, S.C. Study on meteorological and hydrological drought characteristics in the lower reaches of Nujiang River Basin. *Acta Sci. Nat. Univ. Pekin.* **2019**, *55*, 764–772.
3. Xue, H.Z.; Li, Y.Y.; Dong, G.T. Analysis of spatiotemporal variation of meteorological drought in Hexi Corridor based on SPEI index. *Chin. J. Agrometeorol.* **2022**, *43*, 923–934.
4. Zhang, L.; Zhang, Z.; Peng, Z.; Xu, Y.; Zhang, Y.; Mao, J. Linkages between Meteorological and Hydrological Drought in the Jinsha River Basin under a Changing Environment. *Water* **2023**, *15*, 3644. [CrossRef]
5. Jiang, D.B.; Wang, X.X. Interpretation of drought change in IPCC Sixth Assessment Report. *Trans. Atmos. Sci.* **2019**, *44*, 650–653.
6. Zhang, X.; Huang, S.Z.; Guan, Y.H. Research progress, challenges and prospects of drought propagation. *Adv. Earth Sci.* **2023**, *38*, 563–579.
7. Mo, C.; Tang, P.; Huang, K.; Lei, X.; Lai, S.; Deng, J.; Bao, M.; Sun, G.; Xing, Z. Evolution of Drought Trends under Climate Change Scenarios in Karst Basin. *Water* **2023**, *15*, 1934. [CrossRef]
8. Gao, F.; Liu, J.; Chen, W.; Yang, S.K.; Feng, H. Study on meteorological and hydrological drought propagation characteristics and response probability in the upper reaches of Huaihe River Basin. *Res. Soil Water Conserv.* **2023**, *30*, 257–265. [CrossRef]
9. Mukti, R.S.; Wei, M.X.; Christopher, B.E.; Sandra, R.H.; Ming, Y. Tree mortality and biomass loss in drought-affected forests of East Texas, USA. *J. For. Res.* **2021**, *32*, 67–80.
10. Chen, Y.N.; Li, Z.Q.; Xu, J.H.; Shen, Y.J.; Xing, X.X.; Xie, T.; Li, Z.; Yang, L.S.; Xi, H.Y.; Zhu, C.G.; et al. Changes of water resources and ecological environment in arid area of Northwest China and their protection suggestions. *Bull. Chin. Acad. Sci.* **2023**, *38*, 385–393.
11. Yang, D.; Xiong, W.; Xu, Y.L. Impacts of climate change on water resources and agriculture in Africa. *Chin. J. Agrometeorol.* **2016**, *37*, 259–269.

12. Yang, S.K.; Liu, J.; Zhang, T.; Peng, T.; Chang, W.J. Study on the probability of meteorological and hydrological drought *characteristic variables* response in the upper reaches of Hanjiang River. *Water Resour. Prot.* **2023**, *39*, 143–151. [CrossRef]
13. Li, X.; Jiao, Y.; Liu, J. Changes in Drought Characteristics in the Yellow River Basin during the Carbon-Neutral Period under Low-Emission Scenarios. *Water* **2024**, *16*, 1045. [CrossRef]
14. Cao, S.P.; Zhang, L.F.; He, Y.; Zhang, Y.L.; Chen, Y.Y.; Sheng, Y.; Wang, S.Q. Effects and contributions of meteorological drought on agricultural drought under different climatic zones and vegetation types in Northwest China. *Sci. Total Environ.* **2022**, *821*, 153270. [CrossRef] [PubMed]
15. Liu, Q.; Huo, A.; Zhao, Z.; Zhao, X.; Rebouh, N.Y.; Luo, C. Spatial Differentiation and Influencing Factors Analysis of Drought Characteristics Based on the Standardized Precipitation Index: A Case Study of the Yellow River Basin. *Water* **2024**, *16*, 1337. [CrossRef]
16. Xu, Y.R.; Lu, F.; Xie, Z.B.; Zhu, K.; Song, X.Y. Meteorological and hydrological drought characteristics and their response relationship in Chaobai River Basin. *Agric. Res. Arid. Areas* **2019**, *37*, 220–228.
17. Xue, L.Q.; Bai, Q.Y.; Liu, Y.H. The propagation of meteorological drought to hydrological drought in Tarim River Basin under the influence of human activities. *Water Resour. Prot.* **2019**, *39*, 57–72.
18. Meng, D.; Huang, S.Z.; Huang, Q.; Zheng, X.D.; Su, X.L.; Leng, G.Y.; Li, Z.Y.; Fang, W.; Liu, Y.J. Propagation characteristics and mechanism from meteorological to agricultural drought in various seasons. *J. Hydrol.* **2022**, *610*, 127897.
19. Liu, Q.; Yang, Y.T.; Liang, L.Q.; Jun, H.; Yan, D.H.; Wang, X.; Li, C.H.; Sun, T. Thresholds for triggering the propagation of meteorological drought to hydrological drought in water-limited regions of China. *Sci. Total Environ.* **2023**, *876*, 162771. [CrossRef]
20. Qiong, F.L.; Peng, F.H.; Yong, C.H.; Xing, Y.H.; Tian, S.Z.; Guo, B.L.; Hong, J.W. Investigation to the relation between meteorological drought and hydrological drought in the upper Shaying River Basin using wavelet analysis. *Atmos. Res.* **2020**, *234*, 104743.
21. Gao, Q.; Sun, J.W.; Zhao, X.X.; Wu, L.F.; Zhao, J. Multi-scale spatiotemporal characteristics of drought in China based on SPEI. *Water. Saving. Irrig.* **2024**, *6*, 111–120+127.
22. Muhammad, J.; Muhammad, N.S.; Joo-Heon, L.; Tae-Woong, K. Investigating effect of climate change on drought propagation from meteorological to hydrological drought using multi-model ensemble projections. *Stoch. Environ. Res. Risk Assess.* **2020**, *34*, 7–21.
23. Li, S.; Zeng, L.; Zhang, C.J.; Xiao, Q.; Zhang, Q.; Gong, W.T. Temporal and spatial variations of meteorological drought and hydrological drought in the upper reaches of the Yangtze River in recent 120 years and their transmission characteristics. *Clim. Chang. Res.* **2023**, *19*, 263–277.
24. Qin, J.L.; Xue, L.Q. Spatial and temporal variation of vegetation and its relationship with topographic factors in Manas River Basin, arid region of Northwest China. *Ecol. Environ. Sci.* **2019**, *29*, 2179–2188.
25. Chen, F.L.; Yang, K.; Cai, W.J. Research on hydrological drought index based on GAMLSS model: A case study of Manas River Basin. *Geogr. Res.* **2021**, *40*, 2670–2683.
26. Thornthwaite, C.W. An approach toward a rational classification of climate. *Geogr. Rev.* **1948**, *38*, 55–94. [CrossRef]
27. Fu, Y.H.; Shen, X.J.; Li, W.C.; Wu, X.; Zhang, Q.Q. Applicability of crop reference evapotranspiration calculation based on Hargreaves-Samani regression modification. *Arid. Land Geogr.* **2022**, *45*, 1752–1760.
28. Wu, Z.J.; Cui, N.B.; Xu, J.Z.; Cui, Y.L.; Liang, J. Prediction of reference crop evapotranspiration in Guangxi Basin by improved Hargreaves-Samani model. *J. Drain. Irrig. Mach. Eng.* **2022**, *40*, 404–410.
29. Shi, P.; Zhan, H.J.; Qu, S.M.; Feng, J.; Guan, X.X. Correlation analysis of meteorological drought and hydrological drought in the source region of the Yellow Rive. *Water Resour. Prot.* **2022**, *38*, 80–86.
30. Chen, Y.; Shen, H.; Wang, X.H.; Zhao, W.K.; Pan, Z.J.; Wang, J.Y.; Li, S.Y.; Han, D. Urban energy carbon peak assessment method based on Mann-Kendall trend test. *J. Shanghai Jiaotong Univ.* **2023**, *57*, 928–938.
31. Li, C.Y.; Cui, P.; Hao, J.S.; Zhang, G.T.; Wang, J. Variation characteristics of temperature and precipitation in southeast Tibet since 1960. *Plateau Meteorol.* **2019**, *42*, 344–358.
32. Jia, Y.; Cui, N.B.; Wei, X.P.; Gong, D.Z.; Hu, X.T. Applicability evaluation of reference crop evapotranspiration algorithm in Yangtze River Basin based on inverse distance weighting method. *Trans. Chin. Soc. Agric. Eng.* **2016**, *32*, 130–138.
33. Beshr Ashraf, A.A.; Kaloop, M.R. Using modified inverse distance weight and principal component analysis for spatial interpolation of foundation settlement based on geodetic observations. *Open Geosci.* **2022**, *14*, 1310–1323. [CrossRef]
34. Shi, P.; Tang, H.; Qu, S.M. Characteristics of meteorological drought to hydrological drought propagation in Southwest China. *Water Resour. Prot.* **2019**, *39*, 49–56.
35. Yao, J.Q.; Li, M.Y.; Dilinur, T.; Chen, J.; Mao, W.Y. Characteristics of climate warming and humidification at dif-ferent time scales in Xinjiang. *Arid. Zone Res.* **2022**, *39*, 333–346.
36. Zhang, Z.Y.; Li, Z.Q.; He, X.L.; Liu, L.; Wang, P.Y. Research progress of glacier change and water resources in Manas River Basin. *Res. Soil Water Conserv.* **2014**, *21*, 332–337.
37. Wang, Y.J.; Xu, L.P.; Guo, P.; Li, H.T. Trend analysis of ice and snow cover in Mahe River Basin from 1977 to 2013. *Res. Soil Water Conserv.* **2015**, *22*, 208–218.
38. Zhang, Z.Y.; Liu, L.; Tang, X.L. Regional variation and abrupt change of precipitation in the Tianshan Mountains of China in recent 50 years. *J. Arid. Land Resour. Environ.* **2013**, *27*, 85–90.
39. Yao, J.Q.; Li, M.Y.; Dilinur, T. The signal and influence of "wet-dry transition" in Xinjiang. *Acta Geogr. Sin.* **2021**, *76*, 57–72.

40. Zhou, Z.Q.; Shi, H.Y.; Fu, Q.; Ding, Y.B.; Li, T.X.; Wang, Y.; Liu, S.N. Characteristics of Propagation from Mete-orological Drought to Hydrological Drought in the Pearl River Basin. *J. Geophys. Res. Atmos.* **2021**, *126*, e2020JD033959. [CrossRef]
41. Han, R. Impact of Climate Change on Snow Melt in Manas River Basin. Master's Thesis, Shihezi University, Shihezi, China, 2017.
42. Zheng, L.H.; Liu, Y.; Ren, L.L.; Zhu, H.; Yi, H. Temporal and spatial characteristics of meteorological drought and hydrological drought in the Yellow River Basin. *Water Resour. Prot.* **2022**, *38*, 87–95+146.
43. Zhi, X.F.; Tian, Y.T.; Chen, C.C. Research progress and prospects of drought transmission II: Influencing factors and prospects of drought transmission research. *South-North Water Divers. Water Conserv. Sci. Technol.* **2023**, *21*, 654–668.

water

MDPI

Article

Water Governance in the Cambodian Mekong Delta: The Nexus of Farmer Water User Communities (FWUCs), Community Fisheries (CFis), and Community Fish Refuges (CFRs) in the Context of Climate Change

Mak Sithirith [1,*], Sok Sao [1], Sanjiv de Silva [2], Heng Kong [3], Chay Kongkroy [3], Tim Thavrin [3] and Hy Sarun [3]

1 WorldFish, Phnom Penh P.O. Box 1135, Cambodia; soksao1987@yahoo.com
2 International Water Management Institute (IWMI), Battaramulla 10120, Sri Lanka; s.s.desilva@cgiar.org
3 Inland Fishery Research and Development Institute (IFReDI), Fishery Administration (FiA), Phnom Penh 120101, Cambodia; heng.kongspidermans@gmail.com (H.K.)
* Correspondence: maksithirith@gmail.com or m.sithirith@cgiar.org

Abstract: Cambodia faces the challenge of managing excess water during the wet season and insufficient water during the dry season. This harms human life and endangers aquatic and natural resources, agricultural practices, and food security. Water governance is crucial to ensure the well-being of both people and their food security. However, Cambodia's water governance is hindered by various obstacles, including sectoral and centralized influences, top-down and large-scale strategies, weak coordination among relevant agencies, and limited involvement of local communities. This study examines water governance across different sectors, from centralized to community-based natural resources management, and explores the opportunities that can be done to improve water governance. This study undertakes the literature and case studies of farmer water user communities (FWUCs), community fisheries (CFis), and community fish refuges (CFRs) in three Mekong Delta provinces in Cambodia. This study concludes that water governance has been challenged by FWUCs competing for water resources to intensify rice production at the expense of increased pesticides and fertilizer uses, which undermine the fishery productivity, degrade the natural resources in rivers and water bodies, and increase water conflicts among farmers and sectors in the face of climate change. To enhance water governance in Cambodia, it is critical to integrate it at the district level. This will promote sustainable water use and management across the country and pave the way for a brighter future.

Keywords: water governance; rice farming; irrigation; community fisheries; community fish refuge; water conflict

1. Background

Cambodia faces the challenge of managing excess water during the wet season and insufficient water during the dry season. This harms human life and endangers aquatic and natural resources, agricultural practices, and food security [1]. Water governance is crucial to ensure the well-being of both people and their food security. However, Cambodia's water governance is hindered by various obstacles, including sectoral and centralized influences, top-down and large-scale strategies, weak coordination among relevant agencies, and limited involvement of local communities. This study examines water governance across different sectors, from centralized to community-based natural resources management, and explores the opportunities that can be done to improve water governance [1–3].

It is widely believed that effective water management entails the development and management of irrigation systems to store water for rice farming. Despite the Ministry of Water Resources and Meteorology's (MOWRAM) efforts to support rice farming through large-scale irrigations, farmers continue to experience water scarcity issues and annual crop damage from floods. These challenges can be attributed to the centralization of water management through large-scale irrigation development [2–4].

In recent years, Cambodia has adopted a decentralized approach to water management by introducing the Farmer Water User Community (FWUC) system. This system entrusts farmers responsible for managing irrigation systems through contributions in cash and kind [5]. However, the study has revealed that FWUCs tend to be weak and managed centrally, with limited financial and technical support, minimal input and poor ownership from community members, and unclear benefit-sharing mechanisms, contributing to the weak performance of FWUCs. In addition, other community organizations such as community fisheries (CFis) and community fish refuges (CFRs) also compete for water resources to sustain their livelihoods [6–9]. While FWUCs utilize water for rice farming, CFis and CFRs rely on water resources for sustainable fishery productivity. Furthermore, the decentralized water governance is complicated by a centralized system whereby FWUC management and performance are influenced by the decisions and direction of the Ministry of Water Resources and Meteorology (MOWRAM), while CFis/CFRs are influenced by the Fishery Administration (FiA) and the Ministry of Agriculture, Forestry, and Fishery (MAFF). While water is viewed as part of the fishery sector and its management, MOWRAM considers water management as its mandate, and to do so, it is done through irrigation management and development [10]. Nevertheless, MAFF has a role to play when it comes to rice production, and so, water has been strategically planned as part of rice farming planning and programs. These different dimensions and approaches have made the coordination and integration between sectors challenging, implicating decentralized water governance [8,9]. Moreover, the situation is further complicated by climate change, which impacts water availability for all three communities [11,12].

This study examines water governance in FWUCs, CFis/CFRs, and rice farming in three Mekong Delta Provinces in Cambodia and evaluates how water resources are shared among these communities. First, this article reviews the literature concerning water governance and constructs the framework that analyzes water governance in the study sites. Second, this article delves into the subject of water governance, investigating the manner in which water is used by FWUCs, CFis, and CFRs, as well as its impact on water resources. Third, it explores the ramifications of climate change on water usage, including its effects on rice farming seasons and the competition that arises between FWUCs and CFis/CFRs. This study concludes with recommendations for enhancing water governance to promote fishery, rice farming, and livelihood.

2. Conceptual Framework

Water is an essential ingredient for the sustenance of life, the environment, and growth. It naturally flows and is stored in various forms such as rivers, streams, lakes, ponds, and underground reservoirs. These bodies of water provide a habitat for aquatic animals, fish, and plants, while terrestrial plants rely on underground water for growth. The dry season can cause water stress in lakes, rivers, and streams due to evaporation [13]. These are relevant to the Mekong River and Tonle Sap Lake, where Cambodia is part of these river systems, and it experiences heavy floods in the wet season and severe drought in the dry season, raising the need for water governance [3,4]. Water governance is key to the development of Cambodia. In the past, water governance was governed by the open access regime, for instance, Tonle Sap Lake (TSL) and Cambodia's Mekong Delta (CMD) play a critical role in terms of providing natural and cultural capital for numerous communities living around the Lake and the Delta. Hitherto, there have always been relatively plentiful supplies of fish that provide a 'safety net' against famine. Thus, many Cambodians rely on the river and Lake's resources for their living, and they consider these water bodies as a 'social safety net' [14]. However, the growing population and development pressures have increased the demand for water resources, resulting in technical and sectoral systems taking over, including increasing irrigation development, commercial fishing, and industrial uses.

These have made water governance significant challenges with multi-sectoral dimensions. Thus, water governance in Cambodia has evolved and changed over time. The literature on water governance primarily discusses its effects on multiple sectors, including fishery and rice farming. A literature review highlights six dimensions of water governance related to fishery and rice production [15,16].

First, water governance is influenced by different sectors and actors at different levels. Some sectors are considered more economically important than others and, therefore, receive priority in terms of planning and investment aimed at extracting more water to generate income and benefits. Within each sector, actors with power and interests drive decision-making. In the Mekong River Basin, water has been prioritized for hydropower development over supporting the fishery sector, with the goal of securing energy and boosting industrial development in the Mekong countries [17,18]. Hydropower companies are actively involved in driving this development, supported by upstream states of the Mekong River. The industrial sector has also competed for water to support its own development, which has impacted the fishery, agriculture, natural resources, and food security during the production cycle [19–21].

In Cambodia, water governance is characterized by top-down, sectoral, and large-scale approaches, emphasizing the role of the irrigation system as a key sector in managing water resources. As such, it requires a high technical capacity, high costs, and state-driven interventions, which are highly complex, administratively and politically challenging, and have limited capacity of local authorities [22,23]. Consequently, farmers still face water scarcity, affecting agricultural productivity and low yields. In 2000, concerted efforts were made to decentralize water governance by establishing Farmer Water User Communities (FWUCs) to manage water resources at local levels for rice cultivation. FWUCs were formed in accordance with the Water Law (2007) and a subsequent sub-decree, comprising the regulations on water use and fee collection, controlling, and monitoring [10]. However, water governance via FWUCs has been challenged by the lack of focus on roles and responsibilities, particularly with regard to distributing water equitably, effectively, and efficiently to members of FWUCs. There has been little communication and mediation between farmers and the Provincial Department of Water Resources and Meteorology (PDOWRAM) in administrative processes to comply with IWRM procedures and frameworks and even less financial support [24]. At the same time, another form of decentralization of related water governance is the establishment of the community fisheries (CFis) and the community fish refuges (CFRs) to manage fishery resources under the Fishery Administration (FiA), as well as oversight by the Ministry of Agriculture, Forestry, and Fishery (MAFF) [25]. While CFis/CFRs protect water resources and water bodies to improve fishery productivity and fish production, FWUCs extract water to irrigate rice cultivation in the wet and dry seasons. Thus, while decentralized water governance has been constrained by the weak FWUCs, it has also been challenged by competing sectoral interests and weak coordination between sectors at local levels [9,26].

Second, the management of water resources is intricately linked to the unique physical attributes of the Mekong River, lakes, and floodplains, including their overall volume and quality. This issue primarily affects the communities residing upstream and downstream of these vital water sources during both wet and dry seasons, which are the cases in the study areas, where some communities are located upstream of the rivers, lakes, and canals, while others are in the downstream [9,18]. In times of drought in the context of climate change, for instance, upstream communities may consume a disproportionate amount of water, causing a scarcity of this resource for downstream farming and fishing communities. This can result in tensions and disagreements between the various groups. Similarly, during the flood season, the release of excess water by upstream communities can inundate the rice fields of downstream communities, exacerbating existing tensions [17,19,20,24].

Third, water usage, governance, and management across various sectors are guided by institutional frameworks and policies. These policies are formulated by institutions and governing bodies to provide a set of rules and regulations that help these sectors access and govern water resources for their benefit [2,3,27]. These policies are influenced by technical expertise and specialization within institutions, which in turn affect other institutions. The creation of policies is intrinsically linked with power and politics, where power sustains politics and policy and politics involves the processes of achieving, exercising, and resisting power [28]. Politics operates within institutions and sectors, while power as a strategy involves controlling and organizing spaces and resources through forms of territoriality and the classification of precise geographic areas and boundaries. The organization and management of water resources are based on technical, scientific, economic, and political interests [29]. Henri Lefebvre [29] highlights the following:

> Specializations divide space [water resources] among them and act upon its truncated parts, setting up mental barriers and practice-social frontiers. Thus, architects are assigned architectural space as their private property, economists come into possession of economic space, geographers get their own place in the sun, and so on. The ideologically dominant tendency divides space up into parts and parcels in accordance with the social division of labor ([29], 1991: 89–90).

In Cambodia, the Ministry of Water Resources and Meteorology (MOWRAM) has developed and managed the irrigation systems to manage water resources, but very few have enough water for rice farming in the dry season, while rice farming is under the responsibility of MAFF, which is not managing water [10]. Meanwhile, fisheries and water are closely linked, but each management has been separated by sectoral policies and interests; for instance, the irrigation system may be beneficial for the national output of rice, but tensions and conflict over land and water use often arise between local user groups and large-scale commercial actors in the irrigated and intensively cultivated land. Also, the segmentation of traditional rice-field fisheries, which are unaccounted-for trade-offs with inland capture fisheries, remains a critical issue. In the fisheries sector, unregulated fisheries tend to exclude household fisheries, impacting food security and household income.

Fourth, using, sharing, and controlling water for fishery, agriculture, industry, etc., involves decision-making by actors from different sectors at different levels. Dore [30] in the deliberative water governance and Dore et al. [16] in a framework for analyzing transboundary water governance complexes in the Mekong Region suggest that water governance is a social process of dialogue, negotiation, and decision-making in which many different actors from different sectors are dealing with a variety of issues influenced by their individual and shared context: actors from different sectors engage in multiple arenas, depending on the opportunity, necessity, and choice; drivers are what influence and motivate actors in different sectors; actors employ drivers to establish and legitimize their positions, inform debate, and influence negotiations; decisions emerge from the arenas, and the impacts of decisions result in fairness and water allocation [19,27]. The same happens in Cambodia. Along this line, Ratner et al. [21] look at the governance of the aquatic agricultural system in TSL from three governance dimensions: (i) Stakeholder representation—which actors are represented in decision-making and how? (ii) Distribution of authority—how are formal and informal authority distributed concerning decisions over resource access, management, enforcement, dispute resolution, and benefit sharing? (iii) Mechanisms of accountability—how are power-holders held accountable for their decisions and to whom? These form the basis of governance of water, where decision-making by actors from different sectors and levels, is always challenging and dominated by powerful sectors, in this case, the irrigation development more than the fishery sector [31].

Fifth, access to water resources is crucial for the livelihoods of the farmers, fishers, and the rural population. It also plays an essential role in ensuring the well-being of people, reducing crop failures during dry spells, and providing opportunities for farmers to grow two or three rice crops a year in the study areas. However, treating water as a public good and assuming that it is accessible to all can create a few problems. Firstly, powerful farmers with resources and equipment would maximize the extraction of water for their rice farming at the expense of other farmers, leading to potential water conflicts and shortages, which state institutions may be unable to address fairly due to limited financial resources, capacities, rules, and regulations [32]. Secondly, treating water as a public good can lead to wasteful use, as it is free, and wasting it does not incur any cost [33,34]. On the other hand, when people have to pay to use water, such as a member of FWUCs, they tend to use only enough water to satisfy their immediate needs; however, poor farmers may be unable to do so due to small farmlands and low yield with the uncertainty of climate change. Thus, this article examines these dilemmas in the study areas in a great deal [32].

Sixth, water governance is intricately linked to the physical structures that are put in place to regulate, collect, store, move, and distribute it. The design and operation of water infrastructure have a profound impact on the natural and social environments in which they exist. Furthermore, social systems and processes play a crucial role in shaping the physical infrastructure used for water management. To truly understand a water management system, it is essential to recognize the interdependence of social, technological, and biophysical systems. This interconnectedness also highlights the continuous evolution of governance arrangements and processes. Ultimately, decision-making related to water management will involve a diverse group of stakeholders at different levels, utilizing a range of platforms and technologies [15,32]. There has been a variety of infrastructural interventions affecting the natural environmental regime [35]. Chief amongst these are hydropower dams, which have a high disruptive potential, altering flow, quantity, temperature, and flooding patterns, sometimes in combination with large-scale irrigation schemes. Finally, the effects of global climate change are increasingly being felt. Southeast Asia in general, and particularly Cambodia and its water regime, are hypothesized to be one of the most vulnerable areas in the world [36].

3. Materials and Methods

The conceptual framework above has been utilized to analyze water governance in Cambodia's Mekong Delta (CMD) and assess its impact on rice field fishery productivity, rice production, food security, and livelihoods. The CMD is a rice-producing region in Cambodia; it is rich in fishery resources and agricultural products, which produce foods to support millions of people with water at the center of food production. The increased population, the development pressures from the upstream and around the CMD, and climate change underline the capacity of the CMD to produce foods, affecting many people. Thus, the above framework is employed to analyze water governance in the FWUCs, CFis, and CFRs in the CMD under the above changing dynamics aiming at increasing rice farming and fishery management for food production to feed its population.

Empirical research was carried out in two of Cambodia's Mekong Delta Provinces, namely, Prey Veng and Takeo, as well as in Kampong Thom—a province located in the Tonle Sap Lake (TSL) region (Figure 1). This study selected four sites across three provinces: (1) Beung Sneh Lake (Beung Sneh) and (2) Beung Plang in Prey Veng Province, (3) Beung Ream in Kampong Thom Province, and (4) Ta Soung in Takeo Province (Table 1). Beung Sneh, a 3924 ha freshwater lake in Prey Veng Province, is connected to the Mekong River and surrounded by 44 villages located in 8 communes in 4 districts. Hundreds of thousands of people living in these villages are dependent on Beung Sneh for water for both domestic uses and agriculture, fish, and biodiversity. Within Beung Sneh, you can find four CFis, two FWUCs, and a community-based eco-tourism (CBET). Our study examines the Chamcar Kouy Irrigation Scheme (CKIS) and the four CFis situated in Beung Sneh. Beung Plang is a serene freshwater pond nestled in Ampil Krav commune, Sithor Kandal District, Prey Veng Province. The pond is a beloved fixture in the area, surrounded by five vibrant villages and providing a home to 2112 households. It is noteworthy that three of these villages have come together to establish the community fish refuge (CFR), known as a Beung Plang CFR, to protect the last remaining fish refuge areas and link the CFR to rice fields, aiming at increasing fish and rice production for food security. For our study, we directed our focus toward the CFR site and the Vaiko Irrigation Scheme (VIS) located in the Ampil Krav commune.

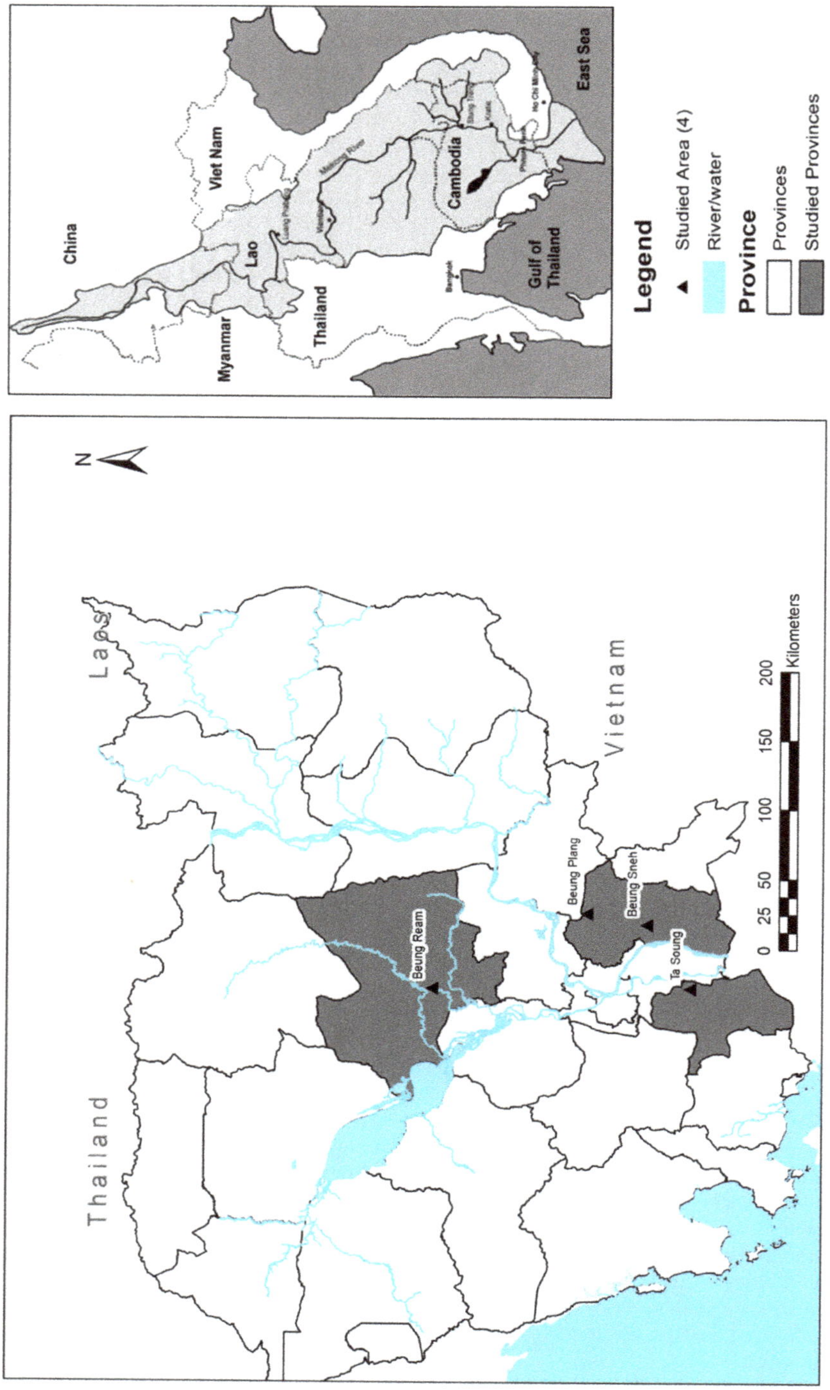

Figure 1. Map of the study provinces.

Table 1. Characteristics of the study sites, FWUCs, CFis, and CFRs.

Studied Sites	FWUC					CFis				CFRs			
	Irrigation System	Irrigation Canal (km)	Irrigated Area (ha)	No. of Villages	Membership (HHs)	No	Areas	No. Villages	Membership	No	Area	No. of Village	Membership
Beung Phlang	Vaiko (no FWUC)	78	153,400	3	93	-	-	-		1	27	3	4981
Ta Soung	Ta Soung Irrigation	50	1511	15	970	4	844,793	15	1016	-	-	-	
Beung Ream	Taing Krasain	22	9869	10	13,058	-	-	-		1	13	2	572
Beung Sneh	Chamcar Kouy	9	2010	6	984	4	85,236	6	11,034	-	-	-	
Total	4	159	166,790	31	15,105	8	930,029	21	12,050	2	40	5	5553

Note: Source: authors.

The Farmers Water User Community (FWUC) of the Ta Soung Irrigation Scheme (TSIS) serves nearly 1000 farming households across 15 villages in Prey Kabbas District, Takeo Province. This community is linked to the Prek Ambel River, which feeds into the Bassac River. It is a rice-producing region where farmers cultivate three rice crops a year using water from the Bassac River, and it has a former fishing lot area, where rice and fish are produced for food consumption and trade. Additionally, this study delves into the connection between the Ta Soung Irrigation Scheme and the four community fisheries (CFis) that were established to manage former fishing lot no. 20 in the Prek Ambel River, Takeo Province.

Beung Ream is a pristine freshwater pond nestled in Kakoh Commune, Santuk District, Kampong Thom Province. It has been designated as Beung Ream CFR and is linked to the Tang Krasaing Irrigation Scheme, which has been established as FWUC. Nestled within Tang Krasaing Irrigation Scheme, Kakoh Commune is home to approximately 3309 households that are spread across 10 villages and organized into Sub-FWUC. This study explores the Beung Ream CFR and the Sub-FWUC of Kakoh Commune as a crucial component of Tang Krasaing Irrigation Scheme.

The researchers utilized both qualitative and quantitative methods to gather primary and secondary data from the studied sites. The team, comprised of members from WorldFish, International Water Management Institute (IWMI), Inland Fishery Research and Development Institute (IFReDI), and Fishery Administration Cantonments (FiACs), and data collection was conducted between December 2022 and June 2023. Secondary data on various factors such as CFis, CFRs, FWUCs, irrigation schemes, rice production, fishery, pesticides and fertilizers, population, and farming lands were collected from the commune database (2021), CFi and CFR databases (2022), and the irrigation database of MOWRAM (2019).

Primary data were obtained through key informant interviews (KIIs) and focus group discussions (FGDs). The KIIs were conducted with Provincial Departments of Water Resources and Meteorology (PDWRAMs), FiACs, District Officers in charge of Agriculture, Environment and Water Resources, Commune Chiefs, and NGOs in the respective sites to obtain their knowledge on sectors, the policy and legal frameworks, the institutional arrangements, roles, responsibilities, activities, challenges and opportunities in carrying out their works. The FGDs were conducted with various groups such as CFis, CFRs, FWUCs, Identity of the Poor 1 and 2 (ID Poor 1 and 2), and non-ID Poor (Table 2). The KIIs and FGDs were conducted concerning the performance of FWUCs, CFis, and CFRs. Specifically, the water usage of FWUCs during three rice farming seasons and its effects on CFis/CFRs were discussed, along with the competition between FWUCs and CFis/CFR for water resources. The roles of local governments in water governance were also examined, and recommendations were made for improving water governance for FWUCs, CFis/CFRs, rice farming, and fishing. Furthermore, FGDs were held with both ID Poor and non-Poor people to discuss the changes in water resources, fishery, agriculture, and food over the past 10–15 years. The impact of these changes on ID Poor 1 and 2 and non-Poor individuals was also explored. The ID Poor is a government system under the Ministry of Planning to classify poor households into four categories to which households can be assigned: Poor 1 (very poor); Poor 2 (poor); at-risk; and non-Poor, which were promulgated by sub-decree no. 291 issued in December 2011 [37]. Finally, the group discussed the positive developments in water resources, rice farming, fishery, foods, and livelihoods that have resulted from the implementation of irrigation systems, FWUCs, CFis, and CFRs in the studied areas. The information gathered from the FGDs, interviews, and secondary sources underwent analysis utilizing an Excel spreadsheet. The data were then transformed into percentages, figures, and tables, with qualitative data included to support the findings. This article is descriptive and based on the data and analysis, and it is structured into first, the introduction; second, the analytical frameworks; third, the results and discussions; and fourth, the conclusion.

Table 2. The primary data collection using KIIs and FGDs.

Data Collection Methods/Sites	KIIs	FGDs with CFis/CFRs	FGDs with FWUCs	FGDs with ID Poor 1 and 2; and the Non-ID Poor
Beung Sneh	• Chief of PDOWRAM and two staff, • FiAC Prey Veng Province, • Commune Chiefs of Prey Kandieng, Theay, Samrong and Damrei Poun Communes	4 FGDs with CFis: (1) Theay, (2) Samrong, (3) Damrei Poun; (4) Prey Kandieng	• 01 FGD with Chamcar Kyou's FWUC • 01 FGD with Private FWUC in Toap Sday Village/Theay Commune	15 FGDS in 5 villages in five Communes around Beung Sneh: 1. Samrong, 2. Prey Khla, 3. Kok Trom, 4. Kampong Sleng, 5. Chamcar Kyou,
Beung Phlang		One FGDs with CFR	n/a	03 FGDs in Peanea Village
Beung Ream	• PDWRAM of Prey Veng Provinve, • District of Officers Santuk District in charge of FWUC, • Kakoh Commune Authority, • Chief of Kakoh's FWUC	02 FGDs	3 FGDs of FWUCs	15 GFDs in five villages in Kakoh Commune: (1) Chey Chomneas, (2) Kiriwone, (3) Samnak. (4) Santuk Krav, (5) Cheay Spai
Ta Soung	• District Officers in charge of irrigation, agriculture and fishery	03 FGDs with 3 CFis	One FGD with FWUC	15 FGDs in five villages three communes: (1) Sethey, (2) Prey Lvea Keut, (3) Pontong, (4) Kampomg Reab, (5) Prey Tapong

Note: Source: authors.

4. Results and Discussion

4.1. Results

4.1.1. Water Resources in Cambodia

Cambodia is situated in the Lower Mekong Basin, spanning an area of 181,035 km^2. A substantial portion of Cambodian land, roughly 86% (156,000 km^2), is within the Mekong catchments. As a downstream and lowland nation, Cambodia is blessed with plentiful water resources. It boasts approximately 1216 km^3 of water within its borders, with an additional 355.5 km^3 flowing into the Mekong River from external sources. Cambodia's estimated annual total renewable water resources are around 476 km^3 [38] (units of volume: 1 km^3 = 1 billion m^3 = 1000 million m^3 = 10^9 m^3). The annual water usage amounts to around 2.18 cubic kilometers, with agriculture being the primary consumer at 94%. The withdrawal of irrigation water alone accounts for about 1.928 million cubic meters on a yearly basis, while the remaining water is allocated for domestic and industrial use. The estimated water withdrawal per individual ranges from 130 to 160 cubic meters per year [38].

With 39 river basins located in five sub-regions, Cambodia is home to a vast array of water resources. The Tonle Sap Lake alone is made up of sixteen sub-river basins, while the Upper Mekong River basin contains five, the 3S basin has three, the Mekong Delta has eight, and the coastal river basin has eight sub-river basins. At Kratie, the Mekong River provides Cambodia with its primary external water resources, with an average discharge of 476 km^3 per year before it flows into the South China Sea [1].

4.1.2. Water Management through Irrigation Development

Irrigation Development

There are 2500 irrigation schemes across Cambodia, which could irrigate 2.32 million ha, among which 65% are located in the Mekong floodplains and Delta and 35% in the Tonle Sap floodplains. We studied three irrigation systems in Cambodia's Mekong Delta in the Prey Veng and Takeo Provinces. Prey Veng province has 177 irrigation systems and Takeo has 136 irrigation schemes. In the Tonle Sap floodplain, this study focuses on Kampong Thom Province, with 258 irrigation schemes [39].

In Cambodia, there are 2500 irrigation schemes that could irrigate 2.32 million hectares of land. Of these, 65% are found in the Mekong floodplains and Delta, while the remaining 35% are situated in the Tonle Sap floodplains. Our investigation specifically examined three irrigation systems in Cambodia's Mekong Delta, which are located in Prey Veng and Takeo Provinces. Prey Veng Province boasts 177 irrigation systems and Takeo has 136. Furthermore, we delved into Kampong Thom Province in the Tonle Sap floodplain, where there are 258 irrigation schemes. These statistics were procured from the CISIS database in 2020 and were authenticated via personal communication in February 2023 [39].

An analysis of four irrigation schemes has been chosen for this study, comprising two in Prey Veng Province (Chamcar KKouy Irrigation Scheme and Vaiko Irrigation Scheme), one in Takeo Province (Ta Soung Irrigation Scheme), and one in Kampong Thom Province (Taing Krasaing Irrigation Scheme). The Chamcar Kouy Irrigation Scheme is primarily supplied with water from Beung Sneh, which provides approximately 85 million cubic meters of water during the wet season and reduces to around 40 million m^3 during the dry season. The Ta Soung Irrigation Scheme uses water from the Prek Ambel River, a tributary of the Bassac River, while the VIS scheme relies on the Mekong River. The Taing Krasaing Irrigation Scheme uses water from the Taing Krasaing and Stung Chinit Rivers. These irrigation schemes cover an area of approximately 63,895 hectares, which accounts for around 3% of the targeted national irrigated area of 2 million hectares by 2023 [40].

The Prey Kabbas District boasts the Ta Soung Irrigation Scheme, which is an irrigation system that comprises two main canals, ten secondary earth canals, and eight secondary concrete canals. The system is equipped with a pumping station that houses five pumping machines and irrigates 1511 hectares of land owned by 970 farming households from 15 villages across four communes. Another notable irrigation scheme in Kampong Thom Province is the Taing Krasaing Irrigation Scheme. The Taing Krasaing Irrigation Schem in Kakoh Commune consists of a main canal, six secondary canals, and sixteen tertiary canals, with two water gates that can cover a distance of 22 km from the Stung Chinit River to the Beung Ream in Kakoh Commune, Santuk District. Part of the Taing Krasaing Irrigation Scheme located in Kakoh Commune can irrigate up to 9869 hectares of land across ten villages in Kakoh Commune (Table 1).

Farmer Water User Community

The Royal Government of Cambodia (RGC) introduced the Water Law in 2005, which allows farmers who utilize the irrigation system to form FWUCs under Article 19 [12]. To better manage water resources, RGC decentralized the implementation and maintenance of irrigation schemes to FWUCs via Prakas 306 in 2006. To date, 544 FWUCs have been established to manage irrigation schemes [41], but in Prey Veng Province, only 38 out of 177 irrigation systems have FWUCs. Meanwhile, Kampong Thom has 258 schemes and 30 FWUCs, but, unfortunately, many FWUCs are inactive, with only 10% currently active. Five of the identified FWUCs were examined in their respective study areas—two in Beung Sneh and one each in Beung Phlang, Beung Ream, and the Soung Irrigation Scheme. These FWUCs cover a total of 9067 hectares of agricultural land and have 15,781 agricultural households as members. The largest of the studied areas is Damrei Puon's FWUC in Beung Sneh, followed by the FWUC in Ta Soung Irrigation Scheme in Takeo Province and Theay's FWUC in Beung Sneh (Table 3).

Table 3. The selected FWUCs for the studies of water governance.

Site/Commune	No. of FWUCs	No. of Sub-FWUC	No. of Villages in FWUC	Total Areas (ha)	No. of Members (HHs)	Year of Establishment
Beung Phlang	1	0	3	107	93	n/a
Beung Sneh (Damrei Puon and Theay)	2	0	10	2350	1660	(2018, 2005)
Boeung Ream (Kakoh)	1	9	10	5099	13,058	2018
Ta Soung	1	15	15	1511	970	2022
Grand Total	5	24	38	9067	15,781	0

Note: Source: authors.

4.1.3. Fishery Resource Management

The areas under study are rich in fishery resources, with connections to a variety of water bodies including rivers, streams, and lakes. The Soung Irrigation Scheme was a former fishing lot n. 20 in Takeo Province, and it is connected to Prek Ambel River, a tributary of the Bassac River. Similarly, the Beung Sneh is linked to the Mekong River, which is a habitat for a diverse array of fish that people from 44 villages depend on for their livelihoods. Beung Ream, located in the Tonle Sap Floodplain, serves as a migration route for fish between Tonle Sap Lake and Beung Ream during the rainy season. Despite this, only 24% of households in the studied areas rely on fishing, with Prey Kabbas District having the highest percentage of fishing households at 33%. Around 30% of households in Kampong Thom's Beung Ream still rely on fishing, while Beung Sneh in Prey Veng Province has a significantly lower percentage of fishing households at just 19%.

Community Fisheries (CFis)

The studied CFis were established after 2000, following the release of commercial fishing areas for public open access to local communities. The prominent roles of CFis are to conserve and protect fishery resources within the CFi territories. All members could fish openly throughout the year using the fishing gear defined in the CFi by-laws. The by-laws allow CFi members to fish with subsistence and not commercial, aiming at conserving fishery resources.

The CFis in Beung Sneh are connected to eight large irrigation systems that utilize water from the CFi areas to irrigate rice fields spanning over 22,899 hectares around the lake. These CFis are established at the commune level, with elected committees from villages in the communes. Their primary objective is to protect and conserve the fishery resources by reserving approximately 40 hectares inside the lake and nine deep water areas in the Beung Sneh as CFi-protected zones (Table 1). To achieve this, CFis maintain the water level in the lake at approximately 4–5 m deep during the dry season, providing fish with shelter. However, their actions often contradict those of Farmer Water User Committees (FWUCs), as farmers need to pump water to irrigate their rice fields. This puts CFis under immense pressure as they lose water to rice cultivation, leading to a decrease in the lake's water level, which impacts both the CFis and fisheries and results in illegal fishing within the protected areas. Additionally, they also face the destruction of flooded forests around the lake and water pumping from the lake.

From 2000 to 2002, four community fishery institutions (CFis) were established in the Prey Kabbas District of Takeo Province, specifically in the areas of TSIS. These CFis boast a total of 1016 members, with 550 of them being female (Table 1). The leadership of the CFis consists of 36 committee members, 4 of whom are female. The establishment of these CFis aimed to protect fishery resources and support the communities. Oxfam Australia partnered with these CFis from 2002 until 2015 to protect fishery resources to support communities. However, since Oxfam phased out of the area, CFis have become inactive due to financial and technical support, staffing, and budget constraints. In 2022 and 2023, the European Union provided a small grant of USD 1000 per year to Kampong Reab, one of the CFis. However, this grant only addresses patrolling, conservation, and signboard for CFi awareness-raising and not the other pressing issues that CFis are facing.

CFis encounter several obstacles, including encroachment in their conservation zones, illegal fishing within CFi core areas, and limited participation from both members and non-members in the management of CFi areas. Moreover, the lack of support from FiA, FiAC, and local government in managing CFi areas, as well as limited financial and technical support from concerned agencies, hinders the protection of fishery resources. Addressing conflicts between CFis and Farmers Water User Committees (FWUCs) over water pumps from CFi areas to irrigate rice fields, overlapping areas between CFis and FWUCs, lack of fishery management within FWUC areas, and the use of pesticides and fertilizers in rice farming that results in the killing of aquatic animals and fish are crucial issues that require attention.

Community Fish Refuges (CFRs)

This study delves into two CFRs, namely the Beung Ream CFR in Kampong Thom Province and the Beung Phlang CFR in Prey Veng Province. The Beung Ream CFR was established in 2021 by two villages in the Kakoh Commune, Santuk District, Kampong Thom Province. It covers a vast area of 13 hectares, with a 2-hectare core protected region and an average water depth of 2.5 m during the dry spell (Table 1). The CFR area is equipped with a water level monitoring system that triggers an alarm to prevent any further water extraction from the lake. It is marked with pillars, a security guard post, and a signboard that explicitly states the prohibition of illegal fishing within the Beung Ream CFR area.

In the vicinity of Beung Ream CFR, three canals are present: O' Praing, Beung Karav, and irrigation canals constructed by MoWRAM. O' Praing underwent rehabilitation through a private company that utilized the soil to construct roads. Since then, it has ensured a year-long water supply to Beung Ream CFR. The Kakoh irrigation canal enters Beung Ream CFR via a water gate that regulates water supply to CFR areas. The Kakoh irrigation system is part of the Taing Krasaing Irrigation Scheme. Farmers have been irrigating 995 hectares of rice fields surrounding Beung Ream CFR with water from the Kakoh irrigation canals and Beung Ream CFR, enabling them to cultivate 2–3 crops of rice. Approximately 572 households are actively involved in fishing and harvesting fish and other aquatic animals from the floodplain area and rice fields surrounding Beung Ream CFR, which gets flooded during the rainy season. In addition, around 294 households fish within the Beung Ream CFR area for approximately five months a year. Fishing benefits approximately 716 families of which 20 percent are impoverished households. The estimated annual fish catch per household is about 88 kg, and the estimated annual catch of other aquatic animals is about 48 kg per household.

The Beung Phlang CFR, situated in the Ampil Krau Commune, is home to 17,572 individuals from 2112 households across five villages. The majority of the population, approximately 85%, is engaged in farming, while about 20% are involved in fishing. Established in 2008 by Peanea, Kbal Beung, and Svay Teap villages, the Beung Phlang CFR was created to protect land, fish, and biodiversity for the community's benefit. Covering an area of roughly 27 hectares, the Beung Phlang CFR boasts a length of 1800 m and a width of 200 m. It holds water all year round, with a depth of 6 m in the wet season and 2 m in the dry season. The core area, spanning 12 hectares, is managed as a conservation zone, with poles marking its boundaries. The release of indigenous fish fingerlings 2–3 times has resulted in an increase in fish stock. While villagers from these villages can fish within the CFR, they are not permitted to do so in the conservation areas. The Vaiko Irrigation Scheme encircles the Beung Phlang CFR and includes two primary canals, one pumping station, two vertical sub-canals, and four horizontal sub-canals. One vertical and one horizontal sub-canal link the Beung Phlang CFR to the main canals. The canal system is observed to be shallow and poorly maintained, and villagers have not utilized much of the water. Nonetheless, the connectivity allows fish to move between the sub-canals and the Beung Phlang CFR to rice fields, with an estimated fish catch at about 25 kg/ha.

4.2. Discussion

4.2.1. Rice Export Policy, Irrigation Development, and Fishery Management

Water is a crucial component for both rice farming and fishery production. With abundant water resources, Cambodia is rich in fishery, standing no. 5 globally after China, India, Bangladesh, and Myanmar in inland fishery production. The Tonle Sap Lake and Mekong Rivers are well known for fishery production. However, between 2000 and 2012, RGC removed the fishing lot systems, a hundred-year-old system, and turned the fishing lot areas into CFis/CFRs, fishery conservations and open access areas. The fishery sector has gradually and economically become inactive [41,42]. In 2015, RGC developed a rice export policy to promote the paddy rice export for one million tons a year. These have led to mushrooming of rice production and the development of irrigation systems to manage water to irrigate rice farming [43].

To promote rice intensification and fulfill the rice export policy, investments are poured into irrigation development and rehabilitation to increase water availability for rice cultivation. MOWRAM has developed the National Irrigation and Water Resources Management Investment Program (NIWRMIP) 2019–2033 to develop and rehabilitate irrigation schemes at a total cost of USD 2.64 billion. FWUCs have been established under the MOWRAM, as a local arm, which is in line with the above policy to promote community participation in water management and uses for rice farming [44,45].

Nevertheless, agriculture is managed under the Ministry of Agriculture, Forestry, and Fishery (MAFF), including rice farming, agricultural extension, agricultural land management, and so on. At the same time, fishery policy supports the establishment of CFis/CFRs to manage the changing fishery management system. The fishery views water as part of the fishery and water bodies as natural habitats for fish and fishery production [46]. However, MOWRAM treats water as a separate sector, which is managed as part of the irrigation system. Thus, the irrigation policy considers water as a valuable resource to be leveraged, utilized, controlled, and managed primarily for agricultural purposes. The irrigation system is designed to retain water and employ it for irrigating rice fields [47,48]. The rice intensification for commercial rice exports requires a larger amount of water. Farmers extract more water from irrigation canals and, in some cases, from rivers, lakes, and ponds to irrigate dry-season rice farming. This often causes conflicts among farmers and between upstream and downstream communities along rivers and around lakes. With the expansion of the rice farming industry, the use of pesticides, fertilizers, and other chemical inputs has increased to improve rice production.

In the name of economic development, policy, actors, power, and interest drive these changes. Even with a common policy for promoting rice production and export, some sectors (agriculture, irrigation) gain momentum, while others (e.g., fishery sector, etc.) are vulnerable in terms of their contribution to the economy and the country's development. Instead, dominant and powerful sectors get leveraged, which to some extent undermines the credibility of other sectors, in this case, the fishery sector (Table 4) [27,28,30].

Table 4. Framework for analyzing water governance in the studied areas.

Water User Community	No. of FWUCs	Source of Water—Upstream vs. Downstream	Policy	Institution	Water Conflict and Cooperation—Power, Politics, and Position	Dialogue/ Negotiation
FWUCs	• 01 FWUCs in TSIS • 01 FWUC in Taing Krasaing Irrigation Scheme • 02 FWUCs in Beung Sneh—One is not active • 01 FWUC in Beung Phlang, not active.	• Beung Sneh, • Prek Ambel River in Takeo, • Taing Krasaing River, and • Mekong River (Vaiko Irrigation Scheme)	• The Constitution 1993—Article 58 and 59. • Water policy (2004), • Water Law (2007). • Sub-decree for effective and sustainable management, protection, and development of surface water and groundwater in 39 river basins (July 2015). • FWUC Sub-decree (2015)	• MOWRAM was established in 1999, managing water and irrigation management—the operation of the irrigation system. • MAFF is responsible for rice farming and seems to have no role in irrigated areas provided by MOWRAM. • The community authority plays a role in water management in their communes only when water conflicts occur. • The District Office of Agriculture, Environment, and Water coordinates activity concerning water uses, agriculture, and the environment	• Water is pumped from water sources through irrigation to irrigate three rice crops annually. Water shortages lead to competition and conflict among farmers between farming, fishing, and domestic uses. • Irrigation is a weapon for fighting water shortage at the community level, and water pumping generators are weapons for fighting water shortage at the household level. • Competing between upstream and downstream communities: communities with and without irrigation schemes. • Rice farming and in-datedness and migration.	• At the regional level, there are regional dialogues on water sharing in the Mekong River Basin. • At the national level, coordination between MOWRAM and MAFF remains challenging over water management and water uses for rice farming. • At the local level, there is a weak conflict between FWUCs and CFis over water use.
CFis	• 04 CFis in Prek Ambel River • 04 CFis in Beung Sneh	• Beung Sneh in Prey Veng. • Prek Ambel River in Takeo.	• Fishery Law • Sub-decree on CFis. • Strategic Planning Framework for Fisheries. • Planning Framework	• FiA/FiAC, • The Fishery Sector has been decentralized, and FiAC is not under the FiA, but under the PDAFF. • FiA/FiACs organize the election to set up the CFi committee to manage CFis. • CFis are weak and lack funding and technical support. Many CFis are not organizing re-elections and developments of the management plans. • Commune Administrations do not have a mandate in fishery management and CFis.	• CFis protect water to keep it at 3–4 m deep to keep the fish habitat productive—anything below that will affect fish—increased fish catch, and encroachment of the wetlands in the lake to cultivate rice farming. • The irrigation scheme has blocked the river flow and fish migration, changing landscapes. The irrigation management has no fishery management system included. • The increased water pumping resulted in lower water, affecting the fishery. • The increased rice farming leads to the use of pesticides and fertilizers, affecting fishery and aquatic animals.	• CFis and FWUCs are in the same villages and communes, but water conflicts remain unresolved. • Communes facilitate the CFis and FWUCs in the monthly commune meeting but cannot resolve the tensions between CFis and FWUCs. • FiA has decentralized to FiACs, but FiACs have limited resources and capacity to resolve these conflicts at local levels. • CFis—no fishery officers at the district level. The fishery has not been integrated into the District Office of Agriculture, Environment, and Water.

Table 4. *Cont.*

Water User Community	No. of FWUCs	Source of Water—Upstream vs. Downstream	Policy	Institution	Water Conflict and Cooperation—Power, Politics, and Position	Dialogue/ Negotiation
CFRs	• 01 Beung Ream CFR • 01 Beung Phlang CFR	• Taing Krasaing River, and • Mekong River (Vaiko Irrigation canals)	• Legal framework is being developed • Sub-decree on CFR is being drafted.	• FiA is the leading institution. • The Department of Aquaculture Department (DAD) is mandated to manage CFRs. • FiA/DAD organized the election to set up the CFR committee to manage CFRs. • Commune Administration has no knowledge and mandates in CFR management. • NGOs and donors support CFRs financially.	• CFRs protect water sources to provide habitats to fishery and aquatic animals to seek refuge in the dry seasons in order for them to keep breeding so that in the wet season these fish and aquatic animals will migrate to rice fields and water bodies. • CFRs are often pumped by farmers, who are members of FWUCs and CFRs, to save their rice farming in the dry season when water is critically low in the CFRs and the is a shortage.	• No mechanism is in place to support the CFRs at the ground, only through donors/NGOs. • Commune Administration often facilitates CFRs and FWUCs in the monthly meetings of the communes.

The rice export policy has triggered the expansion of agricultural land. At a national level, the total rice farming area has expanded to reach 3.34 million hectares by 2023, with 82% dedicated to wet-season rice farming and 18% to dry-season rice farming. Multiple irrigation schemes, including the Vaiko, Taing Krasaing, Ta Soung, and Chamcar Kouy Irrigation Schemes, have undergone rehabilitation to enhance agriculture, specifically rice farming.

In the study areas, the total rice farming area is 37,363.5 hectares, with 71% dedicated to wet-season rice farming and 29% to dry-season rice farming. Agricultural landholdings are relatively small, at approximately 1.95 hectares per household, and the increased rice production primarily relies on water and agricultural inputs. In Beung Sneh, rice farming covers a vast area of 22,899 hectares. The majority of this land, 69%, is dedicated to wet-season rice farming, while the remaining 31% is for dry-season rice farming. The Ta Soung Irrigation Scheme is the second-largest rice farming area, with 40% allocated to wet-season rice farming and 60% to dry-season rice farming. Although the Beung Ream and Beung Plang areas have constructed irrigation canals, farmers currently do not engage in dry-season rice farming. Nonetheless, many agricultural households in Beung Ream have been observed to utilize the Taing Krasaing Irrigation Scheme and the FWUC to manage water supply and transform wet-season rice farming areas into dry-season rice farming areas.

Under the rice export policy and improved irrigation system, farming households have intensified the rice farming industry. They have graduated from one rice crop a year to 2–3 rice crops a year, moving away from rainfed farming to irrigated agriculture, from labor-intensive to mechanization, from farming for subsistence to farming for trading, and from low yield to high yield. Farmers cultivate both wet- and dry-season rice at present. These happen following the increased water availability, irrigation systems, and access to water bodies.

On the other hand, the rice market is influenced by rice traders from Vietnam who buy rice from Cambodian farmers to sell in Vietnam. The Vietnamese rice traders have to introduce the Vietnamese rice varieties to Cambodian farmers if they intend to trade their types of rice with Vietnamese rice traders. For these reasons, Cambodian farmers in the study areas cultivate a high-yield rice variety, including IR 504, IR 5154, and others imported from Vietnam.

About 98% of Cambodian agricultural households own farmlands, of which 99.3% are in Takeo Province, 99.1% in Prey Veng Province, and 98.3% in Kampong Thom Province. Furthermore, about 5% of agricultural households in Cambodia rent farmlands from others for agriculture; perhaps they are landless. In Prey Veng Province, about 4% of agricultural households rent their farmlands or are landless, followed by 3.4% in Kampong Thom Province and 2% in Takeo Province [49]. In the study areas, about 57% of all agricultural households are engaged in agricultural production [50]. The average household landholding is 1.95 ha. About 12.5% of farming households own less than one hectare of farmland, and 10.3% of households are landless. About 24% of the total population is in fishing. In Prey Kabbas District, the fishing population constitutes 33% of the total population. In Beung Ream in Kampong Thom, about 30% of its population is still engaged in fishing. However, the fishing population in Prey Veng, generally, and in Beug Sneh constitute 19%, which is low compared with other provinces.

Cultivating three rice crops per year requires much water. The irrigated command area is about 1.2 million hectares, which represents only 22% of all arable land [51]. In the study area, about 19% of farming households have their farmlands irrigated, of which 26% of farming households are located in the Ta Suong Irrigation Scheme, 18% of farming households around the Beung Sneh in Prey Veng, and 17% of farming households in Kah Koh Commune/Beung Ream. However, 42% of the total dry-season farmlands are irrigated, while 24% of the wet rice farming areas are irrigated in the wet season.

This study found that most farming households cultivate 2–3 rice crops per year. The first rice farming season starts in May and harvests in October, which is called 'wet-season rice farming'; it is when rainfalls and river water flood the floodplains, and when farmers cultivate the wet-season rice in the upper rice fields or 'sreleu' that are not flooded. Farmers cultivate the local varieties, namely, 'neang minh, senkra oup, malis, and others, which yield between 4 and 5 tons/ha. Farmers cultivate the 'wet-season rice' mostly for household consumption and the rice surplus is sold to those in need. However, given the increased rice trades, farmers gradually changed to cultivate the high-yield rice varieties from Vietnam for sale, and some farming households buy local rice production with local varieties for household consumption.

The second rice farming season starts from November to January, which is called the first dry-season rice farming, which is when the flood water recedes the floodplains and farmers start cultivating the recession rice, namely, the 'first dry-season rice farming'. During this period, farmers cultivate high-yield rice varieties, including rice varieties from Vietnam, such as IR 504 and IR 5154 in the middle rice fields, or 'srekandal'. These varieties consume so much chemical agricultural inputs and water, which could yield 5–6 tons/ha for three three-month periods. With the increased rice farming during the first dry-season rice farming, farmers experience water shortage to some degree but in many cases manage to secure water for their rice farming. The rice production during this period is mostly sold to rice traders who trade them to Vietnam (Figure 2).

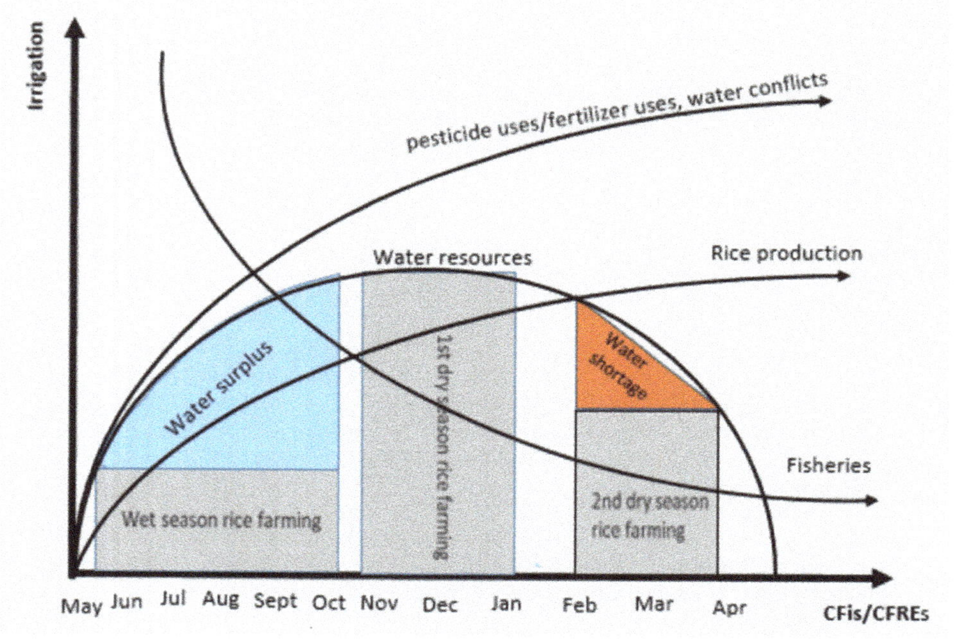

Figure 2. Water governance challenges (source: authors).

Some farmers cultivate the second dry-season rice farming, from February to April, but are at risk of water shortage, which could harm rice farming and increase the cost of production. Due to water shortage, farming households cultivate the second dry-season rice farming in the lower rice fields or 'srekrom' located adjacent to water bodies (lakes, rivers, and streams), where farmers could pump out water to irrigate the rice farming. The second dry-season rice farming is mainly for rice trades to Vietnam, and so, farmers cultivate the high-yield rice varieties from Vietnam, but it may not have a high yield, approximately 4–5 tons/ha, as they could be spoiled due to shortages of water or severe droughts. Water competition during this period is relatively high between farmers and between 'sreleu, srekandal, and srekrom'. Water competition sometimes could lead to water conflicts, which could increase the cost of rice production (Figure 2).

4.2.2. Changing Hydrological Regime and Climate Change

In the last 20 years, changes to hydrological flows in the Mekong River, floodplains, and lakes have occurred. The hydrological flows of the Mekong River to the Tonle Sap Lake and the Mekong Delta have dropped, affecting rice farming, fishery production, food security, and the livelihoods of millions of people downstream. Many scholars blame this on the developments of hydropower, particularly the Chinese hydropower dams [52–56], while others in favor of Chinese dams deny the accusations but claim climate change to be the cause [57,58]. Nevertheless, a third group of scholars came out and asserted neither Chinese dams nor climate change caused the declines in the Mekong flow downstream to Tonle Sap Lake and the Mekong Delta but argued that it was due to local developments, particularly the irrigation developments [59,60]. It is estimated that, on average, approximately 13% of the annual discharge, which is equivalent to around 62 km^3 of water, has been withdrawn from the entire lower MRB, of which Vietnam, Thailand, China, Laos, Cambodia, and Myanmar account for approximately 52%, 29%, 9%, 5%, 3%, and 2%, respectively. The expansion of irrigation and croplands will play a role in decreasing the annual streamflow by 3% over the period of 2036–2065 compared with the period of 1971–2000 [61]. Furthermore, another study has confirmed that hydropower and other infrastructure developments could reduce the water discharge in the Mekong River by 21% at Kratie, 5% at the Kampong Cham, and 8% at Prek Kdam and Chak Tomuk. They also confirm that rainfall in the Cambodian floodplains has remained roughly constant from 1960–2019 and conclude that local anthropogenic factors are likely causing the flow reduction [59].

In the study areas, particularly the Mekong Delta, over the past 10–20 years, the natural flooding events have been altered due to the development of hydropower in the upper Mekong River and the construction of irrigational canals and dyke systems. In the Mekong River at Neak Luong, which is in the Mekong Delta Region in Cambodia, the annual wet season discharge dropped by 10% between 2010 and 2020 [59]. Local villagers reported that the Mekong River has not caused any significant flooding since 2011. Furthermore, the Vaiko Irrigation Scheme has played a role in reducing the Mekong flooding from reaching rice fields in Sithor Kandal District, Prey Veng Province. Farmers in the Ta Soung Irrigation Scheme area have reported no flood events in the past decade. Similarly, the natural Beung Ream Lake has not been affected by floods from the Tonle Sap Lake for the last 15 years.

Climate change has impacted the availability of water for rice farming, causing greater uncertainty of rainfalls. In the Mekong River Basin (MRB), climate change may increase the annual streamflow by 15% over the period of 2036–2065 compared with the period of 1971–2000 [61]. Nevertheless, climate change is also expected to decrease the dry season flow by 2.18 percent. Frequent drought events also occur, offsetting the wet season flows, and resulting in frequent droughts in the lower MRB [62]. In the study areas, local communities report that *'tuk thom'* (the big floods) in the wet season have never occurred in the study areas in the past 10 years. Instead, farmers experience frequent droughts, which cause damage to agriculture and rice farming. It is observed that the frequent droughts make the wet and the dry seasons homogenous, with slight differences in terms of a short wet season with drought-pronged periods and a long dry season duration, particularly between 2014 and 2023 [63]. These have created the farming practices homogenously throughout the year, including using the same rice varieties (IR 504, IR5154) for different rice farming seasons, three-month cultivating periods for rice cultivations, using the same amounts of water quantity to irrigate the same plots of rice cultivating areas, and using the same quantity of agriculture cultural inputs and obtaining the similar rice yields per hectare (4–5 tons/ha) throughout the year. The wet-season rice farming is also irrigated, as is the first and second dry-season rice farming; the only difference is that when the irrigation canals run out of water, particularly during the first and second dry-season rice farming, farmers compete for the remaining water from elsewhere using their extra pumping generators to pump water from rivers, lakes, and ponds near their rice fields (Table 4).

Communities such as Kampong Reap and Pou Rumchak in the Ta Soung Irrigation Scheme previously experienced frequent flooding from the Bassac and Prek Ambel Rivers for half of the year, and waters receded in the dry season, leaving the community areas on the dry land for another six months of the year. However, now they rely on irrigation canals and embankments for protection. Similarly, Penea and Svay Rompea used to be inundated by Mekong floods during the wet season but have not experienced such events since 2015 due to the completion of the Vaiko Irrigation Scheme. These changes have resulted in a shift away from water-based communities to land-based communities reliant on human systems for water supply, such as irrigation systems, wells, and water supply and sanitation.

During the first and second dry-season rice farming periods, the irrigation system serves as a water weapon [64] for the communities, playing a primary role in the fight for water between different communities either upstream or downstream of the Beung Sneh Lake, the Vaiko Irrigation Scheme, the Prek Ambel River, and the Taing Krasaing Irrigation Scheme. Additionally, at the household level, each farming household owns at least one or two water pumping generators, which are considered a water weapon that households use to compete with other agricultural households in the same community to fight for limited water resources during severe droughts [65]. Farmers use water pumping generators to pump water from irrigation canals, rivers, and lakes and distribute it to remote rice fields. There is also competition among farmers with rice fields near the canals, rivers, and lakes and those with remote rice fields. Farmers collaborate at one time to pump water from the canals but compete at other times to pump water to irrigate their rice farming.

Four community fisheries (CFis) in Beung Sneh have experienced low water levels during the dry season between 2020 and 2023, posing a challenge for rice field irrigation. To address this issue, eleven irrigation schemes covering 22,899 hectares have been pumping water from Beung Sneh. In addition, 10,911 households in 44 villages have been utilizing the irrigation canals to irrigate their rice fields. In the Prey Kabbas District, the Ta Soung Irrigation System pumps water from the Prek Ambel River to irrigate 1511 hectares in four communes, while farmers in the Beung Ream in the Taing Krasaing Irrigation Scheme pump water from the Beung Ream CFR to the lowest water level during the dry season of 2023. Despite being from the same communities, these activities have led to tension between the CFis/CFRs and FWUCs [10].

Private individuals in the Torp Sdach village of Theay Commune operate a private water pumping station (PWPS) under permits granted by the district authority. The PWPS operator has a four-year contract (2016–2024) to pump water from Beung Sneh and irrigate 305 hectares of land across five villages, charging water fees ranging from KHR 270,000 to KHR 300,000 per hectare per season. This enables farmers to cultivate 2–3 rice crops annually. Of the 305 hectares, 105 hectares of upland rice fields are not flooded by the rising water level in Beung Sneh, and farmers cultivate three rice crops per year. The remaining 200 hectares are located within Beung Sneh's floodplain, which floods during the wet season, permitting only one dry seasonal rice crop per year (typically from March to May). The PWPS serves approximately 250–300 households, 175 of which do dry seasonal rice farming and own at least one pumping generator per household. In addition, Beung Sneh has three water supply stations operated by private individuals who possess licenses from MOWRAM. These stations pump water from Beung Sneh Lake, filter, clean, and sell it to villagers. Two stations are operational, while one is under construction. It is estimated that 50–60% of the population around Beung Sneh uses water from the water supply system, paying KHR 1800–2000 per cubic meter. On average, a household uses around 10 cubic meters of water per month.

In the region of Beung Ream, the lowest part of the Taing Krasain Irrigation Scheme is known as the Kakoh. During the wet season, farmers from ten villages rely on the Kakoh's canals to cultivate rice, and thanks to the Taing Krasain Irrigation Scheme, they can now grow three crops per year. However, the farmers face challenges as upstream communities tend to use as much water as possible before releasing it downstream, causing delayed water release to the Kakoh's canals. The ten villages also compete for water, especially the upstream and downstream ones. The Chey Chumnas, Kiriwon, and Samnak Villages, located at the lowest reach of the Taing Krasain Irrigation Scheme, have reported water shortages due to upstream villages tapping more water and releasing the leftovers downstream. As a result, the Kakoh Commune experiences water scarcity between January and April, and farmers resort to taking water from the Beung Ream CFR to irrigate their rice fields. Unfortunately, water conflicts have arisen between upstream and downstream villages, as well as between rice farming and CFR.

4.2.3. Water Resources, Water Fees, and Water Conflict

Every year, water resources are limited, as shown in Figure 2, and they have been used throughout for rice farming; they are abundant in the wet-season rice farming, moderate but with a slight shortfall in the first dry-season rice farming, and have a severe shortage in the second dry-season rice farming. Based on the efficiency of the irrigation system utilized in dry-season rice farming, it takes one cubic meter of water to produce 0.11–0.242 kg of paddy rice [66]. The irrigation systems in the studied regions can pump an estimated 245.45 million m^3 of water annually through four primary pumping stations from the Mekong River's connected rivers and lakes. This water is utilized to irrigate dry-season rice farming areas that span at least 10,909 hectares (Figure 3). Nonetheless, the expansion of dry-season rice farming in the region results in the cultivation of more wetlands and the extraction of additional water for irrigation purposes.

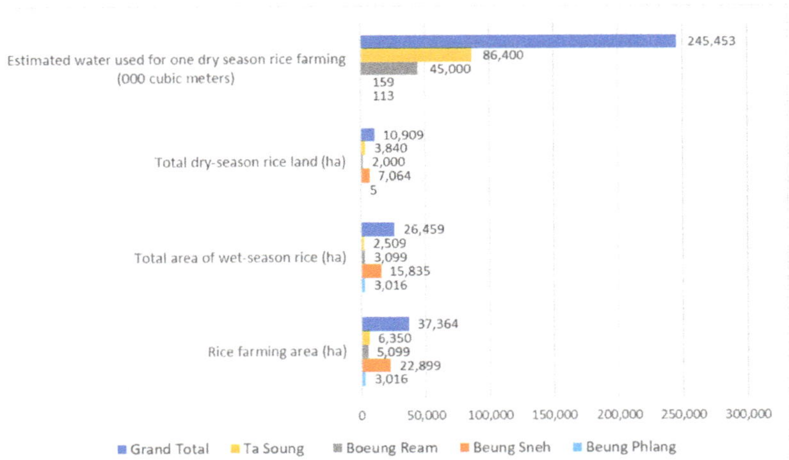

Figure 3. The total area of wet and dry seasonal rice in hectares by targeted communes (source: authors).

The rice farming communities located upstream and downstream of the Prek Ambel River, in Kandal Province and Takeo Province, respectively, compete for water for both their fisheries and rice farming. The Ta Soung Irrigation Scheme has extracted a minimum of 86.4 million cubic meters (MCM) of water from the Prek Ambel River to irrigate approximately 3840 hectares of dry seasonal rice farming in four communes, namely, Ban Kam, Kampong Reab, Pou Rumchak, and Prey Lvea, in Prey Kabbas District, Takeo Province. However, farmers in the Ta Soung Community cultivate three rice farming seasons per year, which means that the amount of water extracted from the Prek Ambel River could be tripled. Additionally, the CFis in the Prek Ambel River, located in Prey Kabbas District, have reported a negative impact of the Ta Soung Irrigation Scheme's water pumping on fishery and fish conservation; this mainly during the dry season when the water level in the Prek Ambel River is low. This results in some areas along the river drying up. Rice farming communities downstream of the Ta Soung Irrigation Scheme, particularly in the Prey Kabbas Commune, have voiced their concerns about the shortage of water for their rice farming during the dry season.

Competitions revolving around water usage for rice farming have been observed among farmers in the vicinity of Beung Sneh. In Damrei Puon Commune, the Chamcar Kouy Irrigation Scheme is responsible for extracting water for rice farming, which puts it in competition with other farmers from different communes. Despite this, the pumping station is capable of extracting a minimum of 4 MCM for dry-season rice farming. Meanwhile, in the Theay Commune, three irrigation schemes—Po Louk, Khse, and Top Sdach—are responsible for extracting approximately 23.35 MCM of water from Beung Sneh. In the Prey Kandieng Commune, three irrigation schemes—Phum Chan, Prey Kandieng, and Russei Muou Kom—use 32.82 MCM of water from the Beung Sneh to irrigate 1459 hectares of dry-season rice. Similarly, in the Ta Kao Commune, 2–3 irrigation canals that date back to the Khmer Rouge period have been revitalized by local communities using their own funds. The farmers in Ta Kao rely on these canals to extract 65.25 MCM of water from the Beung Sneh, competing with other communes to irrigate 2900 hectares of dry-season rice. At the village and household levels, each farmer possesses at least one pumping generator to extract water from the irrigation canals and irrigate their far-off rice fields. In total, around the lake, 158.94 MCM of water is extracted annually from the Beung Sneh to irrigate 7064 hectares. Between 2022 and 2023, the demand for water to irrigate dry-season rice farming areas caused the lake to reach dangerously low levels. Emergency measures were required due to the severity of the situation. The lack of water caused all irrigation canals to dry up, forcing nearby communities to resort to dredging the canals to extract the remaining water from the lake. In March, April, or early May of both years, farmers used water pumping generators to irrigate their rice fields, further exacerbating the already low levels of water in the lake and adversely affecting fishery and aquatic biodiversity.

Given the increased rice farming, farming households in the study areas experience increasing water shortages during the late first dry-season rice farming and the second dry-season rice farming. In response, FWUCs put the prices for water uses for its members to increase efficiency and effectiveness in water governance [9,25]. The fees for water use from the irrigation schemes per farming season range between 270,000 and 300,000 KHR/ha (USD 67–75). The water fee by gravity is around 250,000 KHR/ha and 300,000–350,000 KHR/ha by pumping. In the Ta Sung, FWCU collects water fees based on the costs of electricity usage to pump the water from the Prek Ambil River into the irrigation system, ranging between 200,000 and 250,000 KHR/ha. In Beung Sneh and Beung Ream, farmers pay water fees only for the first dry-season rice farming, which is between November and January, as they receive enough water to irrigate their rice fields, but they do not pay water fees for the wet-season rice farming between May and October, as they cultivate with rainfalls. The dry-season rice farming is between February and April, as there is not enough water to irrigate the dry-season rice farming. Also, about 50% of water users pay water fees. However, due to intensive three rice crops yearly, water use has reached critical levels. In addition, farmers have pumped from nearby water sources to irrigate the dry-season rice farming. Each household owns a water pumping machine/generator. There is competition among farmers over the uses of water from Beung Sneh areas. The water used for rice farming has affected the fisheries [10].

4.2.4. Pesticide and Fertilizer Utilization for Rice Farming

The rise in rice farming has led to an increase in the use of agro-chemical inputs to enhance rice yields. In the areas under study, fertilizers were commonly used by around 73% of households during both wet and dry farming seasons. Prey Veng Province saw the highest percentage of households using fertilizers, with 80% of them using them, compared to 67% in Takeo and 31% in Kampong Thom Provinces. In Prey Veng Province, farming households in Samraong, Tuek Thla, and Damrei Puong communes used fertilizers the most, accounting for 88%, 87%, and 85%, respectively. In Takeo Province, around 82% and 85% of farming households in Kan Kam and Pou Rumchak Communes applied fertilizers to their rice crops. Farmers usually used 5–7 bags (50 kg/bag)/ha of fertilizers for a farming season, from the time they sowed the rice seeds until harvest. They used various types of fertilizers, including DAP, urea, and others, which cost approximately KHR 120,000 (USD 30)/bag. The total cost of fertilizers per hectare ranged between USD 150 and 210. These fertilizers were mainly imported from Vietnam and sold publicly. Based on the interviews conducted, farmers had limited knowledge about fertilizer indications and how to use them.

Pesticides are commonly used by farmers to protect their crops, but this practice can negatively impact rice yield. Research indicates that around 70% of agricultural households in the studied areas use pesticides during the rice farming season. Prey Veng Province has the highest percentage of pesticide use at 78%, followed by Takeo at 69%. Notably, the communes of Damrei Puon, Samraong, Ampil Krav, and Prey Kandieng located far from the BSL, which experiences water shortages during the rainy season, have the highest percentage of pesticide use at 90%, 83%, 81%, and 80%, respectively. In Takeo, Pou Rumchak and Ban Kam Communes have the highest percentages of pesticide use in rice farming, accounting for 83% and 80%, respectively (Table 5).

Table 5. The uses of chemical inputs in rice farming.

Site	Commune	No. of HHs	HHs Using Chemical Fertilizers		HHs Using Organic Fertilizers		HHs Using Pesticides		HHs Using Organic Pesticides (Nature) to Kill Pests and Grass	
			No	%	No	%	No	%	No	%
Beung Phlang	Ampil Krau	1981	1610	81	77	4	1606	81	71	4
Beung Sneh	Theay	2964	2240	76	186	6	2170	73	157	5
	Damrei Puon	2679	2284	85	32	1	2402	90	16	1
	Samraong	2482	2179	88	43	2	2060	83	25	1
	Tuek Thla	2820	2457	87	23	1	2107	75	15	1
	Me Bon	2109	1531	73	67	3	1521	72	0	0
	Baray	1655	1256	76	27	2	1256	76	0	0
	Ta Kao	3739	2752	74	50	1	2754	74	4	0
	Prey Kandieng	2887	2308	80	60	2	2308	80	6	0
Sub-total	Sub-total	23,316	18,617	80	565	2	18,184	78	294	1
Boeung Ream	Kakoh (Sub-total)	3325	1023	31	536	16	609	18	81	2
Ta Soung	Ban Kam	1607	1320	82	111	7	1290	80	0	0
	Kampong Reab	532	220	41	100	19	220	41	100	19
	Pou Rumchak	778	662	85	25	3	648	83	55	7
	Prey Lvea	814	307	38	53	7	416	51	27	3
Sub-total	Sub-total	3731	2509	67	289	8	2574	69	182	5
Grant total	Grant Total	30,372	22,149	73	1390	5	21,367	70	557	2

Note: Source: commune database 2021.

Based on data provided by farmers, each hectare of land requires approximately nine containers of pesticides, each priced at KHR 15,000 (USD 3.75), resulting in a total pesticide cost of USD 33.75. To ensure maximum effectiveness, farmers spray their crops with pesticides 3–4 times per hectare until harvest, at a cost of USD 1.25 (KHR 5000) per spray. As a result, they spend around USD 105–140 per hectare on pesticides alone. On top of this, farmers also use other chemical treatments to combat weeds and invasive species such as snails, which have been causing damage to their rice fields. Unfortunately, the use of pesticides can also have negative effects on aquatic life. The pesticide use kills aquatic animals, including fish, and thus, not many fish are reported by farmers in the rice fields [67]. However, the percentage of households using organic pesticides and fertilizers is relatively low, 2% and 5%, respectively. Farmers no longer use organic fertilizers and pesticides in some villages, such as in the Prey Kandieng Commune (Table 5).

At the market, there was a diverse range of fertilizers offered by different importers and distributors. Urea and Muriate of Potash (KCl) were single-nutrient options, while di-ammonium phosphate (DAP) (18-46-0) and ammonium sulfate (16-20-0) were available as compound nitrogen-based fertilizers. Farmers could also find compound nitrogen, phosphorus, and potassium (NPK) products with ratios of 15-15-15, 16-16-8-(13S), and 20-20-15. Fertilizers could be purchased by the kilogram or in 50 kg bags. Most of the products were labeled in Khmer, with the exception of the 16-16-8-13 fertilizer from the Philippines and the urea from China and Vietnam, which had small Khmer stickers [68].

4.2.5. Impacts on Fishery Resources

The irrigation systems play a crucial role in controlling and regulating water flow between rivers, lakes, and rice fields. It is a type of structure that was built to improve water efficiency for rice farming, which may hinder fish migration patterns, breeding, and feeding grounds between dry and wet seasonal refuges, affecting fishery and agriculture practices [69]. The fishery domains and rice fields' segmentation into different sections can result in a lower fish population in rice fields, as reported by villages in the FGDs and KIIs, especially in the irrigation schemes. Additionally, the irrigation systems prioritize water for rice farming over fisheries, leading to the undermining of fisheries to some extent [70]. For instance, irrigation canals are emptied to get water to rice fields, resulting in the destruction of fishery resources in the canals. Moreover, there is no management system for fisheries in the irrigation systems, so fisheries are being harvested without any restrictions.

CFis and CFRs are established at various water sources, and some specific areas are designated as fishery conservation zones to safeguard rivers, lakes, and other bodies of water, creating a conducive environment for fish and their habitats and ensuring their survival. However, irrigation schemes such as Ta Soung, Tang Krasing, and Chamcar Kouy Irrigation Schemes extract water from the Prek Ambel River, Taign Krasaing River, and Beung Sneh Lake, respectively, in areas where CFis and CFRs are present. Also, FWUCs were established overlapping CFi and CFR areas to decentralize water extraction for rice farming, particularly during the dry season, by agricultural households. Without water, dry-season rice farming would be compromised, resulting in lost income for farmers [25].

The goals of CFis/CFRs and FWUCs/irrigation schemes can sometimes be at odds with each other. While CFis and CFRs aim to protect fishery sources to maintain productivity, FWUCs were established to extract water from sources to improve rice productivity. In dry seasons, these two organizations may compete for access to water resources, despite households being members of both. However, the extraction of water for rice farming can harm fisheries, leading to losses in productivity for sites such as Beung Sneh, Prek Ambel River, and Beung Ream CFR. Farmers in the studied areas face a difficult decision, as prioritizing water for CFis and CFRs may lead to water shortages and a loss of dry-season rice farming while protecting rice farming is necessary to maintain yield and production.

Rice fields are treated as food stocks for rural households in the study areas, including paddy rice, aquatic plants, aquatic animals, and fish. The use of pesticides and fertilizers for rice intensification has harmed and killed the aquatic animals and fisheries, undermined the aquatic plants, and also polluted the water in the canals, lakes, and rivers. Farming households in the study areas have reported a decline in fisheries in the rice fields and water bodies. Also, households indicate that they buy fish from the market to make food for their family members. These issues have affected the food items of rural households.

4.3. Level and Scale of Water Governance

Enhancing rice farming and fishery production requires improving water governance through the promotion and decentralization of integrated approaches to institutions and policies. Improving water governance involves multi-stakeholders at different levels and scales [19,29,30]. In natural resource management, decentralization has been implemented through community fisheries (CFis) and community fisheries refuges (CFRs) to empower local communities to manage their own resources [42,46]. Similarly, farmer water user communities (FWUCs) enable local communities to decentralize water resources. However, centralized control of technical and financial resources presents challenges for sectoral decentralization [5]. To address this, MOWRAM established two FWUCs, while FiA/FiAC established four CFis in Beung Sneh, and MoE established one community-based ecotourism (CBET). Despite these efforts, decentralized practices remain unintegrated as each community organization is institutionalized by their respective line ministries, resulting in unintegrated CFis, CBETs, and FWUCs. In the Ta Soung Irrigation Scheme, village communities are separated by CFis/CFRs or FWUCs, despite being from the same village. CFis are supervised by FiA and FiACs, while FWUCs are supervised by MOWRAM/PDOWRAM (Figure 4).

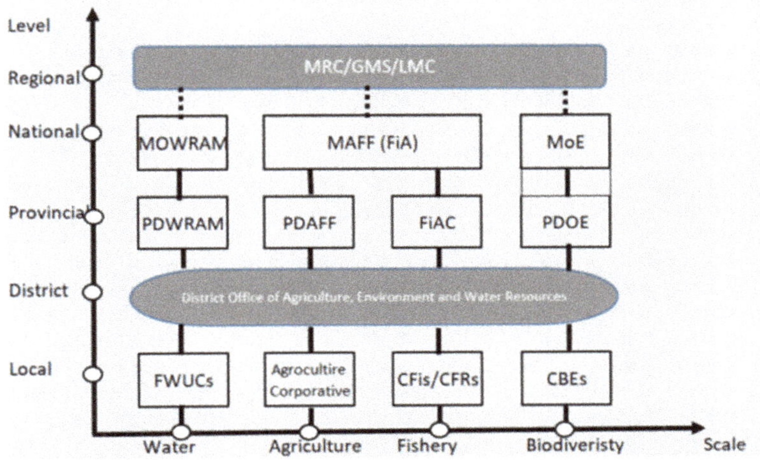

Figure 4. Level and scale of water governance (source: authors).

Furthermore, in Beung Sneh, different communes tend to manage the Beung Sneh from the geographical locations of the communes in the lake and not from integrated approaches. Competitions between communes in the lake have led to the uncertainty of the lake in the future. In the Ta Soung Irrigation Scheme, CFis and FWUCs are two different entities in the same communes, but the FWUC extracts water from the CFis to sustain rice farming and collect water fees from farmers. At the same time, CFis protect the water sources and do not charge any water fees to farmers or fishers to support their protection of the water sources. The lack of integration and connection between FWUCs and CFis leaves them uncertain about the system's future. Also, the Beung Ream CFR and the Kakoh Irrigation Canal are connected in one integrated system, but they operate independently, one under the FiA and another under MOWRAM/PDOWRAM.

However, FWUCs in Taing Krasaing, Ta Soung, and Chamcar Kouy Irrigation Schemes are managed under the District Agriculture, Environment, and Water Resource Office (DAEW) in which district officers in charge of water resources are responsible for managing FWUCs. Nevertheless, CFis/CFRs are not managed under the DAEW due to the fact that the fishery sector is not decentralized to DAEW, and so, its management remains with the FiACs. Thus, there is an urgent need to integrate fisheries into DAEW, so that district agriculture officers are responsible for fishery management and agriculture. This new approach would enable DAEW to coordinate the agriculture, environment, and water management at the district level, and they could report to district governors and the Provincial Departments of Agriculture, Water Resources and Meteorology, and Environment.

Nonetheless, DAEW is still new and has limited capacity and resources to deal with the growing water, fishery, and agriculture issues. Given the limited capacity and staff, they still have not been given the full power to implement their roles and responsibilities. Above all, they need capacity building and orientations to improve water governance, fishery management, and agricultural development. In the future, working with DAEW would address integrating water, fishery, and agriculture and decentralizing natural resource management.

5. Conclusions

Cambodia has abundant water resources in general, but it has little water in the dry season. Following the increased rice export policy in Cambodia in 2015 and the spill-over effects of the rice trade in Vietnam at present, the increased dry-season rice farming in many provinces has led to high water demand for dry seasonal rice farming. These have led to water shortages and conflicts over water among farmers in many provinces in Cambodia and between sectors, for instance, fishery and rice farming.

Irrigation system development and improvement have increased water availability, which improves agricultural development and rice farming. Rice farming areas have been expanded to around 3.34 million ha in 2019, and from one to three rice crops a year, the rice yield has increased from 3 tons/ha to 4–5 tons per ha. Rice production increased to 10.32 million tons in 2019, of which about 7 million tons were surplus for exports [7]. The increased rice production has occurred at the expense of the increased use of pesticides and fertilizers, mechanizations, indebtedness, and migration.

Furthermore, the irrigation system has imposed structures on the physical landscapes; first, it divides the land, wetlands, and water bodies into primary and sub-canals, and second, it blocks the water flows and migration patterns of fish in the floodplains and river systems in order to direct the irrigational flows to the rice fields. Further, it has broken the connectivity of fish migration pathways between the rivers, floodplains, lakes, and rice fields and vice versa. However, no tools and materials are in place to manage fisheries in the irrigation system. In contrast, fisheries management focuses more on central water bodies than rice field fisheries. The fishery is often ignored in irrigation management, as no expertise is involved in irrigation management. Thus, fishery productivity is low in the irrigation system and rice fields, where plenty of water exists. However, stock enhancement is needed to improve rice field fish stock.

The irrigation managements have been decentralized toward FWUCs, promoting the water fee system among members. To do so, FWUCs and irrigation schemes keep pumping water from river and lake systems, where CFis and CFRs are established to protect fisheries and water resources. These two systems are connected by water but are opposite in their approaches. The FWUC and CFis/CFRs often compete for water and conflict over water resources.

On the other hand, fisheries and their productivity are undermined and affected by agricultural practices, particularly the use of agricultural inputs such as pesticides, fertilizers, and chemical inputs to kill pests and herbs. These agricultural inputs are harmful to fishery and aquatic animals. Thus, even though there is water, there are few fish and aquatic animals in the rice fields.

Water governance remains sectoral, technical, and centralized, which has affected the productivity of water, fishery, and agriculture, as well as the cost of production, and it also has induced conflicts between sectors and among farmers. Decentralized water governance has been embedded into the policy, planning, and implementation through FWUCs, Cfis, and CFRs, but they are still sectoral and centralized to some extent, with limited financial and human resources and a lack of capacity.

Water governance can be improved through improving coordination between sectors and agencies at different levels and scales. It also needs an integration of different sectors and agencies and decentralization. Institutional integration should be strengthened to combine water, fishery, agriculture, and water resource management into one management system down at the ground. The district agriculture, water resources, and environment office should be strengthened to manage this integration and promote decentralized water governance.

Author Contributions: Conceptualization, M.S., S.d.S. and H.K.; Methodology, M.S., H.K. and H.S.; Formal analysis, M.S.; Investigation, M.S., S.S., S.d.S., H.K., T.T. and H.S.; Resources, S.d.S.; Data curation, S.S., C.K., T.T. and H.S.; Writing—original draft, M.S., S.S., S.d.S. and H.K.; Writing—review & editing, M.S., S.d.S. and H.K. All authors have read and agreed to the published version of the manuscript.

Funding: This research received funding support from CGIAR through the Asian Mega Delta Project.

Data Availability Statement: Materials are available on request from the corresponding author.

Acknowledgments: The authors thank the teams from WorldFish, IWMI, and IFReDI for their support in conducting this research. Special thanks are extended to Sean Vichet for his time in developing the map for the study.

Conflicts of Interest: The authors declare no conflicts of interest.

References

1. Sithirith, M. Downstream state and water security in the Mekong region: A case of Cambodia between too much and too little water. *Water* **2021**, *13*, 802. [CrossRef]
2. Sithirith, M. Analysing transboundary water governance: A global-regional-national water governance framework of Tonle Sap Lake. *Lakes Reserv. Res. Manag.* **2022**, *27*, e12396. [CrossRef]
3. Sokhem, P.; Sunada, K. The governance of the Tonle Sap Lake, Cambodia: Integration of local, national and international levels. *Int. J. Water Resour. Dev.* **2006**, *22*, 399–416. [CrossRef]
4. Ojendal, J. *Sharing the Good: Modes of Managing Water Resources in the Lower Mekong River Basin*; Göteborg University Department of Peace and Development Research: Goteborg, Sweden, 2000.
5. Royal Government of Cambodia (RGC). *Sub-Decree on Farmer Water User Community*; RGC: Phnom Penh, Cambodia, 2015; no. 73.
6. Center for Agriculture Development Study. *Inventory of Irrigation Schemes and Famer Water User Communities in Cambodia*; CEDAC and Water Program; CEDAC: Phnom Penh, Cambodia, 2009.
7. Cambodian Development Research Institute. *Hydrological Analysis in Support of Irrigation Management: A Case Study of Stung Chrey Bak Catchment, Cambodia*; CDRI Working Paper Series No. 59; CDRI: Phnom Penh, Cambodia, 2011.
8. Nang, P.; Khiev, D.; Hirsch, P.; Whitehead, I. *Improving the Governance of Water Resources in Cambodia: A Stakeholder Analysis—Understanding Stakeholders' Roles, Perceptions and Constraints for Effective Irrigation and Catchment Management and Development*; Working Paper Series No. 54; Cambodia Development Research Institute: Phnom Penh, Cambodia, 2011.
9. Nang, P.; Ouch, C. *Gender and Water Governance: Women's Role in Irrigation Management and Development in the Context of Climate Change*; Working Paper Series No. 89; Cambodia Development Research Institute: Phnom Penh, Cambodia, 2014.
10. Sithirith, M. Water governance in Cambodia: From centralized water governance to farmer water user community. *Resources* **2017**, *6*, 44. [CrossRef]
11. Ly, V. *The Baseline Assessment of the Effectiveness of Community Fisheries in Cambodia*; The Community Fisheries Development Department (CFDD), Fisheries Administration (FiA): Phnom Penh, Cambodia, 2018.
12. Royal Government of Cambodia (RGC). *Sub-Decree on Community Fisheries*; No 25 OrNor Kror BorKor; Royal Government of Cambodia: Phnom Penh, Cambodia, 2005.
13. Ostrom, E. *Governing the Commons: The Evolution of Institutions for Collective Action*; Cambridge University: New York, NY, USA, 1990.
14. Sithirith, M. Political Geographies of the Tonle Sap: Power, Space and Resources. Ph.D. Dissertation, National University of Singapore, Singapore, 2011.
15. Rogers, P.; Alan, W.H. *Effective Water Governance*; TEC Background Papers No. 7; Global Water Partnership/Swedish International Development Agency: Stockholm, Sweden, 2003.
16. Tortajada, C. Water Governance: A Research Agenda. *Int. J. Water Resour. Dev.* **2010**, *26*, 309–316. [CrossRef]
17. Öjendal, J.; Mathur, V.; Sithirith, M. *Environmental Governance in the Mekong. Hydropower Site Selection Processes in the Se Son and Sre Pok Basins*; Stockholm Environment Institute: Bangkok, Thailand, 2002.
18. Arias, M.E.; Piman, T.; Lauri, H.; Cochrane, T.A.; Kummu, M. Dams on Mekong tributaries as significant contributors of hydrological alterations to the Tonle Sap Floodplain in Cambodia. *Hydrol. Earth Syst. Sci.* **2014**, *18*, 5303–5315. [CrossRef]
19. Dore, J.; Lebel, L.; Molle, F. A framework for analysing transboundary water governance complexes, illustrated in the Mekong Region. *J. Hydrol.* **2012**, *466–467*, 23–36. [CrossRef]
20. Grundy-Warr, C.; Sithirith, M.; Li, Y.M. Volumes, fluidity and flows: Rethinking the nature of political geography. *Political Geogr.* **2015**, *45*, 93–95. [CrossRef]
21. Ratner, B.D.; Cohen, P.; Barman, B.; Mam, K.; Nagoli, J.; Allison, E.H. Governance of aquatic agricultural systems: Analyzing representation, power, and accountability. *Ecol. Soc.* **2013**, *18*, 59. [CrossRef]

22. Sreymom, S.; Sokhem, P.; Channimol, K. Governance for Water Security and Climate Resilience in the Tonle Sap Basin. In *Climate Change and Water Governance in Cambodia: Challenge and Perspectives for Water Security and Climate Change in Selected Catchments, Cambodia*; CDRI: Phnom Penh, Cambodia, 2015; pp. 124–153.
23. Seng, S.; Keartha, C.; Phyrom, S.; Sokhem, P. *Institutional Arrangements: Policies and Administrative Mechanisms on Water Governance in the Kingdom of Cambodia*; CPWF Mekong, M-Power, Asian Institute of Technology, CENTDOR, ewarec.: Phnom Penh, Cambodia, 2013.
24. Chea, C.; Nang, P.; Whitehead, I.; Hirsch, P.; Thompson, A. *Decentralised Governance of Irrigation Water in Cambodia: Matching Principles to Local Realities*; CDRI: Phnom Penh, Cambodia, 2011.
25. Fisheries Administration. *Report on Status of Community Fisheries Assessment in 2018*; Community Fisheries Development Department (CFDD), Fisheries Administration (FiA): Phnom Penh, Cambodia, 2022.
26. Chap, S.; Touch, P.; Diepart, J.X. *Fisheries Reforms and Right-Based Fisheries: Insights from Community Fisheries across Cambodia*; The Learning Institute: Phnom Penh, Cambodia, 2016.
27. Lebel, L.; Bastakoti, R.C.; Daniel, R. *CPWF Project Report: Enhancing Multi-Scale Mekong Water Governance*; Project Number PN50; CGIAR Challenge Program on Water and Food: Chiang Mai, Thailand, 2010.
28. Jones, M.; Jones, R.; Woods, M.; Whitehead, M.; Dixon, D.; Hannah, M. *An Introduction to Political Geography: Space, Place and Politics*; Routledge: London, UK, 2014.
29. Lefebvre, H. *The Social Production of Space*; Blackwell Publishing: Malden, MA, USA, 1991.
30. Dore, J. An agenda for deliberative water governance arenas in the Mekong. *Water Policy* **2014**, *16*, 194–214. [CrossRef]
31. Savenije, H.H.G.; van der Zaag, P. Water as an economic good and demand management: Paradigms with pitfalls. *Water Int.* **2002**, *27*, 98–104. [CrossRef]
32. Perry, C.; Rock, M.; Seckler, D. *Water as an Economic Good: A Solution, or a Problem?* International Water Management Institute: Colombo, Sri Lanka, 1997.
33. Asian Development Bank (ADB). *Proposed Revision of the Water Policy of the Asian Development Bank*; Asian Development Bank: Manila, Philippines, 2004.
34. Global Water Partnership. *GWP in Action*; Global Water Partnership: Stockholm, Sweden, 2002.
35. Ratner, B.D.; So, S.; Mam, K.; Oeur, I.; Kim, S. Conflict and Collective Action in Tonle Sap Fisheries: Adapting Governance to Support Community Livelihoods. *Nat. Resour. Forum* **2017**, *41*, 71–82. [CrossRef]
36. Meynell, P.J.; Kong, K.; Sorn, P.; Lou, V. *Climate Change Vulnerability Assessment Boueng Chhmar Ramsar Site, Cambodia*; IUCN: Bangkok, Thailand, 2019.
37. Royal Government of Cambodia (RGC). *Sub-Decree on Identification of Poor Households*; Ministry of Planning: Phnom Penh, Cambodia, 2011.
38. FAO. FAO Database. 2021. Available online: http://www.fao.org/nr/water/aquastat/data/query/results.html (accessed on 12 October 2022).
39. MOWRAM. *Databased on Cambodian Information System on Irrigation Schemes (CISIS)*; MOWRAM: Phnom Penh, Cambodia, 2017.
40. MOWRAM. *Water Strategic Development Plan, 2019–2023*; MOWRAM: Phnom Penh, Cambodia, 2019.
41. Fishery Administration (FiA). *The Strategic Planning Framework for Fisheries: Update for 2015–2024*; Ministry of Agriculture, Forestry and Fishery (MAFF): Phnom Penh, Cambodia, 2015.
42. Royal Government of Cambodia (RGC). *Law on Fisheries Management*; Fishery Administration: Phnom Penh, Cambodia, 2006.
43. World Bank. *Cambodia Rice Sector Review: Turning Cambodian Rice into White Gold*; Technical Working Paper; The International Bank for Reconstruction and Development/The World Bank: Washington, DC, USA, 2015.
44. Ministry of Water Resources and Meteorology (MOWRAM). *The National Irrigation and Water Resources Management Investment Program (NIWRMIP) 2019–2033*; MOWRAM: Phnom Penh, Cambodia, 2020.
45. Royal Government of Cambodia (RGC). *National Water Resources Management and Sustainable Irrigation Road Map and Investment Program 2019–2033*; MOWRAM: Phnom Penh, Cambodia, 2019.
46. Royal Government of Cambodia (RGC). *Sub-Decree on Community Fishery Management*; Fishery Administration: Phnom Penh, Cambodia, 2005.
47. MOWRAM. *Water Law*; MOWRAM: Phnom Penh, Cambodia, 2007.
48. MOWRAM. *Water Policy*; MOWRAM: Phnom Penh, Cambodia, 2005.
49. National Institute of Statistics (NIS). *Cambodia Agriculture Survey (CAS) 2020*; Ministry of Planning: Phnom Penh, Cambodia, 2022.
50. Ministry of Interior. *Commune Database*; Royal Government: Phnom Penh, Cambodia, 2021.
51. Sagara, J. *Surface Water Resources Assessment of the Tonle Sap and Mekong Delta River Basin Groups: Improving Climate Resilience, Productivity, and Sustainability*; Asian Development Bank: Manila, Philippines, 2021.
52. Arias, M.E.; Cochrane, T.A.; Piman, T.; Kummu, M.; Caruso, B.S.; Killeen, T.J. Quantifying changes in flooding and habitats in the Tonle Sap Lake (Cambodia) caused by water infrastructure development and climate change in the Mekong Basin. *J. Environ. Manag.* **2012**, *112*, 53–66. [CrossRef] [PubMed]
53. Chen, A.; Liu, J.; Kummu, M.; Varis, O.; Tang, Q.; Mao, G.; Wang, J.; Chen, D. Multidecadal variability of the Tonle Sap Lake flood pulse regime. *Hydrol. Process.* **2021**, *35*, e14327. [CrossRef]
54. Kallio, M.; Kummu, M. Comment on 'Changes of inundation area and water turbidity of Tonle Sap Lake: Responses to climate changes or upstream dam construction?'. *Environ. Res. Lett.* **2021**, *16*, 058001. [CrossRef]

55. Green, W.N.; Baird, I.G. The contentious politics of hydropower dam development in the Mekong River basin. *Political Geogr.* **2020**, *83*, 102272. [CrossRef]
56. Grumbine, R.E.; Dore, J.; Xu, J. Mekong hydropower Drivers of change and governance challenges. *Front. Ecol. Environ.* **2012**, *10*, 91–98. [CrossRef]
57. Wang, Y.; Feng, L.; Liu, J.; Hou, X.; Chen, D. Changes of inundation area and water turbidity of Tonle Sap Lake: Responses to climate changes or upstream dam construction? *Environ. Res. Lett.* **2020**, *15*, 0940a1. [CrossRef]
58. Frappart, F.; Biancamaria, S.; Normandin, C.; Blarel, F.; Bourrel, L.; Aumont, M.; Azemar, P.; Vu, P.L.; Le Toan, T.; Lubac, B.; et al. Influence of recent climatic events on the surface water storage of the Tonle Sap Lake. *Sci. Total Environ.* **2018**, *636*, 1520–1533. [CrossRef]
59. Chua, S.D.; Lu, X.X.; Oeurng, C.; Sok, T.; Grundy-Warr, C. Drastic decline of flood pulse in the Cambodian floodplains (Mekong River and Tonle Sap system). *Hydrol. Earth Syst. Sci.* **2022**, *26*, 609–625. [CrossRef]
60. Frenken, K. (Ed.) *Irrigation in Southern and Eastern Asia in Figures*; Food and Agriculture Organization of the United Nations: Rome, Italy, 2011.
61. Liu, J.; Chen, D.; Mao, G.; Irannezhad, M.; Pokhrel, Y. Past and future changes in climate and water resources in the lancang–mekong River Basin: Current understanding and future research directions. *Engineering* **2022**, *13*, 144–152. [CrossRef]
62. Hoang, L.P.; van Vliet, M.T.; Kummu, M.; Lauri, H.; Koponen, J.; Supit, I.; Leemans, R.; Kabat, P.; Ludwig, F. The Mekong's future flows under multiple drivers: How climate change, hydropower developments, and irrigation expansions drive hydrological changes. *Sci. Total Environ.* **2019**, *649*, 601–609. [CrossRef] [PubMed]
63. NGO Forum on Cambodia (NGOF). *Impacts of a Low Flow of Water from the Mekong River to Tonle Sap Lake on the Livelihoods of Different Communities and Natural Environments*; NGOF: Phnom Penh, Cambodia, 2023.
64. Zawahri, N. International rivers and national security: The Euphrates, Ganges-Brahmaputra, Indus, Tigris, and Yarmouk rivers. *Nat. Resour. Forum* **2008**, *32*, 280–289. [CrossRef]
65. Sithirith, M. Dams and state security: Damming the 3S rivers as a threat to Cambodian state security. *Asia Pac. Viewp.* **2016**, *57*, 60–75. [CrossRef]
66. Mainuddin, M.; Kirby, M. Agricultural productivity in the lower Mekong basin: Trends and future prospects for food security. *Food Secur.* **2009**, *1*, 71–82. [CrossRef]
67. Sokcheng, S.; Socheat, K.; Sarom, M. *Pesticide Use Practices in Cambodia's Vegetable Farming*; CDRI, Cambodia Development Resource Institute: Phnom Penh, Cambodia, 2021.
68. Vuthy, T. The Supply of Fertilizer for Rice Farming in Takeo. In *White Gold: The Commercialisation of Rice Farming in the Lower Mekong Basin*; Springer: Singapore, 2020; pp. 291–308.
69. Baran, E.; Starr, P.; Kura, Y. *Influence of Built Structures on Tonle Sap Fisheries*; Cambodia National Mekong Committee and the WorldFish Center: Phnom Penh, Cambodia, 2007.
70. Khoa, S.N.; Lorenzen, K.A.I.; Garaway, C.; Chamsinhg, B.; Siebert, D.; Randone, M. Impacts of irrigation on fisheries in rain-fed rice-farming landscapes. *J. Appl. Ecol.* **2005**, *42*, 892–900. [CrossRef]

water

MDPI

Article

Assessing the Impacts of Future Climate and Land-Use Changes on Streamflow under Multiple Scenarios: A Case Study of the Upper Reaches of the Tarim River in Northwest China

Qiang Han [1], Lianqing Xue [1,2,*], Tiansong Qi [3], Yuanhong Liu [1], Mingjie Yang [1], Xinyi Chu [1] and Saihua Liu [1]

1 College of Hydrology and Water Resources, Hohai University, Nanjing 210098, China
2 School of Hydraulic Engineering, Wanjiang University of Technology, Ma'anshan 243000, China
3 Department of Civil, Construction and Environmental Engineering (Dept 2470), North Dakota State University, P.O. Box 6050, Fargo, ND 58108-6050, USA
* Correspondence: lqxue@hhu.edu.cn

Abstract: Climate change and land use/cover change (LUCC) are two major factors that alter hydrological processes. The upper reaches of the Tarim River, situated in the northwest region of China, experience a dry and less rainy climate and are significantly influenced by human activities. This study comprehensively assessed the impacts of individual and combined climate changes and LUCCs on streamflow. Three general circulation models (GCMs) were utilized to predict future climate changes under three shared socioeconomic pathways (SSP119, SSP245, and SSP585). Cellular Automata–Markov (CA–Markov) was employed to predict future LUCC under three scenarios (i.e., ecological protection, historical trend, and farmland development). Streamflow for the period 2021–2050 was simulated using the calibrated MIKE SHE model with multiple scenarios. The results showed that from 2021 to 2050, increments in both average annual precipitation and average annual temperature under the three SSPs were predicted to lead to an increased streamflow. In comparison to the conditions observed in 2000, under three LUCC scenarios for 2030, the grassland area decreased by 1.04% to 1.21%, while the farmland area increased by 1.97% to 2.26%, resulting in reduced streamflow. The related changes analysis indicated that the variation in streamflow during winter is most significant, followed by spring. The study predicted that climate change would increase streamflow, while LUCC would decrease it. Due to the greater impact of LUCC, considering the combined effect of both factors, runoff would decrease. The contribution analysis indicated that climate change contributed between −7.16% and −18.66%, while LUCC contributed between 107.16% and 118.66%.

Keywords: CA–Markov; climate change; land-use change; multiple scenarios; MIKE SHE model

1. Introduction

Climate change and land use/land cover change (LUCC) play a significant role in hydrological variations. Variations in precipitation and temperature due to climate change have impacts on streamflow [1]. LUCC alters streamflow by affecting evapotranspiration, interception, and infiltration. Simultaneously, the interactions between climate and land use make hydrological processes even more complex [2]. The assessment of streamflow responses to climate change and LUCC is crucial for water resources management [3–5].

Many studies have quantified impacts of climate change on streamflow with the aid of General Circulation Model (GCM) [6–8]. Due to the presence of some uncertainty, multiple GCMs are often used in research to reduce errors in simulating scenarios compared to real conditions [9–11]. Currently, there have been various findings on how climate change affects streamflow among different basins. Meanwhile, projections indicated increased streamflow in the future for the Yuan River Basin [12] and Yangtze

River Basin [13] in China in response to escalating temperatures and precipitation levels. Conversely, in the Songkhram River Basin, Thailand [14], and the Amu Darya River Basin [15], streamflow responded differently to climate change. The integration of representative concentration pathways (RCPs) with shared socioeconomic pathways (SSPs) in the latest Coupled Model Intercomparison Project Phase 6 (CMIP6) enhances the realistic simulation of future scenarios [16–18].

At the same time, as human activities continue to intensify, the impacts of LUCC on streamflow have also been the subject of extensive research [19,20]. Zhang, et al. [21] concluded that the continuous expansion of cities would lead to a significant increase in streamflow. Additionally, different land-management measures, such as land reclamation or "Grain for Green" [22], have different impacts on streamflow [23–25]. Therefore, predicting future LUCC under different scenarios and determining its impact on streamflow is important. The CA–Markov model, integrating Cellular Automata (CA) and the Markov Chain, is currently the most commonly used approach in predicting future land use [26–30].

Previous research has predominantly concentrated on the separate effects of climate change or LUCC on runoff, often relying on historical data [31–33]. Research on comprehensive studies about how runoff will change in the future under the combined impacts of climate change and LUCC remains limited. The Tarim River, situated in the arid northwest region of China, is the country's largest inland river [34,35], yet there is still a lack of research on this aspect within its basin. To implement more scientific management of the valuable water resources, it is crucial to precisely assess the impacts of future climate change and LUCC on streamflow.

To fill this research gap, this study employed multi-GCMs to predict future climate change and utilized the CA–Markov model to project future LUCC. Three climate change scenarios (SSP119, SSP245, and SSP585), three LUCC scenarios (i.e., ecological protect, history trend, and farmland development), and nine integrated scenarios were established. The hydrological model was established by coupling MIKE SHE with MIKE 11, followed by separate streamflow simulations for each of the 15 scenarios. Furthermore, the study analyzed the individual and combined influences of climate change and LUCC on streamflow using metrics such as the relative change rate (RCA) and contribution rate (Figure 1).

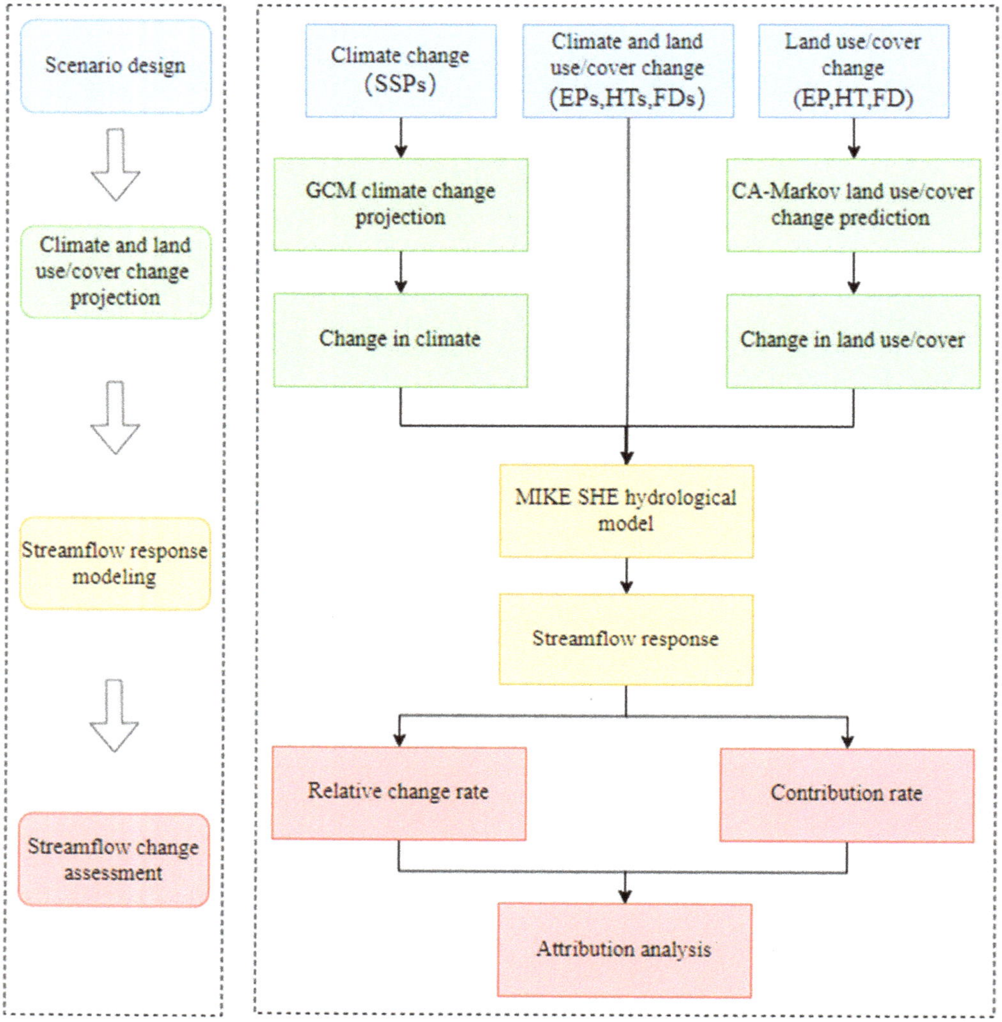

Figure 1. Schematic of methodology.

2. Methodology

2.1. MIKE SHE Model

MIKE SHE is a grid-based model that uses physics-based algorithms to represent distributed processes of the hydrological cycle [36]. MIKE SHE was developed to model water movement, including overland flow, snow melting, rivers and lakes, and unsaturated and saturated flow [37]. The overland flow simulation used a two-dimensional diffusive wave approximation by the finite-difference method. The modified degree–day method was used to simulate snow melting, where the starting times of snowfall and snowmelt were determined by setting threshold temperatures. The characteristic functions of the degree–day method are determined on a time scale based on a day [38]. It has been proved to be accurate of the physically based models [39]. The unsaturated flow was quantified by a two-layer water-balance method. The saturated flow was modeled using a numerical finite-difference scheme.

The rivers and lakes model was based on the one-dimensional unsteady flow Saint–Venant equations. Using MIKE SHE and MIKE 11 [40], the dynamic coupling model of surface water and groundwater was established. In the coupled modeling system, data exchange between two models was achieved through shared storage space, where simulation can be simultaneous in both MIKE SHE and MIKE 11 [41].

2.2. CA–Markov Model

CA–Markov combines the dual advantages of simulating complex spatial dynamic changes through Cellular Automata (CA) models and making long-term predictions using Markov models. The model is widely applied in simulating the dynamic changes in land use and cover structures over long sequences [29,42,43].

The Markov chain is a stochastic probability model characterized by the current state being only related to the preceding state [44]. Additionally, it is discrete in both time series and events [45]. Equation (1) describes the properties of the Markov chain:

$$V_{t+1} = V_t \times P_{ij} \tag{1}$$

$$P_{ij} = \begin{bmatrix} P_{11} & P_{12} & \cdots & P_{1n} \\ P_{21} & P_{22} & \cdots & P_{2n} \\ \cdots & \cdots & \cdots & \cdots \\ P_{n1} & P_{n2} & \cdots & P_{nn} \end{bmatrix} (0 \le P_{ij} \le 1) \tag{2}$$

where V_t and V_{t+1} denote the system state at time t and $t+1$, respectively. Meanwhile, P_{ij} is the probability of transitioning from state i to state j in the subsequent time period.

The CA is a dynamic model that is discrete in both time and space [46]. The principle is to predict the state of the next time step by utilizing specific transformation rules based on the current state of individual cells and their neighboring states in the cellular automaton [47]. The model can be defined as follows:

$$S(t,\ t+1) = f(S(t),\ N) \tag{3}$$

where S denotes a set of cellular states within the field N, while t and $t+1$ are consecutive time periods. The transition rule of the cell is denoted as f.

2.3. Accuracy Evaluation

Calibration and validation are pivotal in enhancing model accuracy, significantly reducing the uncertainty stemming from their intricate iterative equations [48]. To gauge model performance, two widely adopted metrics (the Nash–Sutcliffe efficiency coefficient (NSE) [49–51] and Root Mean Square Error (RMSE) [52]) were utilized. They are defined as follows:

$$NSE = 1 - \frac{\sum_{i=1}^{n}(Q_{obs,i} - Q_{sim,i})^2}{\sum_{i=1}^{n}(Q_{obs,i} - \overline{Q}_{obs})^2} \tag{4}$$

$$RMSE = \sqrt{\frac{1}{n}\sum_{i=1}^{n}(Q_{obs,i} - Q_{sim,i})^2} \tag{5}$$

where $Q_{obs,i}$ and $Q_{sim,i}$ refer to the observed and simulated streamflow on the i-th day (m^3/s), respectively. Meanwhile, Q_{obs} and Q_{sim} are the mean observed and simulated head values throughout the simulation period (m^3/s), with n indicating the total number of time steps [53].

The accuracy assessment of land-use prediction involves testing with indicators. One such indicator is the Kappa index, originally introduced by Cohen [54]. This index assesses the consistency between measured data and simulated result. It is a commonly used criterion for assessing the relationship between simulated data of land use map and actual data [42,55].

$$Kappa = \frac{P_0 - P_c}{P_p - P_c} \quad (6)$$

where P_0 is the observed agreement rate between the reference map and the simulation results. Meanwhile, P_c stands for the expected proportion of correct simulations in a random scenario. On the other hand, P_p denotes the proportion of correct simulations in the ideal classification case, typically considered as 1.

2.4. Separating Effects of Climate Change and LUCC

Following streamflow forecasting, it becomes essential to analyze the subsequent impacts of climate changes and LUCC on streamflow. The quantitative assessment of the contributions made by these changes to streamflow involves calculations using the following formulas:

$$\Delta Q_{L,C} = Q_{L,C} - Q_{baseline} \quad (7)$$

where $\Delta Q_{L,C}$ is the overall change in future streamflow compared to the baseline, $Q_{L,C}$ denotes the streamflow projected under future climate and land-use scenarios, $Q_{baseline}$ stands for the streamflow during the baseline period.

$$\eta_L = \frac{Q_{L,baseline} - Q_{baseline}}{\Delta Q_{L,C}} \quad (8)$$

$$\eta_C = \frac{Q_{baseline,C} - Q_{baseline}}{\Delta Q_{L,C}} \quad (9)$$

where η_L and η_C are the contribution of LUCC and climate changes to streamflow, respectively. $Q_{L,baseline}$ is the streamflow under future land-use scenarios and baseline climate conditions. Similarly, $Q_{baseline,C}$ is the streamflow under the baseline land-use scenario and future climate change conditions.

3. Study Area and Data

3.1. Study Area

The Tarim River (Northwest China, Xinjiang Uygur Autonomous Region) is the longest inland river in the country [34]. Dominated by the typical continental temperate arid climate, the Tarim River Basin (TRB) is a typical water-shortage area with mean annual precipitation of less than 116 mm, an annual mean temperature of about 10 °C, and potential evapotranspiration of up to 2200 mm per year [56,57]. Except for summer precipitation in mountains, seasonal snowmelt and glacier melt dominate the runoff in the TRB. Thus, flow characteristics are sensitive to climate change [58].

As shown in Figure 2, the study area is located in the upper reaches of the Tarim River Basin (URTRB), covering an expanse of 3.43×10^5 km^2. This region comprises the Aksu, Yerqiang, and Hetian Rivers, as well as the mainstream above the Xinquman hydrological station. Among them, the Aksu River is the main source of the TRB, contributing about 73.2% of the water supply [59]. In such an ecological fragile area, with the long-term development and utilization, streamflow processes and land-use patterns have changed greatly [60,61]. The contradiction between economic development and ecological protection poses an unprecedented challenge to the management of the river basin.

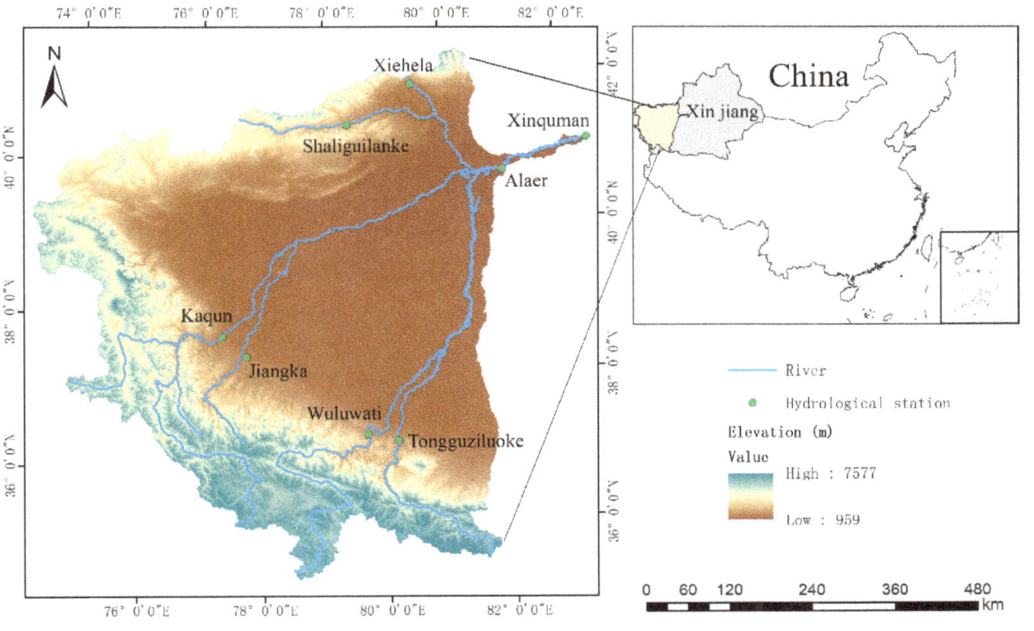

Figure 2. Location of the upper reaches of the Tarim River Basin (URTRB) and hydrological station river network.

3.2. Data

The hydrological data used in this study included: (1) The discharge data collected by six hydrological stations in the source stream (i.e., Xiehela, Shaliguilanke, Kaqun, Jiangka, Wuluwati, and Tongguziluoke) and two hydrological stations in the mainstream (i.e., Alaer and Xinquman); and (2) The water level data of the Xinquman hydrological station in the lower reaches. These hydrological data during ~1985–2014 were monitored and provided by the Tarim River Basin Authority.

All climate data (from 12 meteorological stations during ~1985–2014), including precipitation, temperature, and evaporation, were provided by the China Meteorological Data-Sharing Service System. We used the Thiessen polygon method to deal with the data of each station and obtained the average value of the meteorological data in the study area. The future climate change data were acquired from the National Tibetan Plateau Data Center (http://data.tpdc.ac.cn, accessed on 1 December 2023). The data were downscaled by Peng, et al. [62] using the Delta method [63,64] and have been widely used [65–67]. These monthly scale meteorological data were derived from three climate models: EC–Earth3, GFDL–ESM4, and MRI–ESM2–0 under SSP119, SSP245, and SSP585 scenarios with a spatial resolution of 30 s × 30 s from 2021–2050.

The Digital Elevation Model (DEM) data in this study were obtained from the Landsat series images available on the Geospatial Data Cloud (https://www.gscloud.cn, accessed on 1 December 2023). The DEM, with a resolution of 90 m, served as the primary data source for characterizing the terrain and river network within the catchment. Land-use data for this study were sourced from the Resource and Environment Science and Data Center at the Chinese Academy of Sciences (https://www.resdc.cn, accessed on 1 December 2023). The remote-sensing images were utilized to analyze changes in land-use patterns in the study area. In this study, the categories were reclassified as cropland, forest, grassland, built-up land, unused land, and water bodies. Soil type and soil depth were based on data of Harmonized World Soil Database (HWSD). Using the SPAW hydrology 6.02.75 software, the soil types were reclassified based on soil attributes into 13 different categories.

3.3. Scenario Design

Three LUCC scenarios, representing a spectrum of tradeoffs between ecological preservation and farmland development, were introduced to examine the impact of diverse land-use policies on streamflow. These scenarios include ecological protection (EP), historical trends (HT), and farmland development (FD).

(1) EP Scenario. The scenario enacts ecological conservation, keeping the area of farmland and urban space unchanged while gradually expanding the grassland and forest areas each year.
(2) HT Scenario. This scenario indicates that future land use remains unaffected by any policy influence and continues to develop along historical trends.
(3) FD Scenario. This scenario assumes that human activities are steadily increasing, with farmland land and urban areas experiencing gradual expansion.

A total of nine different scenarios have been created by combining climate and land-use/cover changes. We assemble SSPs with EP as EPs (i.e., EP119, EP245, and EP585); HT as HTs (i.e., HT119, HT245, and HT585); and FD as FDs (i.e., FD119, FD245, and FD585).

4. Result and Discussion

4.1. Climate Change Scenarios

To reduce the uncertainty of future climate data, this study employed the average values of the EC–Earth3, GFDL–ESM4, and MRI–ESM2–0 models as the basis for analyzing future climate data. As shown in Figure 3, the bar chart represents the mean monthly precipitation and temperatures for the period from 2021 to 2050 over URTRB. The error bars depict the range of the multi-model ensemble. The multi-model ensemble range for the monthly temperature is small. In contrast, for monthly precipitation, it is much larger. This indicates that, across various SSP scenarios, monthly temperatures exhibit more similarity, while there is a greater degree of uncertainty in monthly precipitation. Precipitation in the URTRB is extremely unevenly distributed throughout the year, with very little rainfall in the spring and winter, leading to drought conditions. In contrast, precipitation during the summer season is significantly higher, accounting for 49% to 51% of the annual total, and it can easily lead to flooding. At the same time, temperatures vary significantly across different months, with a difference of about 30 °C between the highest and lowest temperatures. January, February, March, November, and December all have temperatures below freezing, which is why precipitation primarily occurs in the form of snow. It is not surprising that the monthly average temperature increases with the increase in radiation intensity.

Figure 4 shows the future precipitation and temperature under various SSP scenarios for the period from 2021–2050. The overall temperature exhibited a notable upward trend. The rising temperature helps to melt snow in mountainous areas, so as the melting snow replenishes runoff, it also helps with vegetation growth. At the same time, there is no significant change in precipitation. The annual average precipitation under SSP245 is 69.48 mm, slightly higher than 66.59 mm under SSP119 and 67.69 mm under SSP585.

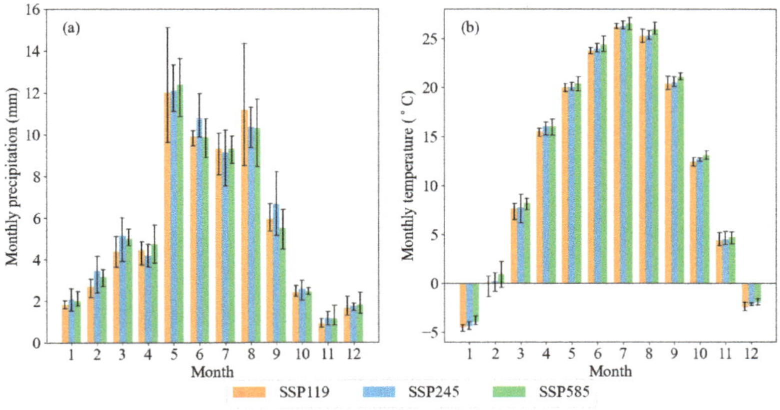

Figure 3. Mean monthly (**a**) precipitation and (**b**) temperature averaged over URTRB under different SSPs from 2021–2050. The error bars indicate the multi-model ensemble range.

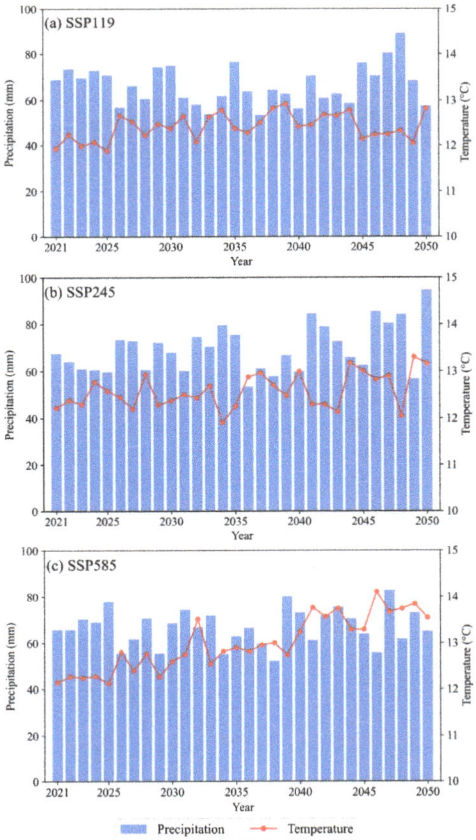

Figure 4. Annual average precipitation and temperature changes from 2021–2050.

4.2. Land Use Change Scenarios

The CA–Markov model was used to simulate 2010 land-use patterns, leveraging data from 2000 as a basis. As shown in Figure 5, the Kappa value is 0.87, indicating that the simulated and measured land-use projections demonstrate satisfactory agreement.

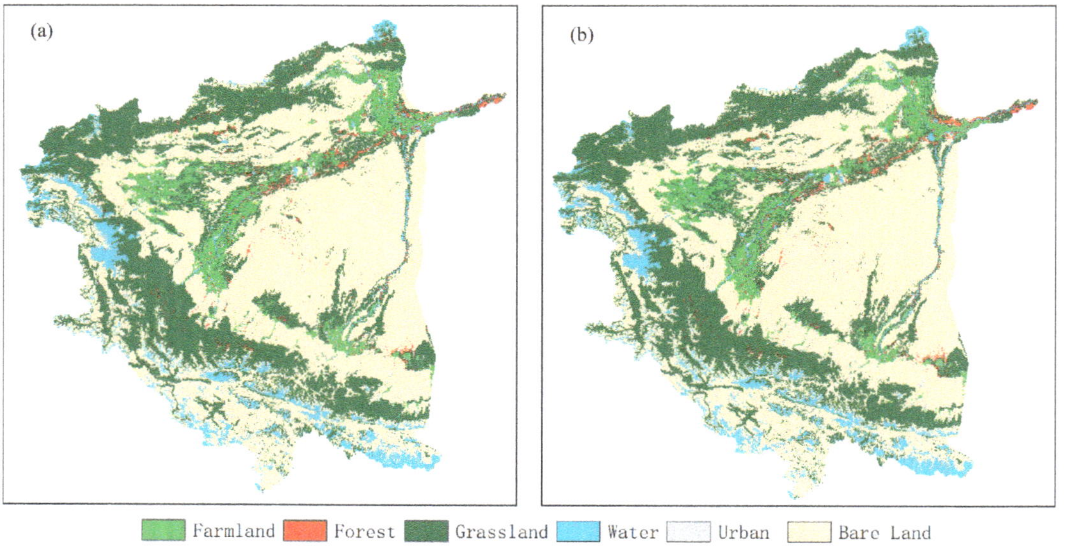

Figure 5. Actual land-use maps of 2010 (**a**) and simulated land-use maps of 2010 (**b**).

As shown in Figure 6, the CA–Markov model was employed to simulate the LUCC under the three scenarios in 2030. Under the influence of different land-use policies, the conversion probabilities between different types of land use have changed. Comparing the land-use/cover simulation result of URTRB in 2030 with those in 2000, we concluded that after 30 years of change, the proportion of farmland in the study area increased by 2.03%, forest decreased by 0.22%, grassland decreased by 1.14%, water bodies increased by 0.02%, urban areas increased by 0.03%, and bare land decreased by 0.66%. Under the influence of human activities, a significant amount of grassland, forest, and bare land has been developed into farmland. When comparing the simulated results of land use/cover across the three scenarios for 2030, noticeable differences in both quantity and spatial distribution among the patterns were evident. Compared to the HT scenario, in the EP scenario, there has been a 1.5% reduction in farmland, a 1.4% increase in grassland, and a 0.1% increase in forest. This indicates that through ecological protection efforts, the expansion of arable land has been reduced, leading to a greater transformation of barren land into grassland and forest. Meanwhile, in the FD scenario, farmland increased by 0.5% compared to the HT scenario while forest decreased by 0.1%, grassland decreased by 0.2%, and water bodies decreased by 0.3%. This indicates that if human activities intensify further, there will be more grassland and forest reclaimed as farmland. This would have a negative impact on the local ecological environment. Therefore, in the EP scenario, the expansion of grasslands and forests has a promoting effect on desert ecology. Moreover, excessive water resource usage will lead to a significant reduction in water bodies.

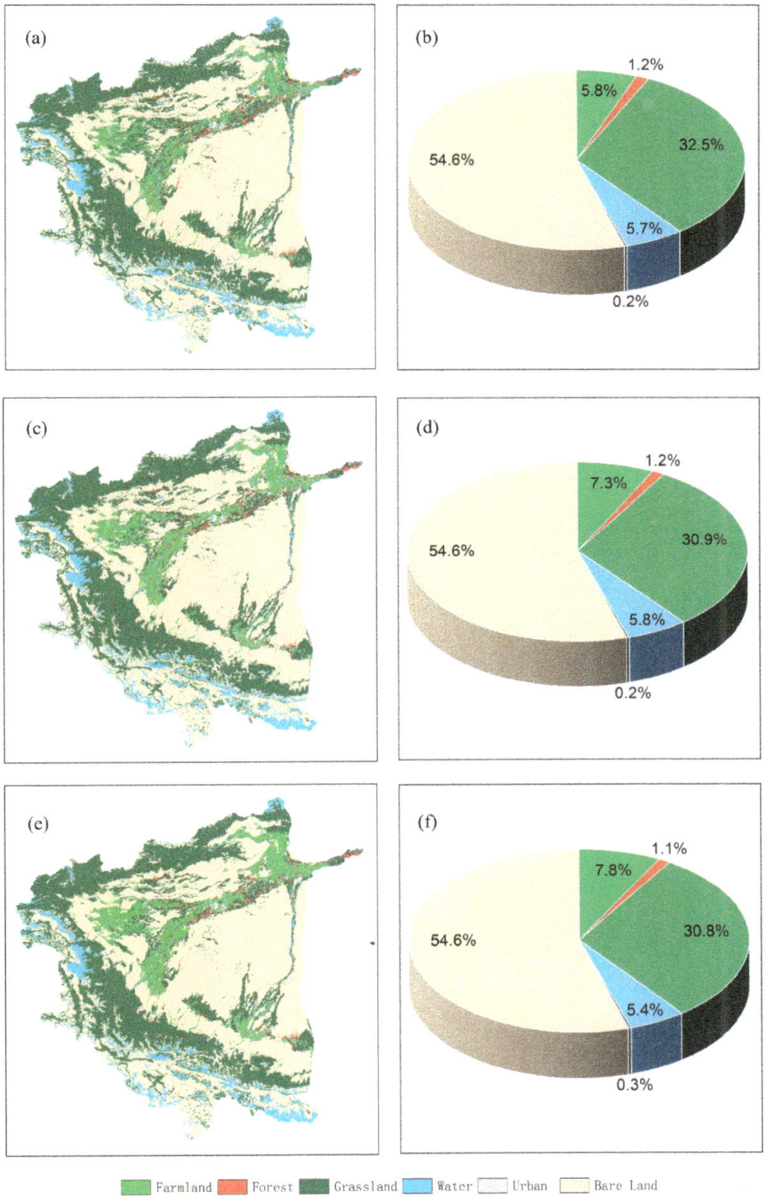

Figure 6. Projected land use maps in 2030 under (**a**) EP, (**c**) HT, and (**e**) FD scenarios, and proportions of area of each land use/cover type under (**b**) EP, (**d**) HT, and (**f**) FD scenarios.

4.3. Calibration and Validation of MIKE SHE Model

The MIKE SHE model was calibrated and validated at a monthly scale for the periods of 1985–1999 and 2000–2014, respectively. Various parameters were calibrated through manual calibration (Table 1). The results indicated a good fit between observed and simulated streamflow, with NSE values of 0.79 and 0.81 and RMSE values of 97.81 m^3/s

and 93.81 m^3/s for the calibration and validation periods, respectively. The hydrological model fits well for both low and peak flow conditions as shown in Figure 7. According to the criteria outlined in Moriasi, Gitau, Pai, and Daggupati [52], the simulation results during both the calibration and validation periods consistently met the "Very Good" standard. They suggested that the MIKE SHE model was deemed suitable for simulating streamflow in the URTRB.

Table 1. Calibration results of model parameters.

Components	Parameters	Units	Calibrated Values
Snow melt	Melting temperature	°C	0
	Degree–day coefficient	mm°C/day	0
Overland flow	Manning number	$m^{1/3}/s$	25
	Detention storage	mm	4
River	Manning number	$m^{1/3}/s$	11
	Leakage coefficient	m/s	1.47×10^{-7}
Saturated zone	Horizontal hydraulic conductivity	m/s	1.25×10^{-3}
	Vertical hydraulic conductivity	m/s	1.25×10^{-4}
	Storage coefficient	m^{-1}	2.22×10^{-5}
	Specific yield	-	0.11

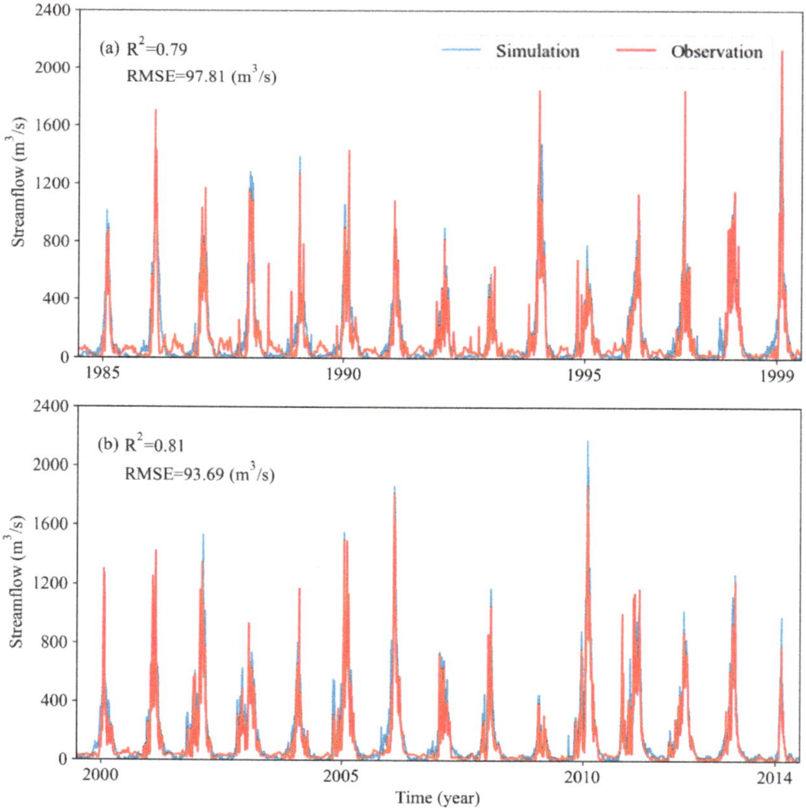

Figure 7. Comparison of observed and simulated streamflow.

4.4. Streamflow Response Modelling under Multiple Scenarios

4.4.1. Under Varying Climate Change Scenarios

We used the calibrated MIKE SHE model to predict long-term (2021–2050) streamflow only under climate change, LUCC, and their combinations. From 1985–2014, the annual streamflow was 133.51 m^3/s. Due to climate influences, the annual streamflow from 2021–2050 ranges from 135.62 to 138.21 m^3/s, representing an increase of 1.6% to 3.5% compared to the 1985–2014 period. In all scenarios, there is a trend of increased streamflow. This outcome is attributed to the fact that the streamflow in URTRB is derived from melting snow, and as a result, streamflow increases with rising precipitation and temperatures.

Figure 8b shows that the monthly streamflow in URTRB is unevenly distributed, with a significant difference in streamflow between the summer flood season and the other three seasons. During the flood season from June to September in the period from 1985–2014, the average streamflow was 312.33 m^3/s, accounting for 79.52% of the total annual streamflow. From 2021 to 2050, the monthly streamflow variation within the year may decrease. The average monthly streamflow in the eight months outside the flood season has increased by 1.19% to 10.08% compared to the years 1976–2005, resulting in the flood season's total streamflow accounting for 76.73% to 78.23% of the total streamflow, which represents a decrease of 1.29% to 2.79%.

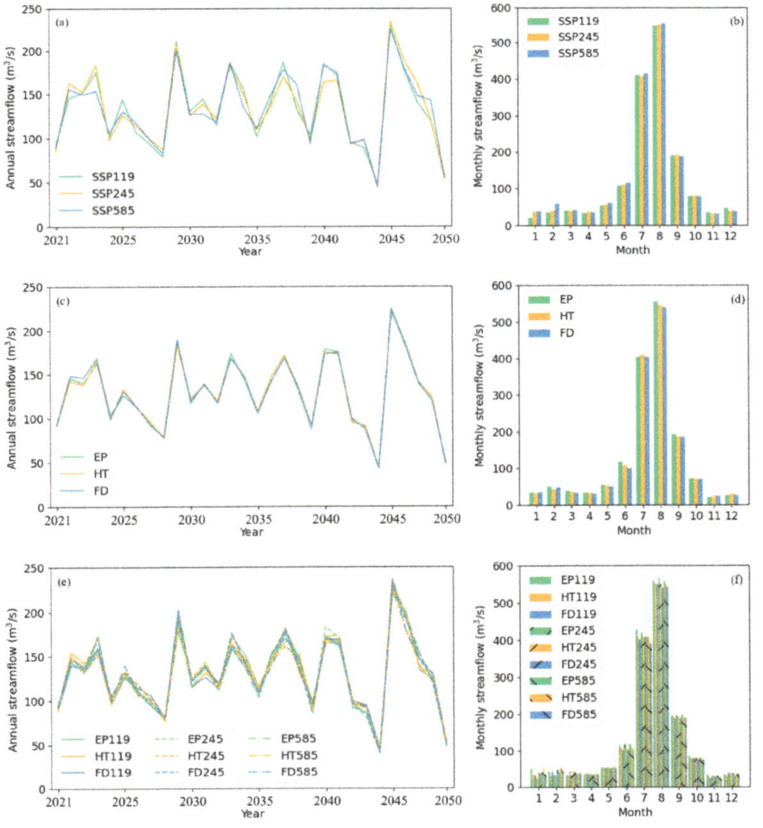

Figure 8. Mean annual and monthly streamflow in URTRB from 2021–2050 under (**a**,**b**) Climate change, (**c**,**d**) LUCC, and (**e**,**f**) Combined climate change and LUCC.

4.4.2. Under Varying Land Use/Cover Change Scenarios

Figure 8c,d illustrated the annual and monthly streamflow variations under LUCC. Under three LUCC scenarios, as grassland and forests are converted to farmland to varying extents and bare land decreases, there is a decreasing trend in streamflow from 2021 to 2050, with a mean annual streamflow ranging from 129.48 to 130.69 m^3/s. The expansion of farmland areas is significantly under the FD scenario. In the period from 2021–2050, the mean annual streamflow shows a decrease of 2.86% compared to the period from 1985–2014. Due to EP's ecological conservation efforts, the degree of conversion from farmland to forests has been reduced, resulting in a higher mean annual streamflow compared to HT and FD, but it is still lower than the baseline period.

Intra-annual monthly streamflow still exhibits the situation where the monthly streamflow during the summer flood season is significantly higher than in other months, but the gap has decreased. Summer season streamflow accounts for approximately 78.24% of the total stream, which is 3.28% less than the baseline period. This may be due to the fact that the growing season of crops in farmlands primarily occurs during the summer, making their impact on summer streamflow more pronounced.

4.4.3. Under Varying Combined Climate and Land Use/Cover Change Scenarios

Figure 8e,f showed the annual and monthly streamflow variations under combined climate change and LUCC scenarios. Over the period from 2021–2050, the streamflow exhibits a decreasing trend in these combined scenarios, with a mean annual streamflow ranging from 129.67 to 130.32 m^3/s. This represents a variation of 2.39–2.87% compared to the period from 1985–2014. The mean annual streamflow is highest under EPs, followed by that under HTs and FDs. Additionally, the alteration characteristics of annual streamflow remain similar under different LUCCs. The annual runoff fluctuation between the comprehensive scenarios is greater than that of the LUCC scenario and smaller than that of the climate change scenario.

5. Discussion

In this study, we analyzed the separate and combined impacts of climate change and LUCC on the alteration of the mean monthly streamflow in the URTRB. We employed multi-GMSs to reduce uncertainties between different SSPs. The RCA of streamflow, attributed to climate change, LUCC, and their combination, was defined as the ratio of the streamflow change under SSPs, LUCC (EP, HT, and FD), and combined conditions (EPs, HTs, and FDs) from 2021–2050 relative to the standard deviation of streamflow from 1985–2014. Refer to Figure 9 for the presented results.

Climate change, a key factor influencing variations in streamflow, results in an augmented streamflow due to the heightened radiation intensity. Its impact on streamflow is mainly manifested in the increase of precipitation and rising temperatures. Under the influence of SSP119, SSP245, and SSP585, the mean RCA is 0.12, 0.11, and 0.18, respectively. During the months of May to September, when runoff is relatively concentrated, the impact of climate change on runoff is smaller, with a mean RCA of only 0.01. However, during the dry season, runoff is more significantly affected by climate change mainly due to the increase in snowmelt following a rise in temperature, which supplements the runoff in the dry season.

In addition, the changes in flow induced by EP, HT, and FD are different, with mean RCAs of 0.28, 0.19, and 0.29, respectively. The RCA under LUCC is larger than that under climate change, which indicates that the impact of LUCC on streamflow is greater than that of climate change. Due to extensive farmland expansion, FD has the greatest land-use change, so its impact on streamflow is greater than EP and HT.

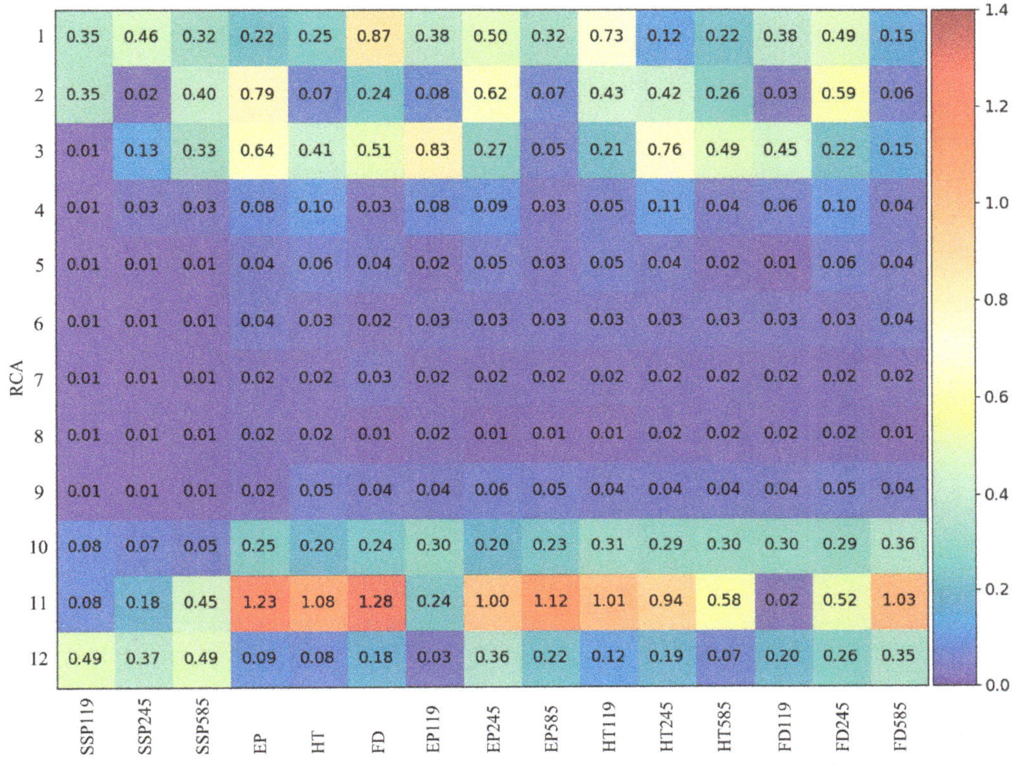

Figure 9. Monthly attribution of streamflow attributed to climate change, LUCC, and combined LUCC and climate change effects in URTRB.

As seen from Figure 9, the streamflow changes in EPs are smaller than those in HTs and FDs. This is because, under ecological protection, the expansion of farmland is significantly restricted and grassland and forests are effectively preserved, resulting in a relatively smaller impact on streamflow compared to the baseline period. Wang, et al. [68] found that the streamflow increases when farmland was converted into forests, which is consistent with the conclusion of this study. The streamflow shows significant changes in both spring and winter. Upon comparing it with Figure 8, it is found that the main reason for this is an increase in spring streamflow and a decrease in winter streamflow.

The impacts of climate change and LUCC on the annual streamflow were then quantified. The results showed that the combined effect of both resulted in a decrease of 3.18–3.84 m^3/s in the mean annual streamflow. The change in annual mean streamflow is primarily driven by LUCC (Table 2). Specifically, climate change increases annual streamflow by 0.27–0.60 m^3/s, contributing from −18.67% to −7.16%, while LUCC leads to a decrease in annual streamflow by 3.78–4.11 m^3/s, with a contribution ranging from 107.16% to 118.67%. In different years and months, the results may vary. The dominant effects of LUCC identified in our study align with findings from prior research in the Yihe River [69], Beiluo River [70], and Luanhe River, China [20]. The opposite was found in Heihe River, China [33,71], where climate change is the dominant factor in runoff variation.

Table 2. Annual contributions of climate change and LUCC in URTRB.

Scenarios	EP119	EP245	EP585	HT119	HT245	HT585	FD119	FD245	FD585
Climate change (%)	7.16	13.67	16.12	11.09	14.48	18.08	11.43	14.94	18.66
LUCC (%)	−107.16	−113.67	−116.12	−111.09	−114.48	−118.08	−11.43	−114.94	−118.66
Total (%)	−100	−100	−100	−100	−100	−100	−100	−100	−100

6. Conclusions

This study projected future climate change and LUCC by designing multiple scenarios. Through the establishment of a hydrology response model, it quantified and characterized the individual and combined impacts of climate change and LUCC on the future streamflow variation in the upper reach of the Tarim River Basin. The main conclusions can be summarized as follows:

(1) Analysis showed that in this study area, compared to the period from 1985 to 2014, the climate from 2021 to 2050 was expected to be warmer and wetter.
(2) From 2021 to 2050, the FD scenario is expected to experience the most pronounced expansion of agricultural land among all scenarios. Similarly, HT, due to consistent human activity trends, also exhibits a significant conversion of grasslands and forests into farmland. Only EP has curbed the extensive expansion of farmland, thereby protecting the ecological environment.
(3) Alterations in mean annual streamflow were primarily influenced by LUCC, while the impact of climate change reduced the influence attributed to LUCC on streamflow. Climate change increases runoff (contribution: −18.67% to −7.16%), while LUCC decreases runoff (contribution: 107.16% to 118.67%), and the combined effect reduces runoff.
(4) In the future, streamflow would shift towards the beginning of the year with increased spring streamflow and decreased winter streamflow, which contributes to the growth of vegetation in the study area.

This study is valuable for predicting climate change and LUCC under various scenarios and understanding the diverse impacts on streamflow in the URTRB. It is beneficial for managers to take proactive measures and devise better responses in the face of environmental changes. The Tarim River is a typical inland river in arid areas. Studying the impact of upstream climate change and LUCC on runoff can serve as a reference for water resource management in other basins with similar conditions.

Author Contributions: Conceptualization, L.X.; Methodology, Q.H., M.Y. and S.L.; Software, Q.H. and S.L.; Validation, L.X.; Formal analysis, X.C.; Investigation, T.Q. and M.Y.; Resources, T.Q., Y.L. and X.C.; Data curation, L.X.; Writing – original draft, Q.H.; Supervision, Y.L.; Funding acquisition, Q.H. and L.X. All authors have read and agreed to the published version of the manuscript.

Funding: This study was supported by the National Key Research and Development Program of China (2023YFC3206800), Xinjiang Production and Construction Corps (No. 2022BC001), National Scientific Foundation of China (No. 51779074), Graduate Research and Innovation Projects of Jiangsu Province (No. SJKY19_0473).

Data Availability Statement: The data presented in this study are available on request from the corresponding author. The data are not publicly available due to data confidentiality.

Conflicts of Interest: The authors declare that they have no known competing financial interests or personal relationships that could have appeared to influence the work reported in this paper.

References

1. Chen, Y.; Shu, L.; Li, H.; Opoku, P.A.; Li, G.; Xu, Z.; Qi, T. Identification of Preferential Recharge Zones in Karst Systems Based on the Correlation between the Spring Level and Precipitation: A Case Study from Jinan Spring Basin. *Water* **2021**, *13*, 48. [CrossRef]
2. Qi, T.; Shu, L.; Li, H.; Wang, X.; Opoku, P.A. Water Distribution from Artificial Recharge via Infiltration Basin under Constant Head Conditions. *Water* **2021**, *13*, 1052. [CrossRef]

3. Zuo, D.; Chen, G.; Wang, G.; Xu, Z.; Han, Y.; Peng, D.; Pang, B.; Abbaspour, K.C.; Yang, H. Assessment of changes in water conservation capacity under land degradation neutrality effects in a typical watershed of Yellow River Basin, China. *Ecol. Indic.* **2023**, *148*, 110145. [CrossRef]
4. Zhai, R.; Tao, F. Climate Change in China Affects Runoff and Terrestrial Ecosystem Water Retention More Than Changes in Leaf Area Index and Land Use/Cover Over the Period 1982–2015. *J. Geophys. Res.-Biogeosci.* **2021**, *126*, e2020JG005902. [CrossRef]
5. Shi, S.; Yu, J.; Wang, F.; Wang, P.; Zhang, Y.; Jin, K. Quantitative contributions of climate change and human activities to vegetation changes over multiple time scales on the Loess Plateau. *Sci. Total Environ.* **2021**, *755 Pt 2*, 142419. [CrossRef]
6. Robertson, D.E.; Chiew, F.H.S.; Potter, N. Adapting rainfall bias-corrections to improve hydrological simulations generated from climate model forcings. *J. Hydrol.* **2023**, *619*, 129322. [CrossRef]
7. Bhadoriya, U.P.S.; Mishra, A.; Singh, R.; Chatterjee, C. Implications of climate change on water storage and filling time of a multipurpose reservoir in India. *J. Hydrol.* **2020**, *590*, 125542. [CrossRef]
8. Gao, J.; Sheshukov, A.Y.; Yen, H.; Douglas-Mankin, K.R.; White, M.J.; Arnold, J.G. Uncertainty of hydrologic processes caused by bias-corrected CMIP5 climate change projections with alternative historical data sources. *J. Hydrol.* **2019**, *568*, 551–561. [CrossRef]
9. Wang, B.; Deveson, E.D.; Waters, C.; Spessa, A.; Lawton, D.; Feng, P.; Liu, D.L. Future climate change likely to reduce the Australian plague locust (*Chortoicetes terminifera*) seasonal outbreaks. *Sci. Total Environ.* **2019**, *668*, 947–957. [CrossRef]
10. Lee, S.; Qi, J.; McCarty, G.W.; Yeo, I.-Y.; Zhang, X.; Moglen, G.E.; Du, L. Uncertainty assessment of multi-parameter, multi-GCM, and multi-RCP simulations for streamflow and non-floodplain wetland (NFW) water storage. *J. Hydrol.* **2021**, *600*, 126564. [CrossRef]
11. Wang, B.; Gray, J.M.; Waters, C.M.; Anwar, M.R.; Orgill, S.E.; Cowie, A.L.; Feng, P.; Liu, D.L. Modelling and mapping soil organic carbon stocks under future climate change in south-eastern Australia. *Geoderma* **2022**, *405*, 115442. [CrossRef]
12. Wen, X.; Liu, Z.; Lei, X.; Lin, R.; Fang, G.; Tan, Q.; Wang, C.; Tian, Y.; Quan, J. Future changes in Yuan River ecohydrology: Individual and cumulative impacts of climates change and cascade hydropower development on runoff and aquatic habitat quality. *Sci. Total Environ.* **2018**, *633*, 1403–1417. [CrossRef] [PubMed]
13. Su, B.; Huang, J.; Zeng, X.; Gao, C.; Jiang, T. Impacts of climate change on streamflow in the upper Yangtze River basin. *Clim. Chang.* **2016**, *141*, 533–546. [CrossRef]
14. Sangam, S.; Binod, B.; Manish, S.; Shrestha, P.K. Integrated assessment of the climate and landuse change impact on hydrology and water quality in the Songkhram River Basin, Thailand. *Sci. Total Environ.* **2018**, *643*, 1610–1622.
15. Xu, Z.P.; Li, Y.P.; Huang, G.H.; Wang, S.G.; Liu, Y.R. A multi-scenario ensemble streamflow forecast method for Amu Darya River Basin under considering climate and land-use changes. *J. Hydrol.* **2021**, *598*, 126276. [CrossRef]
16. Yeh, S.-W.; Wang, G.; Cai, W.; Park, R.J. Diversity of ENSO-Related Surface Temperature Response in Future Projection in CMIP6 Climate Models: Climate Change Scenario Versus ENSO Intensity. *Geophys. Res. Lett.* **2022**, *49*, e2021GL096135. [CrossRef]
17. Sreeparvathy, V.; Srinivas, V.V. Meteorological flash droughts risk projections based on CMIP6 climate change scenarios. *Npj Clim. Atmos. Sci.* **2022**, *5*, 77. [CrossRef]
18. Su, B.; Huang, J.; Mondal, S.K.; Zhai, J.; Wang, Y.; Wen, S.; Gao, M.; Lv, Y.; Jiang, S.; Jiang, T.; et al. Insight from CMIP6 SSP-RCP scenarios for future drought characteristics in China. *Atmos. Res.* **2021**, *250*, 105375. [CrossRef]
19. Zhang, H.; Meng, C.; Wang, Y.; Wang, Y.; Li, M. Comprehensive evaluation of the effects of climate change and land use and land cover change variables on runoff and sediment discharge. *Sci. Total Environ.* **2020**, *702*, 134401. [CrossRef]
20. Wu, L.; Zhang, X.; Hao, F.; Wu, Y.; Li, C.; Xu, Y. Evaluating the contributions of climate change and human activities to runoff in typical semi-arid area, China. *J. Hydrol.* **2020**, *590*, 125555. [CrossRef]
21. Zhang, Y.; Xia, J.; Yu, J.; Randall, M.; Zhang, Y.; Zhao, T.; Pan, X.; Zhai, X.; Shao, Q. Simulation and assessment of urbanization impacts on runoff metrics: Insights from landuse changes. *J. Hydrol.* **2018**, *560*, 247–258. [CrossRef]
22. Xu, Z.G.; Xu, J.T.; Deng, X.Z.; Huang, J.K.; Uchida, E.; Rozelle, S. Grain for green versus grain: Conflict between food security and conservation set-aside in China. *World Dev.* **2006**, *34*, 130–148. [CrossRef]
23. Wang, X.; He, K.; Dong, Z. Effects of climate change and human activities on runoff in the Beichuan River Basin in the northeastern Tibetan Plateau, China. *Catena* **2019**, *176*, 81–93. [CrossRef]
24. Yang, W.; Long, D.; Bai, P. Impacts of future land cover and climate changes on runoff in the mostly afforested river basin in North China. *J. Hydrol.* **2019**, *570*, 201–219. [CrossRef]
25. Ji, L.; Duan, K. What is the main driving force of hydrological cycle variations in the semiarid and semi-humid Weihe River Basin, China? *Sci. Total Environ.* **2019**, *684*, 254–264. [CrossRef] [PubMed]
26. Yin, Z.; Feng, Q.; Zhu, R.; Wang, L.; Chen, Z.; Fang, C.; Lu, R. Analysis and prediction of the impact of land use/cover change on ecosystem services value in Gansu province, China. *Ecol. Indic.* **2023**, *154*, 110868. [CrossRef]
27. Wang, Q.; Xu, Y.; Wang, Y.; Zhang, Y.; Xiang, J.; Xu, Y.; Wang, J. Individual and combined impacts of future land-use and climate conditions on extreme hydrological events in a representative basin of the Yangtze River Delta, China. *Atmos. Res.* **2020**, *236*, 104805. [CrossRef]
28. Wang, Z.; Shu, L.; Xu, P.; Yin, X.; Lu, C.; Liu, B.; Li, Y. Influence of land use changes on the remaining available aquifer storage (RAAS): A case study of the Taoerhe alluvial-proluvial fan. *Sci. Total Environ.* **2022**, *849*, 104805. [CrossRef]
29. Aguejdad, R. The Influence of the Calibration Interval on Simulating Non-Stationary Urban Growth Dynamic Using CA-Markov Model. *Remote Sens.* **2021**, *13*, 468. [CrossRef]

30. Cao, C.; Sun, R.; Wu, Z.; Chen, B.; Yang, C.; Li, Q.; Fraedrich, K. Streamflow Response to Climate and Land-Use Changes in a Tropical Island Basin. *Sustainability* **2023**, *15*, 3941. [CrossRef]
31. Wang, Q.; Cheng, L.; Zhang, L.; Liu, P.; Qin, S.; Liu, L.; Jing, Z. Quantifying the impacts of land-cover changes on global evapotranspiration based on the continuous remote sensing observations during 1982–2016. *J. Hydrol.* **2021**, *598*, 126231. [CrossRef]
32. Farsi, N.; Mahjouri, N. Evaluating the Contribution of the Climate Change and Human Activities to Runoff Change under Uncertainty. *J. Hydrol.* **2019**, *574*, 872–891. [CrossRef]
33. Yang, L.; Feng, Q.; Yin, Z.; Wen, X.; Si, J.; Li, C.; Deo, R.C. Identifying separate impacts of climate and land use/cover change on hydrological processes in upper stream of Heihe River, Northwest China. *Hydrol. Process.* **2017**, *31*, 1100–1112. [CrossRef]
34. Liao, S.; Xue, L.; Dong, Z.; Zhu, B.; Zhang, K.; Wei, Q.; Fu, F.; Wei, G. Cumulative ecohydrological response to hydrological processes in arid basins. *Ecol. Indic.* **2020**, *111*, 106005. [CrossRef]
35. Han, Q.; Xue, L.; Liu, Y.; Yang, M.; Chu, X.; Liu, S. Developing a multi-objective simulation-optimization model for ecological water conveyance in arid inland river basins. *J. Hydrol. Reg. Stud.* **2023**, *50*, 101551. [CrossRef]
36. Abbott, M.B.; Bathurst, J.C.; Cunge, J.A.; O'connell, P.E.; Rasmussen, J. An introduction to the European Hydrological System—Systeme Hydrologique Europeen, "SHE", 2: Structure of a physically-based, distributed modelling system. *J. Hydrol.* **1986**, *87*, 61–77. [CrossRef]
37. Butts, M.B.; Payne, J.T.; Kristensen, M.; Madsen, H. An evaluation of the impact of model structure on hydrological modelling uncertainty for streamflow simulation. *J. Hydrol.* **2004**, *298*, 242–266. [CrossRef]
38. Sengul, S.; Ispirli, M.N. Estimation and analysis of the Antecedent Temperature Index-Melt Rate (ATIMR) function using observed data from the Kirkgoze-Cipak Basin, Turkey. *J. Hydrol.* **2021**, *598*, 126484. [CrossRef]
39. Gan, G.; Wu, J.; Hori, M.; Fan, X.; Liu, Y. Attribution of decadal runoff changes by considering remotely sensed snow/ice melt and actual evapotranspiration in two contrasting watersheds in the Tienshan Mountains. *J. Hydrol.* **2022**, *610*, 127810. [CrossRef]
40. Thompson, J.R.; Sørenson, H.R.; Gavin, H.; Refsgaard, A. Application of the coupled MIKE SHE/MIKE 11 modelling system to a lowland wet grassland in southeast England. *J. Hydrol.* **2004**, *293*, 151–179. [CrossRef]
41. Sahoo, G.B.; Ray, C.; Carlo, E.H.D. Calibration and validation of a physically distributed hydrological model, MIKE SHE, to predict streamflow at high frequency in a flashy mountainous Hawaii stream. *J. Hydrol.* **2006**, *327*, 94–109. [CrossRef]
42. Yu, X.; Xiao, J.; Huang, K.; Li, Y.; Lin, Y.; Qi, G.; Liu, T.; Ren, P. Simulation of Land Use Based on Multiple Models in the Western Sichuan Plateau. *Remote Sens.* **2023**, *15*, 3629. [CrossRef]
43. Lin, Z.; Peng, S. Comparison of multimodel simulations of land use and land cover change considering integrated constraints—A case study of the Fuxian Lake basin. *Ecol. Indic.* **2022**, *142*, 127810. [CrossRef]
44. Iacono, M.; Levinson, D.; El-Geneidy, A.; Wasfi, R. Markov Chain Model of Land Use Change in the Twin Cities. *TeMA J. Land Use Mobil. Environ.* **2012**, *8*. [CrossRef]
45. Zhang, Z.; Hoermann, G.; Huang, J.; Fohrer, N. A Random Forest-Based CA-Markov Model to Examine the Dynamics of Land Use/Cover Change Aided with Remote Sensing and GIS. *Remote Sens.* **2023**, *15*, 2128. [CrossRef]
46. Liu, J.; Zhou, Y.; Wang, L.; Zuo, Q.; Li, Q.; He, N. Spatiotemporal Analysis and Multi-Scenario Prediction of Ecosystem Services Based on Land Use/Cover Change in a Mountain-Watershed Region, China. *Remote Sens.* **2023**, *15*, 2759. [CrossRef]
47. Sanchayeeta, A.; Jane, S. Simulating Forest Cover Changes of Bannerghatta National Park Based on a CA-Markov Model: A Remote Sensing Approach. *Remote Sens.* **2012**, *4*, 3215–3243.
48. Jalil, A.; Akhtar, F.; Awan, U.K. Evaluation of the AquaCrop model for winter wheat under different irrigation optimization strategies at the downstream Kabul River Basin of Afghanistan. *Agric. Water Manag.* **2020**, *240*, 106321. [CrossRef]
49. McMichael, C.E.; Hope, A.S.; Loaiciga, H.A. Distributed hydrological modelling in California semi-arid shrublands: MIKE SHE model calibration and uncertainty estimation. *J. Hydrol.* **2006**, *317*, 307–324. [CrossRef]
50. Qi, T.; Khanaum, M.M.; Boutin, K.; Otte, M.L.; Lin, Z.; Chu, X. Incorporating Wetland Delineation and Impacts in Watershed-Scale Hydrologic Modeling. *Water* **2023**, *15*, 2518. [CrossRef]
51. Khanaum, M.M.; Qi, T.; Boutin, K.D.; Otte, M.L.; Lin, Z.; Chu, X. Assessing the Impacts of Wetlands on Discharge and Nutrient Loading: Insights from Restoring Past Wetlands with GIS-Based Analysis and Modeling. *Wetlands* **2023**, *43*, 103. [CrossRef]
52. Moriasi, D.N.; Gitau, M.W.; Pai, N.; Daggupati, P. HYDROLOGIC AND WATER QUALITY MODELS: PERFORMANCE MEASURES AND EVALUATION CRITERIA. *Trans. Asabe* **2015**, *58*, 1763–1785.
53. Van Liew, M.W.; Garbrecht, J. Hydrologic simulation of the Little Washita River Experimental Watershed using SWAT. *J. Am. Water Resour. Assoc.* **2003**, *39*, 413–426. [CrossRef]
54. Cohen, J. A coefficient of agreement of nominal scales. *Psychol. Bull.* **1960**, *20*, 37–46. [CrossRef]
55. Beroho, M.; Briak, H.; Cherif, E.K.; Boulahfa, I.; Ouallali, A.; Mrabet, R.; Kebede, F.; Bernardino, A.; Aboumaria, K. Future Scenarios of Land Use/Land Cover (LULC) Based on a CA-Markov Simulation Model: Case of a Mediterranean Watershed in Morocco. *Remote Sens.* **2023**, *15*, 1162. [CrossRef]
56. Xue, L.; Bao, R.; Meixner, T.; Yang, G.; Zhang, J. Influences of topographic index distribution on hydrologically sensitive areas in agricultural watershed. *Stoch. Environ. Res. Risk Assess.* **2014**, *28*, 2235–2242. [CrossRef]
57. Xue, L.; Zhang, H.; Yang, C.; Zhang, L.; Sun, C. Quantitative Assessment of Hydrological Alteration Caused by Irrigation Projects in the Tarim River basin, China. *Sci. Rep.* **2017**, *7*, 4291. [CrossRef] [PubMed]

58. Chen, Y.; Li, W.; Deng, H.; Fang, G.; Li, Z. Changes in Central Asia's Water Tower: Past, Present and Future. *Sci. Rep.* **2018**, *6*, 35458. [CrossRef]
59. Xue, L.; Yang, F.; Yang, C.; Chen, X.; Zhang, L.; Chi, Y.; Yang, G. Identification of potential impacts of climate change and anthropogenic activities on streamflow alterations in the Tarim River Basin, China. *Sci. Rep.* **2017**, *7*, 8254. [CrossRef]
60. Zhang, F.; Tiyip, T.; Feng, Z.D.; Kung, H.T.; Johnson, V.C.; Ding, J.L.; Tashpolat, N.; Sawut, M.; Gui, D.W. Spatio-temporal patterns of land use/cover changes over the past 20 years in the middle reaches of the tarim river, Xinjiang, China. *Land Degrad. Dev.* **2015**, *26*, 284–299. [CrossRef]
61. Hou, Y.; Chen, Y.; Li, Z.; Li, Y.; Sun, F.; Zhang, S.; Wang, C.; Feng, M. Land Use Dynamic Changes in an Arid Inland River Basin Based on Multi-Scenario Simulation. *Remote Sens.* **2022**, *14*, 2797. [CrossRef]
62. Peng, S.; Ding, Y.; Wen, Z.; Chen, Y.; Cao, Y.; Ren, J. Spatiotemporal change and trend analysis of potential evapotranspiration over the Loess Plateau of China during 2011–2100. *Agric. For. Meteorol.* **2017**, *233*, 183–194. [CrossRef]
63. Peng, S. 1 km Multi-Scenario and Multi-Model Monthly Precipitation Data for China in 2021–2100. A Big Earth Data Platform for Three Poles. 2022. Available online: http://loess.geodata.cn/data/datadetails.html?dataguid=223722163526125&docid=1 (accessed on 1 December 2023).
64. Ding, Y.; Peng, S. Spatiotemporal change and attribution of potential evapotranspiration over China from 1901 to 2100. *Theor. Appl. Climatol.* **2021**, *145*, 79–94. [CrossRef]
65. Peng, D.; Lyu, J.; Song, Z.; Huang, S.; Zhang, P.; Gao, J.; Zhang, Y. Mercury budgets in the suspended particulate matters of the Yangtze River. *Water Res.* **2023**, *243*, 120390. [CrossRef] [PubMed]
66. Li, X.; Zhang, K.; Li, X.; Verger, A. The Minimum Temperature Outweighed the Maximum Temperature in Determining Plant Growth over the Tibetan Plateau from 1982 to 2017. *Remote Sens.* **2023**, *15*, 4032. [CrossRef]
67. He, Z.-W.; Tang, B.H. Spatiotemporal change patterns and driving factors of land surface temperature in the Yunnan-Kweichow Plateau from 2000 to 2020. *Sci. Total Environ.* **2023**, *896*, 165288. [CrossRef]
68. Wang, H.; Sun, F.; Xia, J.; Liu, W. Impact of LUCC on streamflow based on the SWAT model over the Wei River basin on the Loess Plateau in China. *Hydrol. Earth Syst. Sci.* **2017**, *21*, 1929–1945. [CrossRef]
69. Li, B.; Shi, X.; Lian, L.; Chen, Y.; Chen, Z.; Sun, X. Quantifying the effects of climate variability, direct and indirect land use change, and human activities on runoff. *J. Hydrol.* **2020**, *584*, 124684. [CrossRef]
70. Yang, T.; Yang, X.; Jia, C. Detecting the main driving force of runoff change in the Beiluo River Basin, China. *Environ. Sci. Pollut. Res.* **2023**, *30*, 89823–89837. [CrossRef]
71. Zhang, L.; Nan, Z.; Yu, W.; Zhao, Y.; Xu, Y. Comparison of baseline period choices for separating climate and land use/land cover change impacts on watershed hydrology using distributed hydrological models. *Sci. Total Environ.* **2018**, *622*, 1016–1028. [CrossRef]

water

MDPI

Article

Landscape Ecological Risk Assessment of Kriya River Basin in Xinjiang and Its Multi-Scenario Simulation Analysis

Jinbao Li [1,2], Xuemin He [1,2,3,*], Pengcheng Huang [1,2], Zizheng Wang [1,2] and Ranran Wang [1,2]

[1] College of Ecology and Environment, Xinjiang University, Urumqi 830017, China; 15701919668@163.com (J.L.); wzz952580616@163.com (Z.W.)

[2] Key Laboratory of Oasis Ecology of Education Ministry, Xinjiang University, Urumqi 830017, China

[3] Xinjiang Jinghe Observation and Research Station of Temperate Desert Ecosystem, Ministry of Education, Jinghe 833300, China

* Correspondence: hxm@xju.edu.cn

Abstract: To comprehend the potential impacts of both natural phenomena and human activities on ecological risk, a thorough examination of the spatial and temporal evolution characteristics of Landscape Ecological Risk (LER) in arid river basins is imperative. This investigation holds paramount importance for the proactive prevention and mitigation of LER, as well as for the preservation of ecological security within these basins. In this scholarly inquiry, the Kriya River Basin (KRB) serves as the focal point of analysis. Leveraging three historical land use and land cover (LULC) images and incorporating a diverse array of drivers, encompassing both natural and anthropogenic factors, the study employs the PLUS model to forecast the characteristics of LULC changes within the basin under three distinct scenarios projected for the year 2030. Concurrently, the research quantitatively assesses the ecological risks of the basin through the adoption of the Landscape Ecological Risk Assessment (LERA) methodology and the Spatial Character Analysis (SCA) methodology. The results showed the following: (1) The study area is primarily composed of grassland and unused land, which collectively account for over 97% of the total land. However, there has been a noticeable rise in cropland and considerable deterioration in grassland between 2000 and 2020. The key observed change in LULC involves the transformation of grassland and unused land into cropland, forest, and construction land. (2) The overall LER indices for 2000, 2010, and 2020 are 0.1721, 0.1714, and 0.16696, respectively, showing strong positive spatial correlations and increasing autocorrelations over time. (3) Over time, human activities have come to exert a greater influence on LER compared to natural factors between 2000 and 2020. (4) In the natural development scenario (NDS), cropland protection scenario (CPS), and ecological priority scenario (EPS), the LER of KRB experienced notable variations in the diverse 2030 scenarios. Notably, the CPS exhibited the highest proportion of low-risk areas, whereas Daryaboyi emerged as the focal point of maximum vulnerability. These findings offer theoretical and scientific support for sustainable development planning in the watershed.

Keywords: Landscape Ecological Risk (LER); Kriya River Basin (KRB); PLUS model; multi-scenario simulation; driving factors

1. Introduction

Watersheds are geographical areas with complex structures consisting of multiple systems, including ecological, economic, and social, with different functions, such as maintaining biodiversity and supporting human production, life, and culture [1–3]. Global climate change is becoming more pronounced as human activities intensify, and this, combined with the sensitivity and vulnerability of drylands themselves, makes them one of the most ecologically risky regions [4–6]. On the other hand, the rapid expansion of cities has caused enormous ecological and ecological problems, such as the degradation of land resources, the reduction in regional biodiversity, the reduction in the carrying capacity of the environment and the exacerbation of the problem of ecological security, which is evident in the drylands [7–10]. Arid zones, which are mostly composed of oasis cities and desert ecosystems, have stronger feedback to human activities, which makes them an ideal area for the study of nature–human complex systems [9,11]. In the midst of climate warming and transitional resource exploitation, the high intensity of LULC has brought great pressure on the environment [10]. Consequently, the quantification of ecological risks in a scientific and rational manner, along with the analysis of the drivers responsible for their spatial and temporal variations, has emerged as a prominent focus in ecological and environmental research [12–14].

LERA is an important part of ecological risk assessment, which complements and expands ecological risk assessment [15], and emphasizes the comprehensive analysis of the possible impacts of various large-scale disasters that the regional ecological environment may face [16,17]. LER can be used to evaluate various traditional ecological risk assessment methods, based on the perspective of the coupled association of ecological processes and spatial patterns in landscape ecology, paying more attention to the spatial and temporal heterogeneity of ecological risk and the possible adverse consequences of scale effects, and belonging to an important branch of regional ecological risk assessment [18,19].

The ecological risk assessment method, which is based on the LULC LER, can be used to assess the ecological status of a watershed from both an ecological and landscape perspective [17,20,21]. The method is able to map the interactions between landscape patterns and ecological processes and focuses on describing the spatial and temporal variability of ecological risk, the impact of the spatial distribution of landscape components on ecological risk, and these three areas [22,23]. Thus, from the standpoint of landscape pattern ecological processes, the study of regional ecological security has emerged [24–28]. In the 1990s, Heggem et al. (2000) [29] introduced a landscape pattern analysis approach to assess the impact of human activities on ecological change in watersheds. Later, Kapustka et al. (2001) [30] and other landscape ecology theories introduced ecological risk assessment and used it to propose control strategies. Paukert et al. (2011) [31] evaluated the ecological health of the Colorado River Basin using an LERI constructed using the landscape index method. Research on LER is increasing again both at home and abroad as we enter the 21st century, and most of the evaluation objects are ecologically fragile and sensitive areas as well as areas with high intensity of human activities, which are mainly centered around watersheds, cities, mines, nature reserves, and ecologically fragile areas [23,32–36]. The methods of evaluation include landscape pattern index method [9], entropy value method [37], exposure–response method [38], etc.; additionally, the evaluation scale evolves from a single scale to multiple scales, and numerous researchers have looked into the LER's multi-scale changes [39].

Simulating and forecasting dynamic trends under different conditions is imperative, as is looking into the characteristics of the temporal and spatial evolution of LER in rapidly developing watersheds. Furthermore, in order to support future high-quality economic and social development in watersheds, strategies for optimizing LULC structure in arid watersheds must be proposed [40,41]. Multi-scenario LULC change models can be classified into quantitative predictive models, spatial predictive models and coupled models [36]. Currently, commonly used quantitative forecasting models include Markov, system dynamics (SDs), grey forecasting models (GMs), and artificial neural network (ANN) models [42,43]. Spatial prediction models include the CA model, the CLUE model, and the FLUS model [44–46]. By combining data prediction models and spatial prediction models into a coupled model, the requirements of quantitative and spatial accuracy can be highly met. Liang et al. developed a preamble model called the Patch-Level LULC Simulation (PLUS) model was created [47]. It proposes a rule mining framework based on the Land Expansion Analysis Strategy (LEAS) and CA based on muti-type random patch seeds (CARSs) that can be used to explore the drivers of multiple types of land expansion to determine and predict the patch-level development of the LULC landscape, leading to more accurate LULC simulation results [9,35].

The Keriya River Basin (KRB) is an area with important ecological functions in the north-west arid zone, which is of great significance for regional ecological security as well as water resource protection. The basin is located in the hinterland of the Eurasian continent, far from the sea, with the Taklamakan Desert in the north and the Kunlun Mountains in the south, which makes the basin's climate arid due to its unique geographical location. In recent years, with the increase in population and the demand for economic development, a large amount of land has been reclaimed in the middle and lower plains of the basin, resulting in the evolution from desert land to oases. The main water resources in the basin are composed of surface water and groundwater, but the spatial and temporal distribution of water resources is extremely uneven, and seasonal water shortages often occur in the oases, which leads to constraints on agricultural production [48]. In this context, it is particularly important to effectively assess ecological risks. Based on the LULC of the KRB in 2000, 2010, and 2020, this study analyzes the changing characteristics of landscape types, examines the spatial and temporal evolution of LER, and simulates the development trend of LER of the KRB in three different scenarios using the PLUS model and presents the scenarios of ecological environmental protection, aiming to solve the three practical problems of development and ecological protection of KRB: (1) How did the spatial pattern of LULC in the KRB change from 2000 to 2020? (2) What were the characteristics of the spatial distribution and changes in the LER in the basin, and what are the dominant factors leading to the changes of LER? (3) Which development scenarios optimize ecological risk in catchments?

2. Materials and Methods

2.1. Study Area

The Keriya River Basin (KRB), located in the heart of the Eurasian continent, is flanked by the Taklamakan Desert to the north and the Kunlun Mountains to the south (refer to Figure 1). Its geographical coordinates range from 80°09′ to 83°51′ E longitude and 35°14′ to 39°29′ N latitude, encompassing a land area of 70.5×10^4 km^2. The basin is characterized by a warm temperate inland arid desert climate, marked by an average annual temperature of 11.6 °C, a scant annual rainfall of 44.7 mm, and considerable evaporation totaling 2500 mm [49]. The elevation variance from north to south is approximately 5000 m, creating a sloping terrain with higher elevations in the south and lower elevations in the north. Geological tectonics have shaped five distinct landforms in the region: mid-altitude mountains in the upper reaches, pre-mountainous hills in the upper and middle reaches, pre-mountainous sloping plains in the middle and lower reaches, alluvial plains in the lower reaches, and the desert area. These have given rise to a natural desert oasis known as the Daryabuyi Oasis. The area exhibits a typical arid continental climate, with scanty annual precipitation of 44.7 mm, high evaporation of 2500 mm, and predominantly drought-resistant vegetation, Poor vegetation conditions. The central plain serves as the primary agricultural zone, cultivating crops such as cotton, maize, and wheat.

2.2. Data Acquisition

LULC data were obtained from the Globeland30 global land cover database (http://globeland30.org, accessed on 23 April 2023). The dataset encompasses three distinct time periods, namely 2000, 2010, and 2020, and possesses a spatial resolution of 30 m × 30 m [50] that reclassified the LULC into six categories, including: cropland, forest, grassland, water, unused land, and construction land. Data on soil type, GDP, population, temperature, and precipitation are available in the Scientific Data Center for Resources and Environment of Chinese Science (https://www.resdc.cn/, accessed on 16 May 2023). The distance variable was retrieved from OpenStreetMap (https://www.openstreetmap.org/, accessed on 24 May 2023). The patterns of the distance variables were also ascertained. The Euclidean distance analysis also revealed the patterns of the distance variables The DEM data were obtained from the geospatial data cloud (https://www.gscloud.cn/, accessed on 3 May

2023) at a resolution of 30 m. The slope, slope direction, and topographic relief data were calculated through the ArcGIS 10.7 platform.

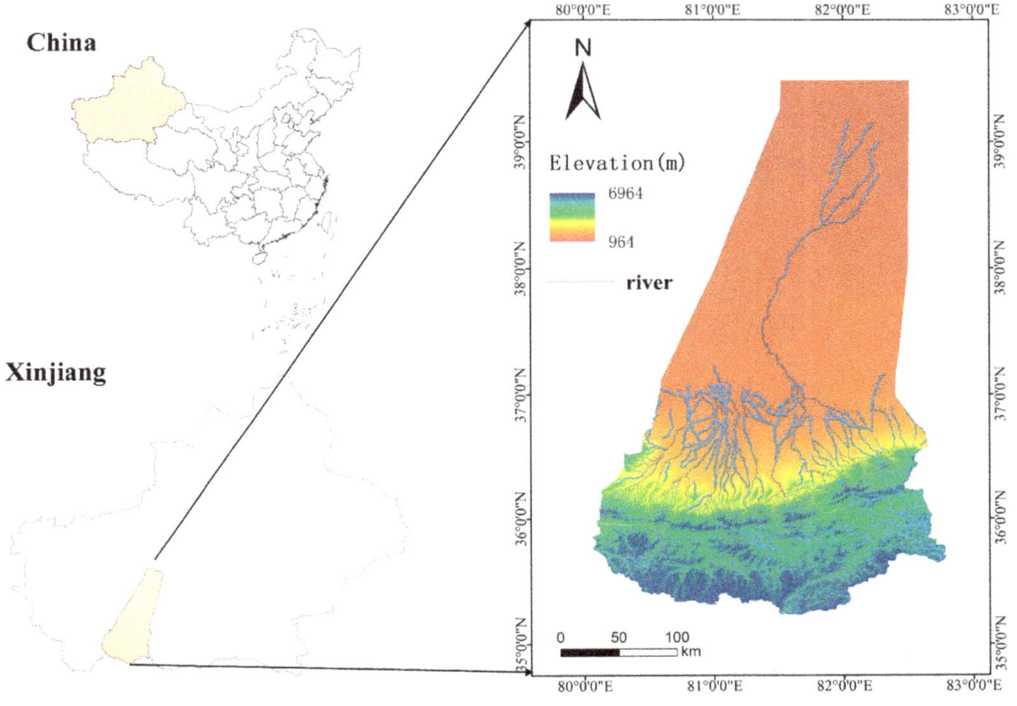

Figure 1. Area of study.

2.3. Landscape Ecological Risk Index

This study investigates the spatial and temporal changes in LULC within the KRB region from 2000 to 2020. The analysis combines the PLUS model with five natural and socioeconomic drivers to simulate different scenarios for 2030, including natural development, agricultural land protection, and ecological priority. The regional ecological risk level was quantified using the LERI. The study's methodology and framework are illustrated in Figure 2.

Building upon previous research, this study develops a rating system for the LERI by incorporating landscape disturbance and vulnerability factors [9], and the landscape type index was determined using Fragstats 4.2 software [51]. The LERI was calculated as follows:

$$ERI_i = \sum_{i=1}^{n} \frac{A_{ki}}{A_k} R_i \quad (1)$$

ERI_i is LERI within the ith sampling unit; A_{ki} is the area of landscape type i in the kth sampling unit; A_K is the area of the i-th sampling unit; R_i is the landscape loss degree index.

$$R_i = \sqrt{E_i \times F_i} \quad (2)$$

$$E_i = aC_i + bS_i + cD_i \quad (3)$$

E_i represents the landscape disturbance index, which is constructed by the landscape fragmentation C_i, landscape separation S_i, and landscape dominance D_i, while a, b, and c denote the weights of the corresponding landscape indexes, and a + b + c = 1. These weights, according to existing research and the actual situation [17,18,52], are assigned as 0.5, 0.3, and 0.2, respectively. Additionally, F_i represents the landscape vulnerability index, and for the six land types of (cropland forest, grassland, watershed, construction land, and unused land), the assigned weights are 0.1905, 0.1429, 0.0952, 0.2381, 0.0476, and 0.2857, based on the characteristics of the study area and the results of previous research [53,54].

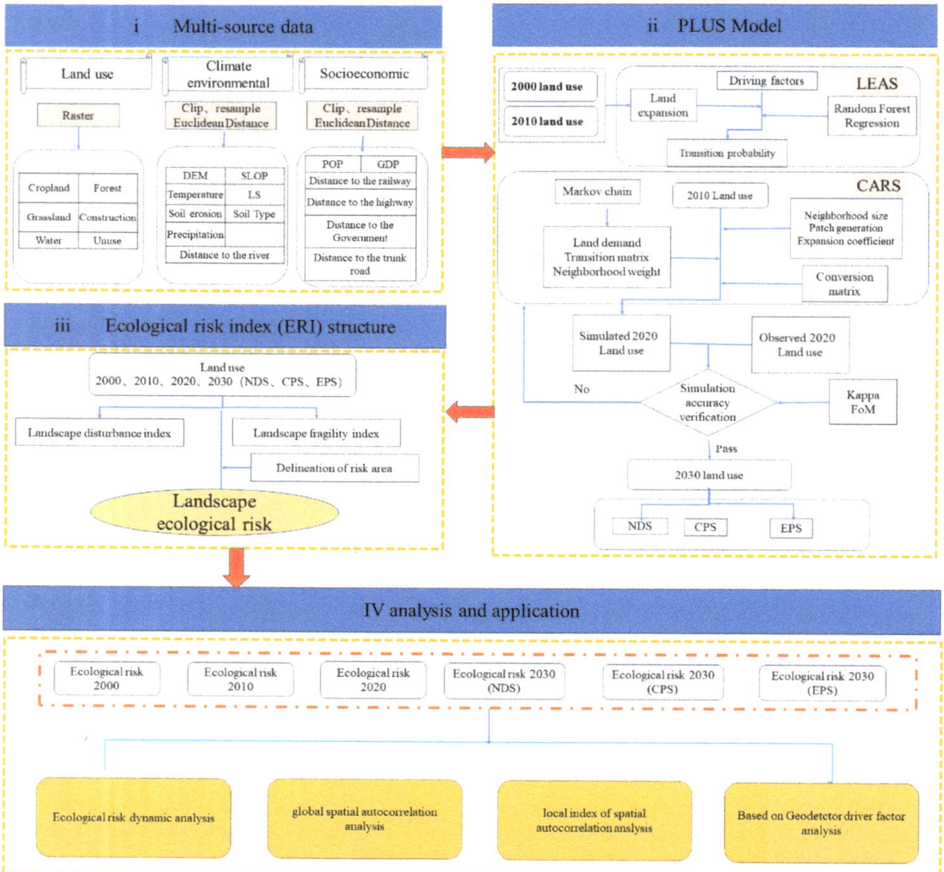

Figure 2. Ecological risk assessment framework. PLUS Model: patch-generating land use simulation.

In order to visually analyze the characteristics of spatial distribution of ecological risks, ordinary Kriging interpolation in the geostatistics module of GIS was used to obtain the LER distribution of KRB. The natural breakpoint method was also used to divide the 2020 Landscape Ecological Risk values into five classes: lowest risk (LER < 0.1517), lower risk ($0.1517 \leq$ LER < 0.2039), medium risk ($0.2039 \leq$ LER < 0.2846), high risk ($0.2846 \leq$ LER < 0.3764), and highest risk (0.3764 < LER). The data for the remaining periods were standardized using the 2020 assessment intervals for better comparability.

2.4. Analysis of Spatial Autocorrelation for the LER

Spatial autocorrelation analysis is employed to examine the presence of significant correlations in the spatial distribution of LER. This analysis involves two types of indices: the global Moran's I index evaluates the spatial correlation of attribute values across the entire study area, while the local Moran's I index assesses the correlation between LER and neighboring spatial units. Moran's I shows a positive correlation when its value is positive, with increasing significance as the value increases. The LISA map is utilized to identify distinct patterns of high–high and low–low clustering of LER within local areas [17,55].

2.5. Analysis of the LER Driving Mechanism GeoDetector

Geographic probes are statistical methods that reveal spatial heterogeneity and, consequently, its driving factors [56]. With reference to related studies, the topographic and climatic characteristics and socioeconomic development of the KRB were also combined to analyze the driving forces of LER changes in the KRB by selecting the following 10 influencing factors from natural and social factors: slope, elevation, climate factors include soil type, annual precipitation, average temperature, distance from a river, GDP, population density, and distance from residential areas [35].

$$q = 1 - \frac{\sum_{h=1}^{L} A_h \sigma_h^2}{A_\sigma^2} \tag{4}$$

In the equation, the q value represents the influence of the driving factor on the LER of the KRB. It ranges between 0 and 1, where a larger q-value indicates a stronger explanation of the spatial distribution of the factor on LER. The variable h (1, 2, . . ., L) represents the number of subregions of the detection factor X. A_h represents the unit count of layer h, and A represents the total area σ_h^2 denotes the variance of the h layer, while σ^2 represents the overall variance of the entire region.

2.6. Multi-Scenario Ecological Risk Prediction Using the Markov-PLUS Model

The PLUS model uses a number of steps to analyze changes in LULC. The LULC data are initially transformed into the appropriate format. The expansion proportion of LULC is then extracted using the LEAS module in two stages. Natural and socioeconomic drivers are integrated into the model, and the random forest classification algorithm is employed to determine the driving contribution rate of each driver. This information is then used to calculate LULC change and determine the expansion potential for each LULC type [47]; the Markov model is used to predict the demand of each LULC in 2030, and the panels are automatically generated to acquire the future LULC simulation map in the CARS module [57].

To further investigate the alterations in ecological risk due to different developmental trends in future, this research presents three scenarios: NDS, CPS, and EPS. NDS: The rate of change between 2010 and 2020 is used as a reference to forecast land usage demand in 2030. CPS: Protect cropland by preventing it from being converted to any other LULC type, except for construction land, all other LULC can be converted to cropland. EPS: Preserve forested land and watersheds by prohibiting their conversion to any other LULC. The cost settings for every situation are illustrated in Table 1 underneath.

Table 1. Cost matrix for land use conversion in each scenario.

	NDS						CPS						EPS					
2000–2030	a	b	c	d	e	f	a	b	c	d	e	f	a	b	c	d	e	f
a	1	1	1	1	1	1	1	0	0	0	0	0	1	1	1	1	1	1
b	1	1	1	1	1	1	1	1	1	0	1	1	0	1	0	0	0	0
c	1	1	1	1	1	1	1	1	1	1	1	1	0	1	1	1	0	0
d	1	1	1	1	1	1	1	0	1	1	1	1	0	0	0	1	0	0
e	0	0	0	0	1	0	0	0	0	0	0	0	0	0	0	0	1	0
f	1	1	1	1	1	1	1	1	1	1	1	1	1	1	1	1	1	1

Notes: a, b, c, d, e, and f represent cropland, forest, grassland, water, construction land, and unused, respectively, and 0 means no conversion was allowed and 1 means conversion was allowed.

Accuracy Verification

This study uses the 2010 LULC, driving factor data, and LULC extension analysis data as its raw inputs in order to keep the simulation error at a level that is reassuringly acceptable. By contrasting it with the predicted outcomes from the 2020 PLUS model, the accuracy of the spatial distribution of LULC in the study area for the year 2020 is assessed. The assessment of simulation accuracy in this article employs the Kappa coefficient and the Fom coefficient [35,58]. The kappa coefficient, which ranges from 0 to 1, is used to assess the consistency and accuracy of the simulation results. A value greater than 0.7 indicates a higher level of consistency and accuracy. Conversely, the FOM coefficient is computed as the ratio of the intersection of the projected and actual land changes to the total of the two. Higher values of this coefficient, which likewise has a range of 0 to 1, denote increased simulation accuracy [36,59].

3. Results

3.1. Spatiotemporal LULC Change

Figure 3 and Table 2 show the spatial and temporal changes in the composition of the six LULC types in the KRB. The overall change in the area of LULC types in the KRB from 2000 to 2020 is more obvious, with unused land and grassland being the most important land types, accounting for more than 78% and 19% of the total land area at all times, respectively, showing a gradually decreasing trend. Arable land, water catchment areas, forest areas, and construction areas generally show increasing trends. In 2000–2020, cropland area continued to increase, with the cropland area increasing by 354.88 km^2 and the unused area increasing by 45.51 km^2. During 2010–2020, construction land grew the fastest, with an expansion rate of 172%. The distribution of construction land is usually surrounded by cultivated land, which represents the main area of human activity and is subject to the most human intervention. The amount of undeveloped land and grassland will drop by 425.42 km^2 and 51.43 km^2, respectively, between 2000 and 2020, while the area of water will hardly change.

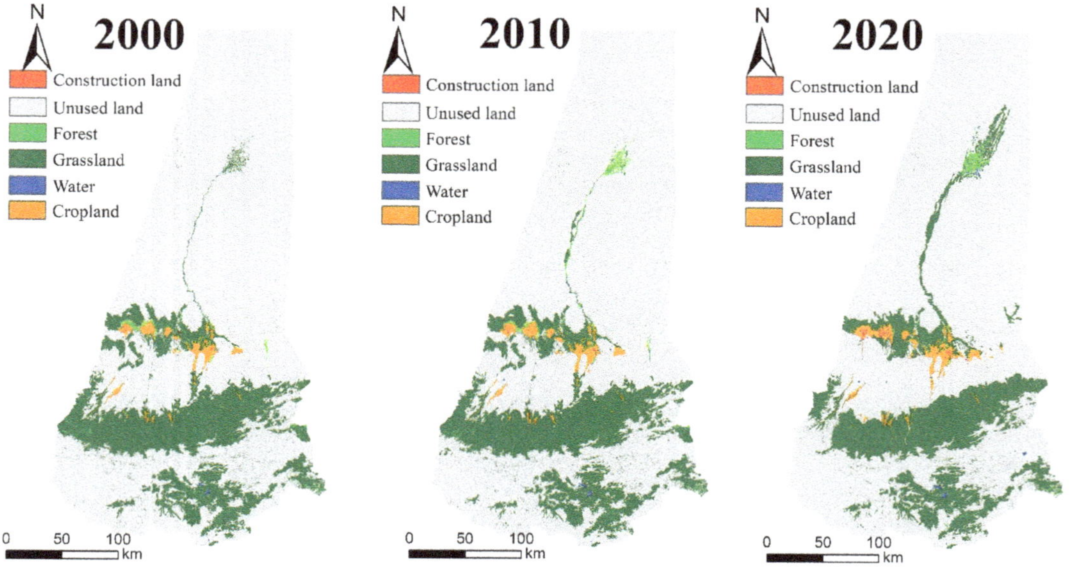

Figure 3. Map of LULC in KRB.

Table 2. LULC in KRB by area and percentages in 2000, 2010, and 2020.

Land Use Type	2000		2010		2020	
	Area (km^2)	Proportion of Total Area (%)	Area (km^2)	Proportion of Total Area (%)	Area (km^2)	Proportion of Total Area (%)
Cropland	872.8434	1.24	967.9833	1.37	1227.725	1.74
Forest	214.4655	0.30	364.6845	0.52	265.6377	0.38
Grassland	13,505.01	19.13	13,482.52	19.10	13,453.58	19.06
Water	79.9893	0.11	86.0904	0.12	105.2874	0.15
Construction	19.9017	0.03	24.0372	0.03	65.4102	0.09
Unused	55,902.14	79.19	55,669.04	78.86	55,476.72	78.59

Analysis of LULC Structure Change

Figure 4 shows the conversion relationship of the LULC for the three phases from 2000 to 2020, and it is represented visually by a chord diagram. The expansion of cropland and forest areas was obvious between 2000 and 2010. The expansion of cropland came mainly from the development of grassland and unused land, with a total increase of 95.14 km^2, while the forest area increased by 150.21 km^2, mainly from the evolution of grassland, and it was the only increase period of the three phases. During the time period of 2010–2020, the area of grassland transferred in was 2449.11 km^2 and the area transferred out was 2420.17 km^2. The overall area had increased, and the development of unused land had become an important way for grassland expansion to occur. The area of construction land doubled during this period, expanding much faster than in the previous period, while cropland was the main contributor to urban expansion, providing a total of 38.82 km^2 or 93.83% of the total. The area of forest shrank severely, with 64.62 km^2 of forest converted to unused land and 131.9 km^2 converted to grassland. From 2000 to 2020 as a whole, the main use types transferred out from the watershed were grassland and unused land, with 51.4323 km^2 and 425.428 km^2, respectively, being transferred out, and the use types transferred in were cropland, forest, and construction land, with 354.88 km^2, 51.17 km^2, and 45.50 km^2, respectively, being transferred in.

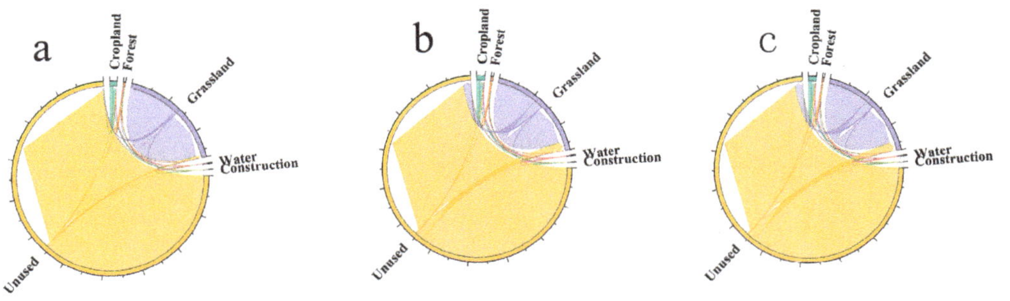

Figure 4. LULC transfer map of the KRB from 2000 to 2020. (**a**) 2000–2010; (**b**) 2010–2020; (**c**) 2000–2020.

3.2. Spatiotemporal Variations in the LER in the KRB

The LERI for the watershed had mean average values of 0.1721, 0.1714, and 0.1669 in 2000, 2010, and 2020, respectively. These values indicated a slight downward trend. Figure 5 shows that the LERI of the watershed had significant spatial and temporal variability. In the spatial distribution, the entire watershed was dominated by low and lower risks, while high and higher risks were mainly distributed along the Kriya River system, especially in the Daryaboyi, which are deep in the desert with sparse surrounding vegetation and poor natural conditions. The combined percentage of the study area's total area that was made up of the lowest risk and the lower risk was approximately constant, making up 98.69%,

97.88%, and 99.18% in 2000, 2010, and 2020, respectively. In 2000, 2010, and 2020, the total of the highest and highest risks is 0.29%, 0.58%, and 0.44%, respectively, with a slightly expanding and fluctuating trend towards the Northern Desert Region.

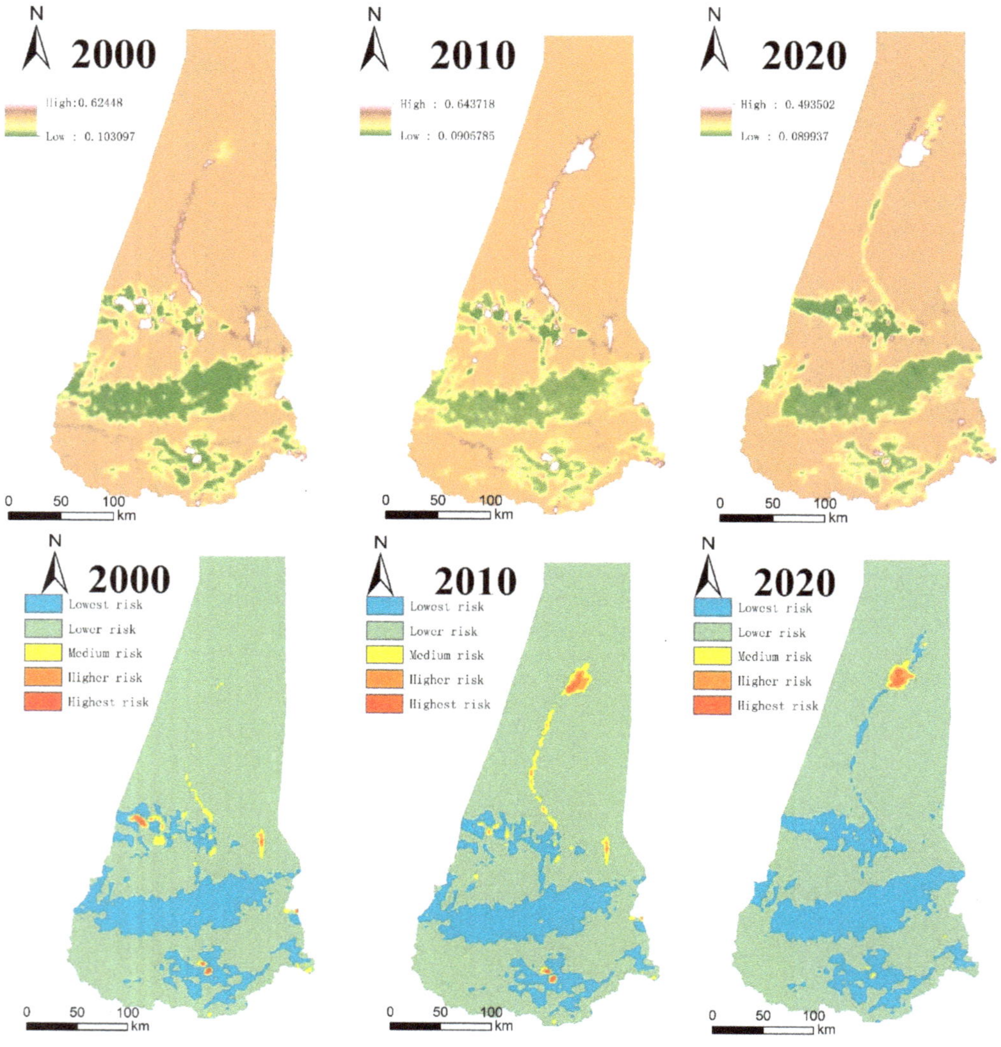

Figure 5. The spatially variable distribution of ERI in the KRB.

3.2.1. Defining the Temporal and Spatial Distribution of KRB

LER Moran's I values are, respectively, 0.818, 0.847, and 0.920 from 2000 to 2020 (Figure 6), all greater than 0.5. This indicated that the LERI of the KRB over the past 20 years had a positive and significant spatial correlation; in addition, the scattered points were distributed close to the regression line, suggests that the distribution of LER is spatially clustered and that this spatial clustering increases over time.

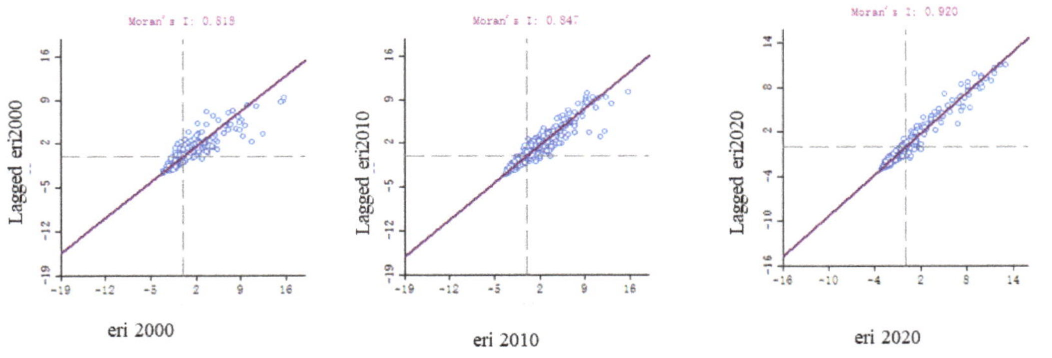

Figure 6. The ERI values in the KRB from 2000 to 2020 are shown in a scatter distribution of Moran's I index.

The trend of the LISA map for the period 2000 to 2020 (Figure 7) shows that the LER in the catchment area mainly has an aggregated "high–high" and "low–low" distribution. High-risk areas within the KRB exhibit concentration in the transition zone between desert and green land, particularly along the lower course of the Kriya River. These areas are characterized by pronounced landscape fragmentation. Additionally, sporadic high-risk areas are observed around the city. The high vegetation cover contributes to the relatively stable ecological quality, while low value zones are typically confined to grassland and cropland regions.

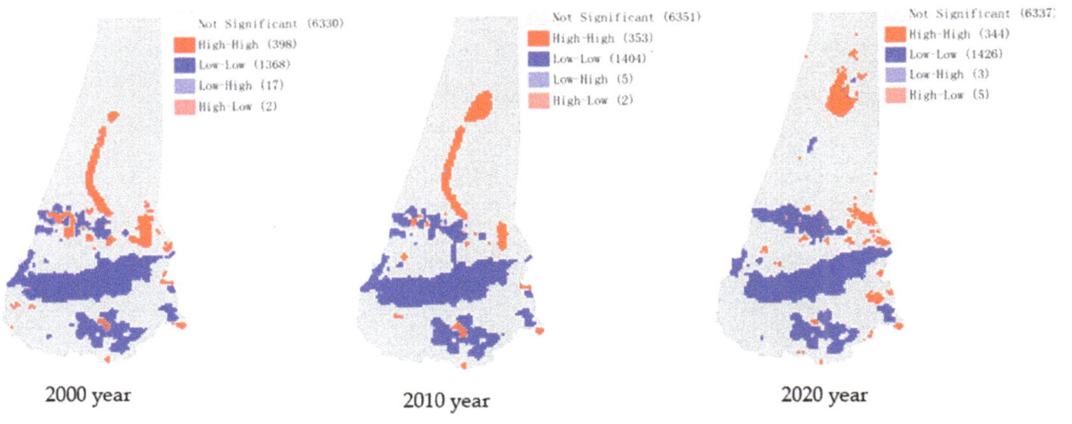

Figure 7. Map of the LISA clusters of the ERI in the KRB.

3.2.2. Analysis of Deriving Factor on the LER by the Geodetector Model

Based on factor detection, the LER of the KRB was examined with respect to the drivers for three periods from 2000 to 2020, and all factors passed the significance test ($p < 0.01$). Figure 8 displays the results of factor detection for each year. The LER of KRB is primarily driven by socioeconomic and natural condition factors, and each driving factor has a different contribution rate. For example, in 2000, soil conditions, annual temperature, precipitation, and DEM were the main influencing factors; in 2010, urban settlements, annual temperature, precipitation, and DEM were the dominant factors; and in 2020, GDP, population, annual temperature, and precipitation were the main influencing factors. According to the findings, environmental drivers will continue to have the greatest influence on the spatial distribution of LER through the year 2020, while social drivers' influence will grow as society advances.

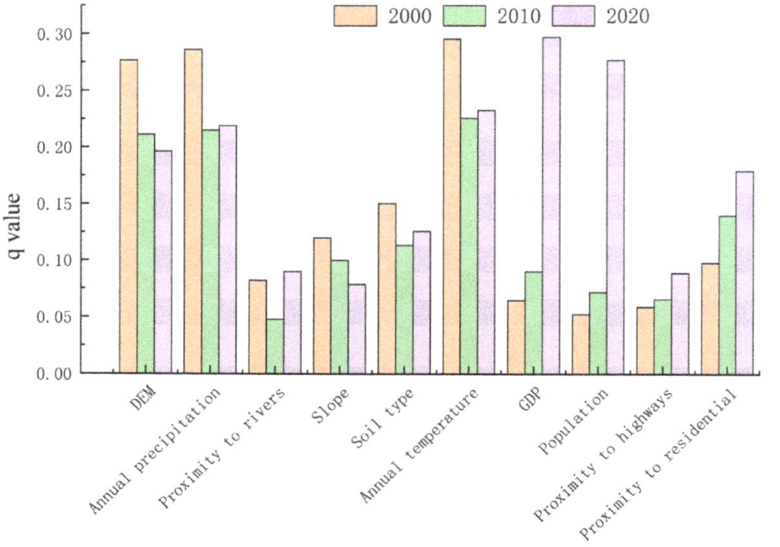

Figure 8. q-statistics of the factors influencing the changes in the ERI.

3.3. Multi-Scenario LULC and Multi-Scenario Modelling, 2030

A comparison between the simulated and real LULC in 2020, together with the validation results and the spatial distribution of the PLUS model simulation error (Figure 9). With a Kappa coefficient of 0.818 and an FOM coefficient of 0.253, the PLUS model is generally more accurate. With its ability to accurately simulate variations in LULC demand within the KRB, the PLUS model shows a high degree of accuracy. This model provides a reliable foundation for future LER simulation predictions, enabling a more precise analysis of LER dynamics in response to LULC demand changes [13,58,59].

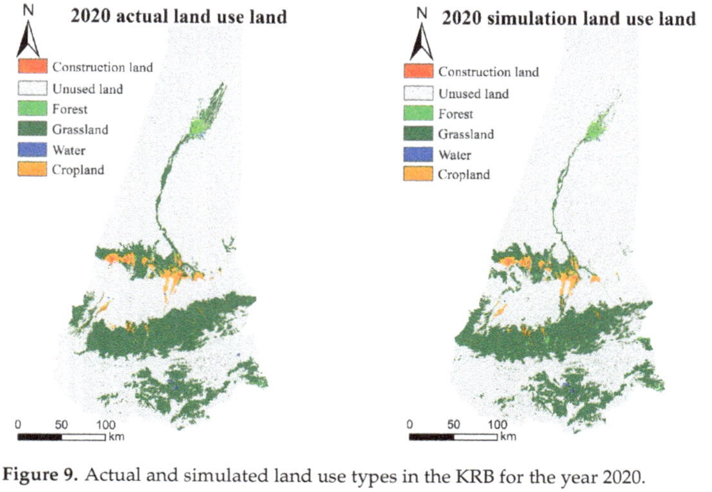

Figure 9. Actual and simulated land use types in the KRB for the year 2020.

3.3.1. Analysis of KRB Land Use While Modeling Multiple Scenarios

Figure 10 and Table 3 show the simulated LULC under different development scenarios for the year 2030, and they show different trends for each scenario for 2030 compared to the LULC for 2020. In accordance with Table 3, we can derive the following: (1) According to the NDS, compared to 2020, cropland, forest land, water, and construction land all increased. Forest land saw the largest increase, with a 92.83 km^2 increase, while grassland and unused land areas decreased. The construction land change rate was 60.37%. (2) Within the framework of CPS, the primary focus was on the protection of cropland. In this context, cropland witnessed a substantial increase in area, with a growth of 156.02 km^2 or 12.71%, representing the largest expansion among all types of land. The expansion of construction land, on the other hand, primarily resulted from the conversion of grassland and unused land. In comparison to the 2020 figures, the area of grassland decreased by 26.81 km^2, while the amount of construction land remained relatively stable. (3) Under the EPS: all ecological land areas increased to different degrees, including forest land, which increased by 66.27 km^2, and grassland, which increased by 132.48 km^2, and the total area of ecological land reached 13,918 km^2, which was the maximum area of ecological land under all scenarios.

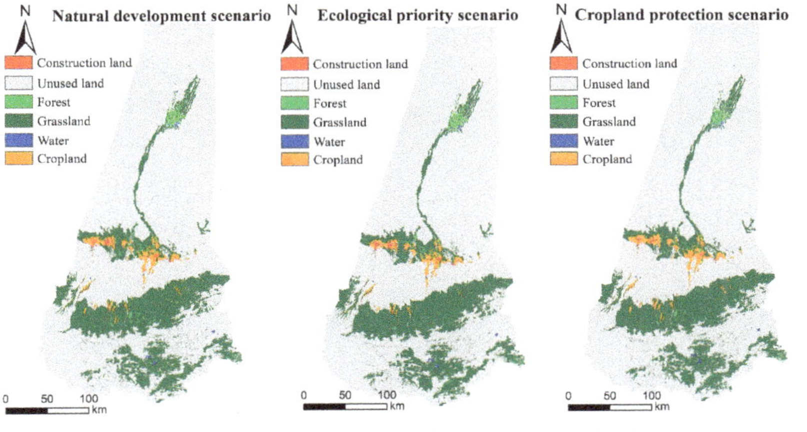

Figure 10. LULC under the different scenarios for the year 2030 in the KRB.

Table 3. Status of 2020 and Multi-scenario simulation of LULC change in 2030.

	Scenario Type	Cropland/km^2	Forest/km^2	Grassland/km^2	Water/km^2	Construction Land/km^2	Unused/km^2
2020		1227.725	265.6377	13,453.58	105.2874	65.4102	55,476.72
NDS	NDS	1295.027	358.4763	13,449.61	111.4929	104.8995	55,274.85
CPS	CPS	1383.749	329.0373	13,426.77	111.4929	68.4603	55,274.85
EPS	EPS	1185.139	331.9137	13,586.06	111.4929	104.8995	55,274.85
2020–2030	NDS	5.48%	34.95%	−0.03%	5.89%	60.37%	−0.36%
2020–2031	CPS	12.71%	23.87%	−0.20%	5.89%	4.66%	−0.36%
2020–2032	EPS	−3.47%	24.95%	0.98%	5.89%	60.37%	−0.36%

3.3.2. Comparative Analysis of LER in Three Scenario Watersheds

Figure 11 depicts the spatial distribution of LER in the catchment in 2030 under various scenarios. Under NDS, CPS, and EPS conditions, respectively, the simulation yields an overall LERI for the catchment of 0.1682, 0.0933, and 0.0903, showing notable variations from those of 2020 (0.1669). The pattern distribution of LERI in the NDS scenario remains largely consistent with 2020. LERs decreased for both CPS and EPS. Comparing the overall spatial pattern distribution of LER, the similarity between the NDS and EPS was extremely high compared to that of 2020, while the lower risk occupied a very large area in the arable

CPS, and the lower risk dominated. In addition, the map clearly showed that ecological land was located in areas with low levels of LER, such as grassland and forest areas. High-risk areas in Daryabuyi under different scenarios in 2030. Combined with Figure 12, the ecological risk ratio of the landscape for 2020 and 2030 under each scenario showed that low risk and lower risk were always in the dominant position, but the area of each ecological risk zone under the different scenarios was still significantly different, especially in the 2030 arable land protection scenario where the low-risk zone accounted for 99.1%. It is also worth mentioning that the largest area of high-risk zone was in 2020, reaching 116.899 km^2, which indicated that there were many ecological risks in the current LULC.

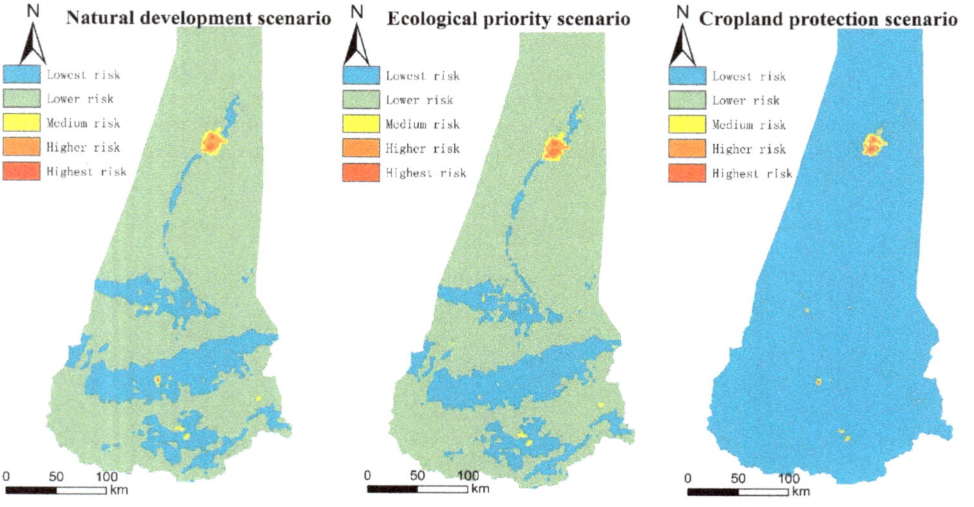

Figure 11. Spatial patterns of the LER rankings in the KRB under the different scenarios for the year 2030.

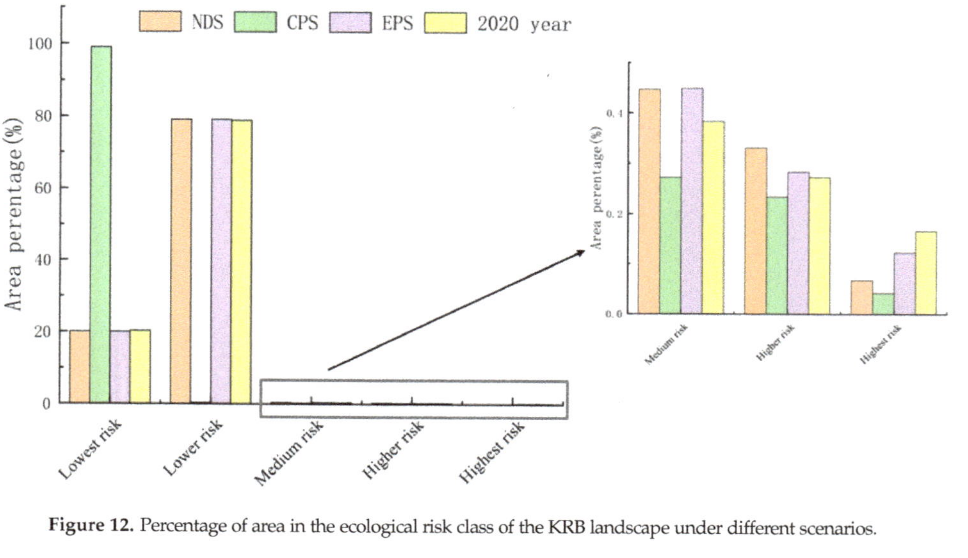

Figure 12. Percentage of area in the ecological risk class of the KRB landscape under different scenarios.

4. Discussion

4.1. The KRB Shows a Spatial Distribution of ERI

Spatial and temporal variations in ecological risk in arid zone watersheds are negative manifestations of natural and social impacts, and LERI can reflect the degree to which watershed ecosystems are threatened by climate change and human activities, thereby revealing mechanisms and trends in ecological processes [60,61]. With LULC serving as a driving force, we found a connection between LER patterns, offering a way to examine the temporal and spatial dynamics of LER. It was discovered that during the period of 2000–2020, the spatial and temporal distribution characteristics of LER in the KRB changed more, primarily in the following ways: Prior to 2010, the area of cultivated land in the central part of the basin was primarily where the high and higher risks were concentrated. This is mainly because the study area is constrained by climatic and environmental factors, the KRB is located in the Gobi Desert, which is arid with low rainfall, and water resource constraints are obvious. This, coupled with the rapid development of agriculture which has led to a large amount of water resource depletion, and because the arable land has been distributed in a fragmented manner, has resulted in poorer landscape connectivity and a relatively high level of ecological risk. The overall LER in the catchment shows a decreasing trend after 2010, with a significant decrease in high-risk areas and an increase in the proportion of low- and medium-risk areas, although there has been an increase in ecologically risky areas in Dariyaboi due to the displacement of downstream water use by upstream water use for irrigated agriculture. In an effort to address the declining ecological environment, the Chinese government has started to restore farmland to forests and grasslands [53,62,63], which to some extent improves the ecological quality and reduces the probability of land degradation, as evidenced by the area of forest the past 20 years.

4.2. Impacts of the Driving Factors on the Pattern of the LER

Taking into account the unique natural geographical, climatic, and social conditions of the KRB, this study explored the intrinsic mechanisms of the LER changes from the perspective of socioeconomic and natural factors. The geodetector results showed significant differences in the main drivers of the LER changes for the three phases from 2000 to 2020. The 20-year period from the initial temperature, precipitation, and topography-driven factors to the GDP, population, temperature, and residence area factors in 2020 showed a very high influence; this indicates, to some extent, that human activities have become an important part of influencing the LER. The period 2000–2020 was a period of rapid economic development from the perspective of the GDP. The 11-fold increase in GDP in the KRB during the two periods was directly manifested by the large expansion in land for construction and was based on encroachment on arable land. However, the increase in the population increased the exploitation and excavation of resources, which was clearly expressed through the expansion in arable land [17,64]. Agriculture and animal husbandry were the main sources of GDP in the basin, and during the study period, regarding the population and socioeconomic development, people increased the area of arable land, which resulted in a large amount of resource plundering in rivers as well as surrounding water bodies, leading to the degradation of the original grassland and artificially changing the spatial configuration of water resources in the basin, which seriously damaged the ecological water use and degraded the grassland vegetation without any resupply [9,65]. Combined with the spatial characteristics of the LER, the locations of the grassland had substantial impacts on the LERs of the watershed areas, and the total area of grassland in 2020 did not change much compared to 2010, but the area of grassland on both sides of the Kriya channel in 2020 increased significantly; that is, this change made the LER of the whole watershed decrease, and the landscape pattern played a key role in relation to the LER.

4.3. Creation of Future LULC Policies and LER Administration

In the KRB region, both natural and societal factors predominantly dictate changes in land use resources, with the natural environment forming the foundational basis that invariably experiences impacts during economic development [66]. Concurrently, amidst the backdrop of severe global climatic alterations, arid zone climates have transitioned from "warm–humid" to "warm–dry" [67]. This shift, coupled with heightened evaporation from river basins, has strained the balance between water resource supply and demand. The land use/cover and landscape patterns in the KRB have drastically transformed over the past two decades [49,68,69]. As the climate in the watershed has evolved, there has been an upsurge in glacial snowmelt runoff upstream, thereby enhancing the accessibility of water resources for agricultural production in the midstream region [70]. However, the expanding cultivated land area in the midstream region, driven by agricultural economic activities, has considerably elevated the demand for agricultural water. Concurrently, the large-scale development of arable land has substantially altered the landscape pattern of natural vegetation in the mid and downstream areas of the river. This alteration has influenced the ecological water demand of the vegetation, leading to a more noticeable trend of vegetation degradation. The rising demand for agricultural water and the decreasing ecological water demand of vegetation have significantly altered the water demand structure of the river basin, as evident in the spatial distribution of the high-risk area of LER from 2000 to 2020. The key ecological LULC in the KRB region is grassland, which, compared to other areas, displays heightened susceptibility and plays a critical role in mitigating ecological risks [9,36].

To safeguard grasslands, moderate grazing intensities can enhance their resilience and bolster their habitat quality stability in riverine ecosystems. Reducing watershed LER necessitates diverse strategies, contingent on the specific region: (1) In high-risk areas, meticulous consideration of natural conditions and actual economic needs is crucial. One strategy involves accommodating the demand for ecological land through natural restoration, encompassing comprehensive protection of forests, grasslands, and arable land, and strategic planning to resolve the conflict between ecological and economic water usage. This includes proactive ecological restoration efforts and the conversion of farmland back to forests and grasslands. Simultaneously, it is essential to develop the agricultural economy in alignment with local conditions, advocate for energy- and water-efficient agricultural management models, and encourage agricultural restructuring. (2) Recognizing the evident spatial variation of LER, adjustments to the spatial distribution pattern of LULC can be employed as a component of ecological risk management. In areas characterized by concentrated ecological risks, it is crucial to address landscape fragmentation and employ spatial governance strategies at the national level. This should entail implementing regionally specific control measures that are tailored to the local natural conditions.

4.4. Limitations and Future Work

In this investigation, an analysis of the Land Ecological Risk (LER) status within the Kriya River Basin (KRB) over the past two decades was undertaken, employing a landscape ecology perspective. The study aimed to elucidate the spatial and temporal distribution of LER while investigating the mechanisms of influencing factors on ecological risk. Furthermore, the research extended to designing and simulating ecological risks under different development scenarios projected for the year 2030. The overarching objective was to provide a theoretical foundation for future ecological risk control measures. Arid zone watersheds, characterized by ecologically sensitive attributes, have witnessed a surge in ecological issues in recent years [2,11]. These concerns have manifested as typical ecological and environmental problems observed globally, garnering extensive attention from various sectors of society. Consequently, the development of a rational and effective ecological assessment model for watersheds, especially those with unique natural geographical conditions like the KRB, assumes critical importance. In this paper, a landscape pattern index was formulated based on land use (LULC) data to assess the KRB. This method proves feasible for managing land resources within the basin and addressing regional ecological and environmental challenges. However, the study acknowledges certain limitations owing to the intricate nature of data sources and ecological risks in the area. The uncertainties introduced as a result of these complexities impact the outcomes of the comprehensive assessment. Notably, the risk assessment method grounded in landscape ecology exhibits a discernible scale effect, where the size of the scale influences the calculation results of the landscape pattern index, leading to potential bias [17,35]. Additionally, the driving factors behind changes in land use landscape patterns within the KRB are exceedingly intricate. The selection of some natural and anthropogenic factors in this study, while omitting others challenging to quantify, hinders the precise prediction of the future spatial pattern of land use in the simulation. Subsequent research endeavors should consider a multi-scale comprehensive assessment, emphasizing the scale effects of diverse influencing factors for a more accurate evaluation. Furthermore, exploring additional quantitative driving analysis indicators in conjunction with the specific conditions of the KRB will contribute to enhancing the simulation's accuracy.

5. Conclusions

This study employs the Landscape Ecological Risk Assessment (LERA) model to conduct a comprehensive examination of the spatial and temporal dynamics within the Kriya River Basin (KRB) from 2000 to 2020. Additionally, the investigation integrates the PLUS model to simulate and forecast the spatial distribution of LER, projecting potential trends in the KRB under diverse scenarios by the year 2030. (1) The KRB predominantly encompasses grassland and unused land, constituting over 97% of its total expanse. Over the two-decade timeframe, notable expansions occur in cropland, water, and construction land, while unused land and grassland witness contractions. Spatial alterations notably reveal that the proliferation of construction land predominantly encroaches upon cropland. (2) The period from 2000 to 2020 manifests a discernible spatial clustering pattern of Ecological Risk Index (ERI) values in the KRB, with an observed escalation in the degree of clustering. Dominating the LER spectrum are low-ecological-risk areas and lower-ecological-risk areas. (3) ERI, as an index, reflects the confluence of socioeconomic and natural conditions, and its principal determinants undergo a nuanced transition over the two decades. In the early 2000s, natural variables including temperature, precipitation, topography, and soil type wield substantial influence, whereas by 2020, socioeconomic factors such as GDP, population, temperature, topography, and proximity to settlements emerge as predominant factors shaping LER. Human activities progressively assert themselves as the paramount catalyst for LER variations. (4) The anticipated landscape configurations under three distinct KRB scenarios in 2030 portray varied degrees of transformation, wherein low-risk and high-risk areas predominate within the NDS, CPS, and EPS scenarios. Spatially, high-risk LERs exhibit notable concentration, particularly evident in the Daryabuyi site across divergent scenarios. The conclusions of the study can provide a decision-making basis for ecological risk early warning for the ecological protection of the Kriya River Basin, enhancement of the ecological security level as well as giving full play to the role of ecological functional zones of inland river basins in arid zones.

Author Contributions: Conceptualization, J.L.; methodology, J.L.; software, X.H. validation, R.W.; writing—original draft, J.L.; resources, P.H., Z.W. and R.W.; project management, X.H.; Funding acquisition, X.H. All authors have read and agreed to the published version of the manuscript.

Funding: This research was funded by the National Science Foundation of China (grant no. 41975115) and research on ecological dispatch and ecological response in the Kriya River Basin (grant no. 2020.B-003).

Data Availability Statement: No new data were created or analyzed in this study. Data sharing is not applicable to this article.

Acknowledgments: We are very reviewers for their comments on the paper revision. We would also like to thank Li Xu and Yanqiu Chen for their comments on the model and paper revisions in.

Conflicts of Interest: The authors declare no conflict of interest.

References

1. He, S.; Wang, D.; Zhao, P.; Li, Y.; Lan, H.; Chen, W.; Chen, X. Quantification of basin-scale multiple ecosystem services in ecologically fragile areas. *Catena* **2021**, *202*, 105247. [CrossRef]
2. Wang, J.; Wu, Y.; Hu, Z.; Zhang, J. Remote Sensing of Watershed: Towards a New Research Paradigm. *Remote Sens.* **2023**, *15*, 2569. [CrossRef]
3. Zhang, H.; Xue, L.; Wei, G.; Dong, Z.; Meng, X. Assessing Vegetation Dynamics and Landscape Ecological Risk on the Mainstream of Tarim River, China. *Water* **2020**, *12*, 2156. [CrossRef]
4. Wang, J.; Zhen, J.; Hu, W.; Chen, S.; Lizaga, I.; Zeraatpisheh, M.; Yang, X. Remote sensing of soil degradation: Progress and perspective. *Int. Soil Water Conserv. Res.* **2023**, *11*, 429–454. [CrossRef]
5. Li, S.; Zhang, J.; Guo, E.; Zhang, F.; Ma, Q.; Mu, G. Dynamics and ecological risk assessment of chromophoric dissolved organic matter in the Yinma River Watershed: Rivers, reservoirs, and urban waters. *Environ. Res.* **2017**, *158*, 245–254. [CrossRef]
6. Ferreira, A.R.L.; Sanches Fernandes, L.F.; Cortes, R.M.V.; Pacheco, F.A.L. Assessing anthropogenic impacts on riverine ecosystems using nested partial least squares regression. *Sci. Total Environ.* **2017**, *583*, 466–477. [CrossRef]
7. Wang, J.; Ding, J.; Yu, D.; Ma, X.; Zhang, Z.; Ge, X.; Teng, D.; Li, X.; Liang, J.; Lizaga, I.; et al. Capability of Sentinel-2 MSI data for monitoring and mapping of soil salinity in dry and wet seasons in the Ebinur Lake region, Xinjiang, China. *Geoderma* **2019**, *353*, 172–187. [CrossRef]
8. Dupras, J.; Marull, J.; Parcerisas, L.; Coll, F.; Gonzalez, A.; Girard, M.; Tello, E. The impacts of urban sprawl on ecological connectivity in the Montreal Metropolitan Region. *Environ. Sci. Policy* **2016**, *58*, 61–73. [CrossRef]
9. Gan, L.; Halik, Ü.; Shi, L.; Welp, M. Ecological risk assessment and multi-scenario dynamic prediction of the arid oasis cities in northwest China from 1990 to 2030. *Stoch. Environ. Res. Risk Assess.* **2023**, *37*, 3099–3115. [CrossRef]
10. Aguilera, M.A.; González, M.G. Urban infrastructure expansion and artificial light pollution degrade coastal ecosystems, increasing natural-to-urban structural connectivity. *Landsc. Urban Plan.* **2023**, *229*, 104609. [CrossRef]
11. Wang, J.; Ding, J.; Li, G.; Liang, J.; Yu, D.; Aishan, T.; Zhang, F.; Yang, J.; Abulimiti, A.; Liu, J. Dynamic detection of water surface area of Ebinur Lake using multi-source satellite data (Landsat and Sentinel-1A) and its responses to changing environment. *Catena* **2019**, *177*, 189–201. [CrossRef]
12. Hou, M.; Ge, J.; Gao, J.; Meng, B.; Li, Y.; Yin, J.; Liu, J.; Feng, Q.; Liang, T. Ecological Risk Assessment and Impact Factor Analysis of Alpine Wetland Ecosystem Based on LUCC and Boosted Regression Tree on the Zoige Plateau, China. *Remote Sens.* **2020**, *12*, 368. [CrossRef]
13. Li, W.; Lin, Q.; Hao, J.; Wu, X.; Zhou, Z.; Lou, P.; Liu, Y. Landscape Ecological Risk Assessment and Analysis of Influencing Factors in Selenga River Basin. *Remote Sens.* **2023**, *15*, 4262. [CrossRef]
14. Malekmohammadi, B.; Rahimi Blouchi, L. Ecological risk assessment of wetland ecosystems using Multi Criteria Decision Making and Geographic Information System. *Ecol. Indic.* **2014**, *41*, 133–144. [CrossRef]
15. Mo, W.; Wang, Y.; Zhang, Y.; Zhuang, D. Impacts of road network expansion on landscape ecological risk in a megacity, China: A case study of Beijing. *Sci. Total Environ.* **2017**, *574*, 1000–1011. [CrossRef] [PubMed]
16. Liu, J.; Kuang, W.; Zhang, Z.; Xu, X.; Qin, Y.; Ning, J.; Zhou, W.; Zhang, S.; Li, R.; Yan, C.; et al. Spatiotemporal characteristics, patterns, and causes of land-use changes in China since the late 1980s. *J. Geogr. Sci.* **2014**, *24*, 195–210. [CrossRef]
17. Du, L.; Dong, C.; Kang, X.; Qian, X.; Gu, L. Spatiotemporal evolution of land cover changes and landscape ecological risk assessment in the Yellow River Basin, 2015–2020. *J. Environ. Manag.* **2023**, *332*, 117149. [CrossRef]
18. Song, Q.; Hu, B.; Peng, J.; Bourennane, H.; Biswas, A.; Opitz, T.; Shi, Z. Spatio-temporal variation and dynamic scenario simulation of ecological risk in a typical artificial oasis in northwestern China. *J. Clean. Prod.* **2022**, *369*, 133302. [CrossRef]
19. Ma, J.; Yu, Q.; Wang, H.; Yang, L.; Wang, R.; Fang, M. Construction and Optimization of Wetland Landscape Ecological Network in Dongying City, China. *Land* **2022**, *11*, 1226. [CrossRef]
20. Qian, Y.; Dong, Z.; Yan, Y.; Tang, L. Ecological risk assessment models for simulating impacts of land use and landscape pattern on ecosystem services. *Sci. Total Environ.* **2022**, *833*, 155218. [CrossRef]
21. Lan, J.; Chai, Z.; Tang, X.; Wang, X. Landscape Ecological Risk Assessment and Driving Force Analysis of the Heihe River Basin in the Zhangye Area of China. *Water* **2023**, *15*, 3588. [CrossRef]
22. Li, W.; Wang, Y.; Xie, S.; Sun, R.; Cheng, X. Impacts of landscape multifunctionality change on landscape ecological risk in a megacity, China: A case study of Beijing. *Ecol. Indic.* **2020**, *117*, 106681. [CrossRef]
23. Li, S.; He, W.; Wang, L.; Zhang, Z.; Chen, X.; Lei, T.; Wang, S.; Wang, Z. Optimization of landscape pattern in China Luojiang Xiaoxi basin based on landscape ecological risk assessment. *Ecol. Indic.* **2023**, *146*, 109887. [CrossRef]
24. Zhao, Y.; Tao, Z.; Wang, M.; Chen, Y.; Wu, R.; Guo, L. Landscape Ecological Risk Assessment and Planning Enlightenment of Songhua River Basin Based on Multi-Source Heterogeneous Data Fusion. *Water* **2022**, *14*, 4060. [CrossRef]
25. Li, H.; Su, F.; Guo, C.; Dong, L.; Song, F.; Wei, C.; Zheng, Y. Landscape ecological risk assessment and driving mechanism of coastal estuarine tidal flats—A case study of the liaohe estuary wetlands. *Front. Environ. Sci.* **2022**, *10*, 2417. [CrossRef]

26. Wang, G.; Ran, G.; Chen, Y.; Zhang, Z. Landscape Ecological Risk Assessment for the Tarim River Basin on the Basis of Land-Use Change. *Remote Sens.* **2023**, *15*, 4173. [CrossRef]
27. Li, J.; Pu, R.; Gong, H.; Luo, X.; Ye, M.; Feng, B. Evolution Characteristics of Landscape Ecological Risk Patterns in Coastal Zones in Zhejiang Province, China. *Sustainability* **2017**, *9*, 584. [CrossRef]
28. De Montis, A.; Caschili, S.; Mulas, M.; Modica, G.; Ganciu, A.; Bardi, A.; Ledda, A.; Dessena, L.; Laudari, L.; Fichera, C.R. Urban–rural ecological networks for landscape planning. *Land Use Policy* **2016**, *50*, 312–327. [CrossRef]
29. Heggem, D.T.; Edmonds, C.M.; Neale, A.C.; Bice, L.; Jones, K.B. A Landscape Ecology Assessment of the Tensas River Basin. *Environ. Monit. Assess.* **2000**, *64*, 41–54. [CrossRef]
30. Kapustka, L.A.; Galbraith, H.; Luxon, B.M.; Yocum, J. Using landscape ecology to focus ecological risk assessment and guide risk management decision-making. *Toxicol. Ind. Health* **2001**, *17*, 236–246. [CrossRef]
31. Paukert, C.P.; Pitts, K.L.; Whittier, J.B.; Olden, J.D. Development and assessment of a landscape-scale ecological threat index for the Lower Colorado River Basin. *Ecol. Indic.* **2011**, *11*, 304–310. [CrossRef]
32. Karimian, H.; Zou, W.; Chen, Y.; Xia, J.; Wang, Z. Landscape ecological risk assessment and driving factor analysis in Dongjiang river watershed. *Chemosphere* **2022**, *307*, 135835. [CrossRef] [PubMed]
33. Zhou, Z.; Zhao, W.; Lv, S.; Huang, D.; Zhao, Z.; Sun, Y. Spatiotemporal Transfer of Source-Sink Landscape Ecological Risk in a Karst Lake Watershed Based on Sub-Watersheds. *Land* **2023**, *12*, 1330. [CrossRef]
34. Pan, N.; Guan, Q.; Wang, Q.; Sun, Y.; Li, H.; Ma, Y. Spatial Differentiation and Driving Mechanisms in Ecosystem Service Value of Arid Region: A case study in the middle and lower reaches of Shule River Basin, NW China. *J. Clean. Prod.* **2021**, *319*, 128718. [CrossRef]
35. Lin, X.; Wang, Z. Landscape ecological risk assessment and its driving factors of multi-mountainous city. *Ecol. Indic.* **2023**, *146*, 109823. [CrossRef]
36. Hou, Y.; Chen, Y.; Li, Z.; Li, Y.; Sun, F.; Zhang, S.; Wang, C.; Feng, M. Land Use Dynamic Changes in an Arid Inland River Basin Based on Multi-Scenario Simulation. *Remote Sens.* **2022**, *14*, 2797. [CrossRef]
37. Gong, J.; Zhao, C.-X.; Xie, Y.-C.; Gao, Y.-J. Ecological risk assessment and its management of Bailongjiang watershed, southern Gansu based on landscape pattern. *Yingyong Shengtai Xuebao* **2014**, *25*, 2041–2048.
38. Shi, Y.; Wang, R.; Lu, Y.; Song, S.; Johnson, A.C.; Sweetman, A.; Jones, K. Regional multi-compartment ecological risk assessment: Establishing cadmium pollution risk in the northern Bohai Rim, China. *Environ. Int.* **2016**, *94*, 283–291. [CrossRef]
39. Wang, W.; Wang, H.; Zhou, X. Ecological risk assessment of watershed economic zones on the landscape scale: A case study of the Yangtze River Economic Belt in China. *Reg. Environ. Chang.* **2023**, *23*, 105. [CrossRef]
40. Islam, K.; Rahman, M.F.; Jashimuddin, M. Modeling land use change using Cellular Automata and Artificial Neural Network: The case of Chunati Wildlife Sanctuary, Bangladesh. *Ecol. Indic.* **2018**, *88*, 439–453. [CrossRef]
41. Albert, C.H.; Hervé, M.; Fader, M.; Bondeau, A.; Leriche, A.; Monnet, A.-C.; Cramer, W. What ecologists should know before using land use/cover change projections for biodiversity and ecosystem service assessments. *Reg. Environ. Chang.* **2020**, *20*, 106. [CrossRef]
42. Zhou, M.; Ma, Y.; Tu, J.; Wang, M. SDG-oriented multi-scenario sustainable land-use simulation under the background of urban expansion. *Env. Sci. Pollut. Res. Int.* **2022**, *29*, 72797–72818. [CrossRef] [PubMed]
43. Gao, B.; Wu, Y.; Li, C.; Zheng, K.; Wu, Y.; Wang, M.; Fan, X.; Ou, S. Multi-Scenario Prediction of Landscape Ecological Risk in the Sichuan-Yunnan Ecological Barrier Based on Terrain Gradients. *Land* **2022**, *11*, 2079. [CrossRef]
44. Darvishi, A.; Yousefi, M.; Marull, J. Modelling landscape ecological assessments of land use and cover change scenarios. Application to the Bojnourd Metropolitan Area (NE Iran). *Land Use Policy* **2020**, *99*, 105098. [CrossRef]
45. Huang, D.; Huang, J.; Liu, T. Delimiting urban growth boundaries using the CLUE-S model with village administrative boundaries. *Land Use Policy* **2019**, *82*, 422–435. [CrossRef]
46. Liu, X.; Liang, X.; Li, X.; Xu, X.; Ou, J.; Chen, Y.; Li, S.; Wang, S.; Pei, F. A future land use simulation model (FLUS) for simulating multiple land use scenarios by coupling human and natural effects. *Landsc. Urban Plan.* **2017**, *168*, 94–116. [CrossRef]
47. Liang, X.; Guan, Q.; Clarke, K.C.; Liu, S.; Wang, B.; Yao, Y. Understanding the drivers of sustainable land expansion using a patch-generating land use simulation (PLUS) model: A case study in Wuhan, China. *Comput. Environ. Urban Syst.* **2021**, *85*, 101569. [CrossRef]
48. Wang, J.; Zhang, F.; Luo, G.; Guo, Y.; Zheng, J.; Wu, S.; Wang, D.; Liu, S.; Shi, Q. Factors Influencing Seasonal Changes in Inundation of the Daliyaboyi Oasis, Lower Keriya River Valley, Central Tarim Basin, China. *Remote Sens.* **2022**, *14*, 5050. [CrossRef]
49. Muyibul, Z.; Jianxin, X.; Muhtar, P.; Qingdong, S.; Run, Z. Spatiotemporal changes of land use/cover from 1995 to 2015 in an oasis in the middle reaches of the Keriya River, southern Tarim Basin, Northwest China. *Catena* **2018**, *171*, 416–425. [CrossRef]
50. Sun, N.S.; Chen, Q.; Liu, F.G.; Zhou, Q.; He, W.X.; Guo, Y.Y. Land Use Simulation and Landscape Ecological Risk Assessment on the Qinghai-Tibet Plateau. *Land* **2023**, *12*, 923. [CrossRef]
51. Lin, Y.; Hu, X.; Zheng, X.; Hou, X.; Zhang, Z.; Zhou, X.; Qiu, R.; Lin, J. Spatial variations in the relationships between road network and landscape ecological risks in the highest forest coverage region of China. *Ecol. Indic.* **2019**, *96*, 392–403. [CrossRef]
52. Li, C.; Chen, J.; Liao, M.; Chen, G.; Zhou, Q. Ecological Risk Assessment of Shan Xin Mining Area Based on Remote Sensing and Geography Information System Technology. *J. Geogr. Inf. Syst.* **2018**, *10*, 234–246. [CrossRef]

53. Zhang, T.; Du, Z.; Yang, J.; Yao, X.; Ou, C.; Niu, B.; Yan, S. Land Cover Mapping and Ecological Risk Assessment in the Context of Recent Ecological Migration. *Remote Sens.* **2021**, *13*, 1381. [CrossRef]
54. Wang, S.; Tan, X.; Fan, F. Landscape Ecological Risk Assessment and Impact Factor Analysis of the Qinghai–Tibetan Plateau. *Remote Sens.* **2022**, *14*, 4726. [CrossRef]
55. Huang, X.; Wang, X.; Zhang, X.; Zhou, C.; Ma, J.; Feng, X. Ecological risk assessment and identification of risk control priority areas based on degradation of ecosystem services: A case study in the Tibetan Plateau. *Ecol. Indic.* **2022**, *141*, 109078. [CrossRef]
56. Wang, J.-F.; Hu, Y. Environmental health risk detection with GeogDetector. *Environ. Model. Softw.* **2012**, *33*, 114–115. [CrossRef]
57. Fu, F.; Deng, S.; Wu, D.; Liu, W.; Bai, Z. Research on the spatiotemporal evolution of land use landscape pattern in a county area based on CA-Markov model. *Sustain. Cities Soc.* **2022**, *80*, 103760. [CrossRef]
58. Zhang, Z.; Hu, B.; Jiang, W.; Qiu, H. Identification and scenario prediction of degree of wetland damage in Guangxi based on the CA-Markov model. *Ecol. Indic.* **2021**, *127*, 107764. [CrossRef]
59. Zhang, S.; Chen, C.; Yang, Y.; Huang, C.; Wang, M.; Tan, W. Coordination of economic development and ecological conservation during spatiotemporal evolution of land use/cover in eco-fragile areas. *Catena* **2023**, *226*, 107097. [CrossRef]
60. Preuss, T.G.; Hommen, U.; Alix, A.; Ashauer, R.; van den Brink, P.; Chapman, P.; Ducrot, V.; Forbes, V.; Grimm, V.; Schafer, D.; et al. Mechanistic effect models for ecological risk assessment of chemicals (MEMoRisk)-a new SETAC-Europe Advisory Group. *Environ. Sci. Pollut. Res.* **2009**, *16*, 250–252. [CrossRef]
61. Chen, L.; Sun, R.; Lu, Y. A conceptual model for a process-oriented landscape pattern analysis. *Sci. China Earth Sci.* **2019**, *62*, 2050–2057. [CrossRef]
62. Bennett, M.T. China's sloping land conversion program: Institutional innovation or business as usual? *Ecol. Econ.* **2008**, *65*, 699–711. [CrossRef]
63. Xu, J.; Yin, R.; Li, Z.; Liu, C. China's ecological rehabilitation: Unprecedented efforts, dramatic impacts, and requisite policies. *Ecol. Econ.* **2006**, *57*, 595–607. [CrossRef]
64. Liu, Z.; Liu, Y.; Wang, J. A global analysis of agricultural productivity and water resource consumption changes over cropland expansion regions. *Agric. Ecosyst. Environ.* **2021**, *321*, 107630. [CrossRef]
65. Yu, Y.; Yu, R.; Chen, X.; Yu, G.; Gan, M.; Disse, M. Agricultural water allocation strategies along the oasis of Tarim River in Northwest China. *Agric. Water Manag.* **2017**, *187*, 24–36. [CrossRef]
66. Ponce-Campos, G.E.; Moran, M.S.; Huete, A.; Zhang, Y.; Bresloff, C.; Huxman, T.E.; Eamus, D.; Bosch, D.D.; Buda, A.R.; Gunter, S.A.; et al. Ecosystem resilience despite large-scale altered hydroclimatic conditions. *Nature* **2013**, *494*, 349–352. [CrossRef] [PubMed]
67. Yao, J.; Chen, Y.; Guan, X.; Zhao, Y.; Chen, J.; Mao, W. Recent climate and hydrological changes in a mountain–basin system in Xinjiang, China. *Earth-Sci. Rev.* **2022**, *226*, 103957. [CrossRef]
68. Wang, R.; Zayit, A.; He, X.; Han, D.; Yang, G.; Lv, G. Ecological Water Requirement of Vegetation and Water Stress Assessment in the Middle Reaches of the Keriya River Basin. *Remote Sens.* **2023**, *15*, 4638. [CrossRef]
69. Yan, W.; Wang, Y.; Ma, X.; Liu, M.; Yan, J.; Tan, Y.; Liu, S. Snow Cover and Climate Change and Their Coupling Effects on Runoff in the Keriya River Basin during 2001–2020. *Remote Sens.* **2023**, *15*, 3435. [CrossRef]
70. Jiang, N.; Zhang, Q.; Zhang, S.; Zhao, X.; Cheng, H. Spatial and temporal evolutions of vegetation coverage in the Tarim River Basin and their responses to phenology. *Catena* **2022**, *217*, 106489. [CrossRef]

water

MDPI

Article

Hydro-Climatic and Vegetation Dynamics Spatial-Temporal Changes in the Great Lakes Depression Region of Mongolia

Batsuren Dorjsuren [1,2,*], Valerii A. Zemtsov [2,*], Nyamdavaa Batsaikhan [3], Denghua Yan [4], Hongfei Zhou [5] and Sandelger Dorligjav [3]

1 Department of Environment and Forest Engineering, National University of Mongolia, Ulaanbaatar 210646, Mongolia
2 Department of Hydrology, National Research Tomsk State University, Tomsk 634050, Russia
3 Department of Geography, School of Art & Sciences, National University of Mongolia, Ulaanbaatar 210646, Mongolia; nyamdavaab@num.edu.mn (N.B.); d.sandelger@num.edu.mn (S.D.)
4 State Key Laboratory of Simulation and Regulation of Water Cycle in River Basin, China Institute of Water Resources and Hydropower Research (IWHR), Beijing 100038, China; yandh@iwhr.com
5 Xinjiang Institute of Ecology and Geography, CAS, Urumqi 830011, China; zhouhf@ms.xjb.ac.cn
* Correspondence: batsuren@num.edu.mn (B.D.); zemtsov_v@mail.ru (V.A.Z.)

Abstract: The Great Lakes Depression region basin is among the most sensitive regions to vegetation change due to climate change. This study estimated spatial-temporal changes and relationships in hydro-climate and vegetation dynamics in the basin. Studying the spatial-temporal variation between vegetation dynamics and hydro-climate in this basin is essential for assessing climate change and sustainability. This research involved an examination of the mean yearly air temperature, overall annual rainfall, fluctuations in river discharge, vegetation cover, and alterations in vegetation types within the selected basin stations. This was accomplished through the utilization of hydro-meteorological analysis, satellite assessment, land cover determination, and statistical analysis. Over the course of the study, it was observed that the average annual air temperature increased at all stations (with a positive change of Z = +1.16). The amount of precipitation decreased (Z = −0.79), especially from 2000 to 2014, and its statistical significance decreased. During the study period, average river discharge significantly decreased (Z = −3.51). Due to these combined factors, the lake's water level also decreased (Z = −2.03). Vegetation cover change varied in high mountains, near river and lake water surfaces, and in arid regions. Changes in air temperature and precipitation in the current year determine vegetation cover. Because of the large amount of precipitation in the summer months from 2000 to 2010 and 2020, the growth of vegetation cover during that period was relatively good. This study was conducted in arid and semi-arid regions of Central Asia and demonstrates the impact of climate change on changes in vegetation cover.

Keywords: Central Asia; arid regions; semi-arid region; vegetation cover; water; climate change

1. Introduction

Vegetation plays many vital roles in nature and society. It performs a crucial function in controlling the carbon cycle, influencing climate patterns, and facilitating the transfer of substances and energy among the atmosphere, land surface, hydrological processes, food chains, and soil-dwelling organisms [1]. In society, vegetation is the primary source of raw materials for food, medicine, medical equipment, beauty, industrial raw materials, and an important component of material wealth [2]. Therefore, it is very important to determine how vegetation cover is changing with climate change and the main factors influencing this change. Looking at the interrelationships of vegetation changes and climate change in areas with the most significant vegetation change, and also vulnerable and sensitive areas, can be useful for predicting the changes in other areas and vegetation communities [3]. A classic example of such a place is Central Asia's arid and semi-arid regions [4–6]. Central

Asia's arid and semi-arid regions are among the most sensitive vegetative regions to climate change. It is vital to estimate the spatial and temporal variations in climate changes and the associated changes in water and vegetation in these sensitive areas, such as the Mongolian Plateau.

Changes in vegetation cover in the Mongolia Plateau region are closely related to water and climate variables [7]. The study of vulnerable ecosystems about hydro-climatic change has enormous implications for the ability of a region to maintain its unique characteristics [8]. This mid-latitude region in the northern part of the world has four seasons, where sensitive plants are dominant, seasonal and winter temperature differences are significant, and plant growth dynamics differ between seasons [6,9]. The region's most favorable climate and weather period for vegetation growth is 5–8 months when the vegetation cover grows best during an active balance of air temperature and precipitation [10]. It is the region where the influence of hydro-climate on vegetation growth is most pronounced and a region where it is changing the most.

Another critical factor affecting vegetation cover is the human impact. In particular, the use of surface water significantly impacts changes in vegetation cover [11,12]. Human factors affecting vegetation cover include surface water use, which can influence agricultural practices, soil and plant cover use, and the grazing capacity of hydrological networks [13,14]. In particular, the impact of grazing on water points is one of the major causes of vegetation cover degradation and erosion. Therefore, studying the process of degradation of vegetation cover due to the effects of hydro-climate on the one hand and livestock grazing on the other hand may be very important to estimate the combined effects in the semi-arid and arid regions of the Mongolian Plateau.

In the study of water and climate change, it can be very important to compare the changes in the main climate parameters with the surface water changes and to calculate the direct and indirect correlations to detect the subsequent changes due to the main parameters [15,16]. It is important to estimate the impact of changes in air temperature in the basin on precipitation and then study the impact of precipitation on surface water [17]. It may be possible to calculate the interrelationship between these changes and the changes and effects of other factors affecting the vegetation cover.

The relationship between these changes may vary from place to place depending on the characteristics of the place. For example, in the arid and semi-arid highlands of Central Asia, surface water mostly flows from high mountain snow and glaciers and precipitation water has the greatest impact on lowland vegetation and surface water [18]. Therefore, it is important to calculate in detail how changes in hydro-climate, such as changes in air temperature, precipitation, and river discharge in high regions, will affect vegetation cover and the rate and extent of spatial and temporal changes.

The purpose of this study is to study the interrelationship between hydro-climatic change and vegetation cover in the Great Lakes Depression region in western Mongolia. In order to achieve the objectives of this study, the following objectives were set, including (i) determining the hydro-climatic changes; (ii) determining temporal and spatial changes in vegetation cover; and (iii) establishing the relationship between hydro-climate and vegetation cover.

2. Materials and Methods

2.1. Study Area

The Great Lakes Depression is in the western part of Mongolia, surrounded by the Mongolian Altai Mountains in the west, the Sayan mountain system in the north, the Khangai Mountains in the east, and the Mongolian Altai and Khangai Mountains in the south (45°51′26″–51°07′03″ N, 87°44′58″–99°03′56″ E), with a small part of the northern area crossing into the Russian Federation (Figure 1). The total land area is 268,309.52 km^2, of which 9% is located in the territory of the Russian Federation, while 91% of the total area, or 244,182.13 km^2, is located in the territory of Mongolia. The region has a fragile ecosystem with a variety of natural typologies, including the desert steppe, gobi, steppe, forest-steppe zones, and high mountain belts with an elevation of 744–4306 m above sea level. Due to the diversity of its nature, the area has its own unique flora. Desert is found in the central part of the Great Lakes basin (*Nanophyton erinaceum, Anabasis brevifolia, Reaumurua soongorica*) *"A desert is a dry area with sparse or no vegetation."*. Gobi (*Artemisia xerophytica, Stipa glareosa, Asteroflamus hereopappoides*), *"The gobi is a large desert and semi-desert with region sparse or little vegetation in Central Asia"* and steppe (*Allium eduardi, Festuca valesiaca, Caragana pygmaea*) *"A steppe is a natural grassy area with dense or large vegetation area."* vegetation communities are formed around large rivers and lakes in basins (*Populus laurifolia, Cleistogenes squarrosa, Agropyron nevskii*). A wide variety of plants [19] are present across the basin. Further, mountain and high-altitude plants (*Adonis apennina* L., *Allium altaicum, Swertia banzragczii*) grow in the Mongolian Altai, Khangai, and Kharhiraa Turgeni Mountains located at the outermost edge of this region [20].

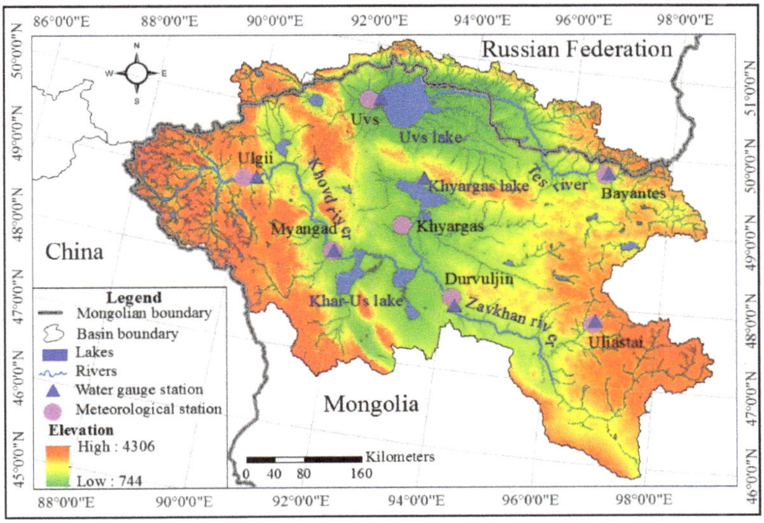

Figure 1. Location of the area and hydro-climate stations in the Great Lakes Depression region of Mongolia.

2.2. Data Sources

To study the interrelationship between hydro-climatic change and vegetation cover, a number of different data sets were collected. Climate data, such as air temperature and precipitation, for the Great Lakes Depression region of Mongolia were obtained from National Centers for Environmental Information, i.e., NOAA's National Centers for Environmental Information (NCEI) "(https://ngdc.noaa.gov/ accessed on 23 September 2022)" and Information and Research Institute of Meteorology, Hydrology, and Environment (IRIMHE) in Mongolia. River discharge and lake water level information were also obtained from the Hydrology and Measurement department of the IRIMHE "(http://irimhe.namem.gov.mn/ accessed on 20 November 2022)". Global SRTM 90 m resolution Digital Elevation Model (DEM) data were obtained from the CGIAR-CSI GeoPortal "(https://cmr.earthdata.nasa.gov/ accessed on 7 August 2022)". The normalized difference vegetation index (NDVI) and enhanced vegetation index (EVI) data were extracted from the MODIS NDVI product (MOD13Q1) obtained from the land processes distributed active archive center, NASA "(https://lpdaac.usgs.gov/ accessed on 10 January 2023)". The land cover data of Landsat resolution of 30 m from 1990 to 2020 were used for land cover classification. The detailed global land cover classification system (containing 16 global and 14 regional land cover types) and 30 m spatial resolution images (GLC_FCS30-2015) were used [21]. The locations of the meteorological and water gauge stations used in the study are shown in Figure 1. The following factors were considered when selecting the climate and water gauge stations: (1) spatial distribution of stations, (2) the amount of data from the stations and the time of measurement, and (3) whether the stations are near water systems.

2.3. Methods

This study used the Mann–Kendall (MK) test method to detect hydro-climate time series data trends. Results were compared with the Innovative Trend Analysis Method (ITAM) and Sen's Slope Estimator Test (SSET) to assess Mann–Kendall (MK) reliability and observe changes in trend. Trend analysis significance levels of 10%, 5%, and 1% were adopted to evaluate the hydro-climatic time series data. A digital elevation model (DEM) calculated the study area and baseline. The relationship between hydro-climatic data, NDVI, EVI, and land cover was calculated to estimate changes in plant cover according to the following general figure (Figure 2).

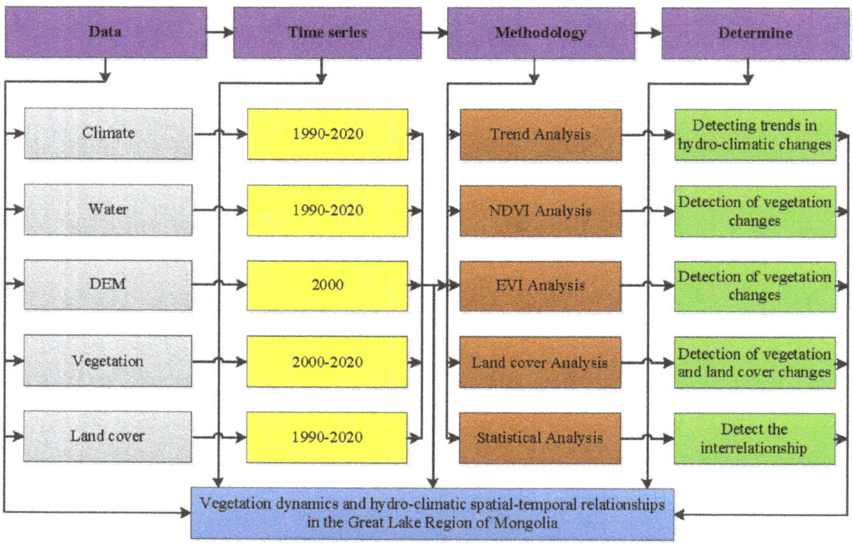

Figure 2. Diagram on detecting spatio-temporal relationships between vegetation dynamics and hydro-climate.

2.3.1. Hydro-Climatic Trend Analysis

The MK test method indicates statistically significant upward and downward trends [22,23]. Annual hydro-climatical data series were used in this study and analysis to detect changing trends. Annual precipitation, air temperature, river discharge, and lake water surface level trends were also analyzed separately. Individual hydro-meteorological time series data were compared with all-time series data for the same year.

The MK statistics are the cumulative results of all the data values. The MK test statistics "S" are then equated as

$$S = \sum_{i=1}^{n-1} \sum_{j=i+1}^{n} sgn(x_j - x_i) \quad (1)$$

The trend test is applied to x_i data values $(i = 1, 2, \ldots n-1)$ and $x_j(j = i+1, 2, \ldots n)$. The data value of each x_i is used as a reference point to compare with the data value of x_j, which is given as

$$sgn(x_j - x_i) = \begin{cases} +1 \text{ if } (x_j - x_i) > 0 \\ 0 \text{ if } (x_j - x_i) = 0 \\ -1 \text{ if } (x_j - x_i) < 0 \end{cases} \quad (2)$$

where sgn is the sign function, and x_j and x_i are the values in period j and i. When the number of data series is greater than or equal to ten $(n \geq 10)$, MK test is then characterized by a normal distribution with the mean $E(S) = 0$, and variance $Var(S)$ is equated as

$$E(S) = 0 \quad (3)$$

$$Var(S) = \frac{n(n-1)(2n+5) - \sum_{k=1}^{m} t_k(t_k-1)(2t_k+5)}{18} \quad (4)$$

where m is the number of the tied groups in the time series, and t_k is the number of ties in the kth tied group.

The test statistics Z are as follows:

$$Z = \begin{cases} \frac{s-1}{\delta} & \text{if } S > 0 \\ 0, & \text{if } S = 0 \\ \frac{s+1}{\delta} & \text{if } S < 0 \end{cases} \quad (5)$$

When Z is greater than zero, δ is the variance of S, indicating an increasing trend, and when Z is less than zero, it is a decreasing trend.

In time sequence, the statistics are defined independently:

$$UF_k = \frac{d_k - E(d_k)}{\sqrt{var(d_k)}}(k = 1, 2, \ldots, n) \quad (6)$$

Firstly, given the confidence level α, if the $UF_k > UF_\alpha/2$, indicating that the sequence has a significant trend. Then, the time sequence of changes is represented in reverse order. The following equation is used to express the inverse relationship according to the calculation of the equation:

$$UB_k = -UF_k \quad (7)$$

$$K = n + 1 - k \quad (8)$$

where UF_k and UB_k are the statistical variables, two statistical order curves UB_k and UF_k are drawn as UB and UF curves, and two critical value lines are drawn on the graph to detect MK change points. UB and UF of these two crucial values indicate a significant upward or downward trend when it exceeds the line, and the range beyond the essential value line is defined as the time interval of the change. In this case, the intersection of the UB and UF curves indicates the start of the change, which is the point of difference [24].

ITAM has been used in several studies to detect changes in hydro-climate trends [23,25]. The ITAM divides a time series into two equal parts, and it sorts both sub-series in ascending order. Then after, the two halves are placed on a coordinate system $(x_i : i = 1, 2, 3, \ldots n/2)$ on X-axis and $(x_j : j = n/2+1, n/2+2, \ldots n)$ on Y-axis. If the time series data on a scattered plot are collected on the 1:1 (45°) straight line, it indicates no trend. However, the trend increases when data points accumulate above the 1:1 straight line, and the trend decreases when data points accumulate below the 1:1 straight line.

The trend indicator is given as

$$\varphi = \frac{1}{n}\sum_{i=1}^{n} \frac{10(x_j - x_i)}{\mu} \quad (9)$$

where φ = trend indicator, n = number of observations on the sub-series, x_i = data series in the first half sub-series class, x_j = data series in the second half sub-series part, and μ = mean of data series in the first half sub-series part.

A positive value of φ indicates an increasing trend. However, a negative value of φ indicates a decreasing trend. However, when the scatter points are closest around the 1:1 straight line, it implies the non-existence of a significant trend.

In SSET, the trend magnitude is calculated using [26] slope estimator methods. The slope Q_i between two data points is given by the equation

$$Q_i = \frac{x_j - x_k}{j - k}, \text{for } i = 1, 2, \ldots N \tag{10}$$

where x_j and x_k are data points at time j and $(j > k)$, respectively. When there is only single datum in each time, then $N = \frac{n(n-1)}{2}$; n is number of time periods. However, if the amount of data in each year is high, then $N < \frac{n(n-1)}{2}$; n is the total number of observations. The N values of slope estimator are arranged from smallest to biggest. Then, the median of slope (β) is computed as

$$\beta = \begin{cases} Q[(N+1)/2] & \textit{when N is odd} \\ Q[(N/2) + Q(N+2)/(2)/(2)] & \textit{when N is even} \end{cases} \tag{11}$$

The sign of β shows whether the trend is increasing or decreasing.

2.3.2. Vegetation Analysis

NDVI is one of the most widely used vegetation indices of plant biomass and vegetation activity [27], which indicates the change in an area's vegetative greenness [28].

$$NDVI = \frac{(NIR - RED)}{(NIR + RED)} \tag{12}$$

where *NIR* and *RED* are the near-infrared and red channels of the electromagnetic spectrum, respectively, corresponding to bands 2 and 1 of the MODIS (MOD13Q1) product. The second vegetation layer is the EVI, which has improved sensitivity for high biomass regions.

$$EVI = \frac{G * (P_{NIR} - P_{Red})}{(P_{NIR} + C_1 * P_{Red} - C_2 * P_{Blue} + L)} \tag{13}$$

where EVI is the enhanced vegetation index, G is the gain factor (=2.5), P_{NIR} is the near-infrared reflectance, P_{Red} is the red reflectance, P_{Blue} is the blue reflectance, C_1 is the atmosphere resistance red correction coefficients (=6), C_2 is the atmosphere resistance blue correction coefficients (=7.5), and L is the canopy background brightness correction factor (=1) [25].

2.3.3. Land Cover Analysis

Analysis of vegetation cover changes in land cover categories allowed for greater control over temporal and spatial changes in vegetation. Therefore, spatial analysis was performed in duplicate to detect differences in land cover change over distinct periods and to establish transitions and correlations between land cover changes. By overlapping the land cover maps from 1990, 2000, 2010, and 2020, we derived a land cover transformation map, which was then employed for further analysis using a transformation matrix. The extent of changes in land cover was subsequently quantified as [25].

$$CA = TA(t_2) - TA(t_1) \tag{14}$$

$$CE = [CA/TA(t_1)] * 100 \tag{15}$$

where TA, CA, and CE stand for the total area, changed the area, and the extent of change, respectively, t_1 and t_2 are the beginning and ending times. An external Kappa coefficient (KC) was calculated to confirm the transition of land cover change. KC is a measure of agreement between predefined producer ratings and user-assigned ratings. The calculation is based on the difference between how much agreement is present ("observed" agreement) and how much agreement would be expected to be present by chance alone ("expected" agreement) [29].

$$K = P(A) - P(E)/1 - P(E) \tag{16}$$

$$P(A) = \frac{(A + D)}{N} \tag{17}$$

$$P(E) = \left(\frac{A1}{N}\right) * \left(\frac{B1}{N}\right) + \left(\frac{A2}{N}\right) * \left(\frac{B2}{N}\right) \tag{18}$$

where K is the Kappa coefficient, $P(A)$ is the number of times the K rates agree, $P(E)$ is the number of times the K rates are expected to agree only by chance, A and D are unchanged categories, $A1$ and $B1$ are subject's categories, and N is the change in results.

2.3.4. Statistical Analysis

Correlation analysis was used to check whether there is a linear relationship between vegetation cover and water climate, and the strength of the relationship was expressed as a correlation coefficient [30,31]. The correlation coefficient is expressed by Equation (19).

$$r = \frac{\Sigma(x - \bar{x})(y - \bar{y})}{\sqrt{\Sigma(x - \bar{x})^2 \Sigma(y - \bar{y})^2}} \tag{19}$$

Here, r: correlation coefficient, x—variable, $\bar{x}$—standard deviation of x variable, y—variable, and $\bar{y}$—standard deviation of y variable.

3. Results

3.1. Hydro-Climatic Analysis

When calculating the water–climate relationships, the parameters that most affect the hydrological process are selected. These parameters are average annual air temperature, total annual precipitation, river discharge, and lake water level. In the last 30 years, the air temperature in the basin has warmed by 1.09 °C, i.e., from −0.5 °C to + 0.5 °C (Figure 3).

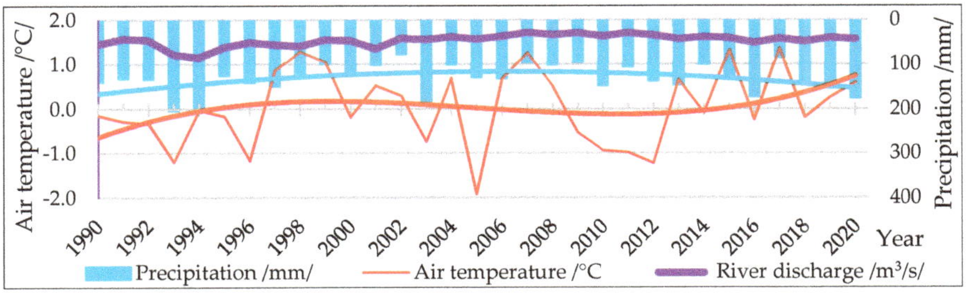

Figure 3. Hydro—climate change.

The amount of total annual precipitation decreased sharply from 1994 to 2010, and especially from 2000 to 2014, decreasing with statistical significance. Overall, this region receives little precipitation. Most of the precipitation falls during the summer months. In I confirm the winter months, the snow water remains stable for a long time. The average annual precipitation here is 134.6 mm. This is indicative of the semi-arid and arid regions of Central Asia.

In this region, Mongolia has the largest rivers in Central Asia, such as the Khovd River, the Zavkhan River, and the Tesin River, all flowing into terminal large lakes. Data from water monitors on these rivers were processed, and the average river discharge of these rivers decreased from 60 to 40 m^3/s in the last 30 years. This decline continued from 1994 to 2012, with a slight upward trend since 2013. However, since 2005, it has continuously decreased with statistical significance.

The water level measurements of the two largest lakes in the study area, Uvs Lake and Khyargas Lake, show that the water level was high from 1995 to 1997, but the water level of the lake has continuously decreased since 1998. Especially since 2011, the decrease has been statistically significant (Figure 4).

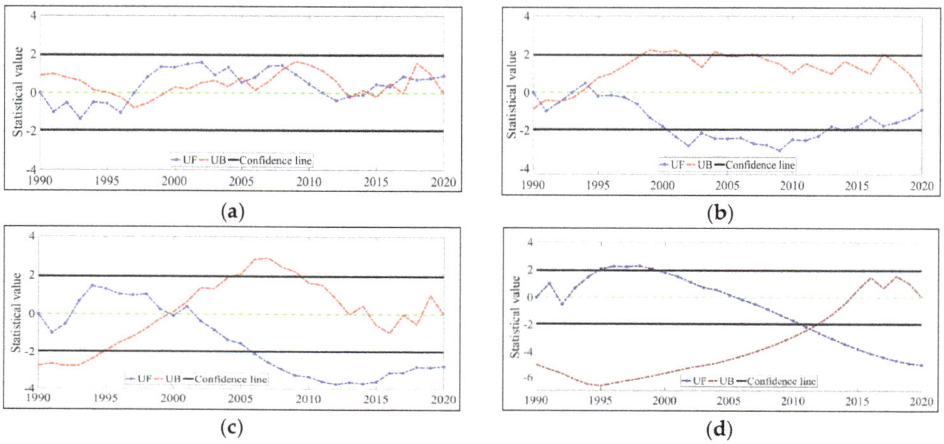

Figure 4. Trends in water climate change in the Great Lakes Depression region: (**a**) annual mean air temperature trends; (**b**) trends in changes in total annual precipitation; (**c**) changes in the mean river discharge; and (**d**) changes in the water level of the Great Lakes.

Hydro-climatic parameters were calculated using MK, and a change analysis was performed using ITAM and SSET to confirm the calculation. The average air temperature increased (Z = 1.16), while precipitation slightly decreased (Z = −0.79). In arid and semi-arid regions, river discharge is directly related to changes in precipitation. However, despite the slight decrease in precipitation in this area, the river discharge (Z = −3.51) significantly decreased. The rate of decline in river discharge is twice as high as the rate of decline in precipitation, suggesting that there may be other factors influencing river discharge. In this region, the flow of river water, which is the main source of lake water, decreased due to this, and the lake's water level (Z = −2.03) decreased significantly. This is directly related to the decrease in river discharge, the primary source (Table 1).

Table 1. Results for hydro-climate trends: MK (Z), ITAM (φ), and SSET (β).

№	Indicator	Z	φ	β	Rate of Change
1	Air temperature	1.16 *	1.86 *	0.01	0.10
2	Precipitation	−0.79	0.07	0.59	−0.16
3	River discharge	−3.51 ***	−2.66 **	−1.36 *	−0.61
4	Lake water level	−2.03 **	−2.05 **	−3.39 ***	−5.9 ***

Notes: * Trends at 0.1 significance level; ** trends at 0.05 significance level; and *** trends at 0.01 significance level.

3.2. Vegetation Analysis

3.2.1. NDVI Analysis

The Great Lakes basin region is located amid a diverse natural landscape with diverse vegetation associated with different ecosystems and climates. There are several large lakes in the central part of the basin, but there is an arid and semi-arid zone around the lake. Depending on the annual air temperature and precipitation, the most active vegetation period is between 5 and 9 months in this region. The MODIS NDVI product (MOD13Q1) was used to calculate the vegetation cover NDVI during the study period from August 12 to 27, 2000, 2010, and 2020. In arid and semi-arid regions of Mongolia, the most active summer vegetation period is from May 20 to August 10. But from September, plant growth stops and starts to wither due to the cold. Therefore, the period between August 12 and 27 was chosen when the vegetation in the region was most stable. NDVI had different values in high-altitude areas where surface water and vegetation cover, such as rivers and lakes, were greatest (Figure 5).

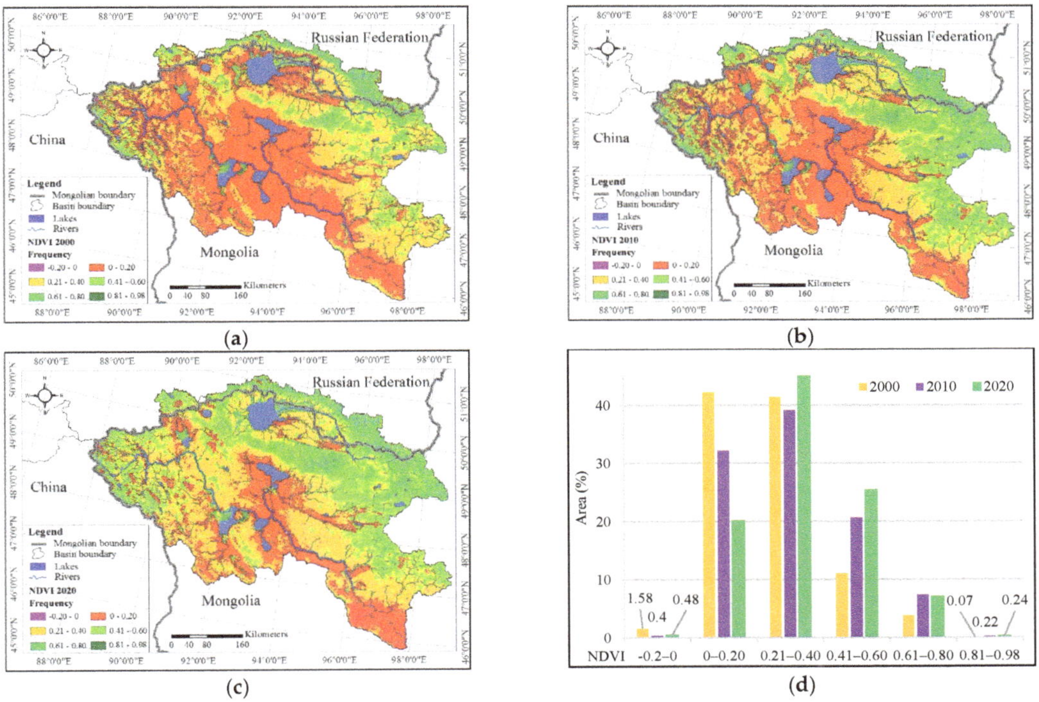

Figure 5. NDVI changes in Great Lakes Depression region of Mongolia. (**a**) NDVI in 2000; (**b**) NDVI in 2010; (**c**) NDVI in 2020; and (**d**) percentage of change in NDVI area.

Considering the number of plants, in August 2000, 43.77% of the total area had a very poor yield or −0.20–0.20, while 41.34% of the total area had a poor yield. However, the rest of the land with relatively good yield accounted for 14.89% of the total area. As of 2010, 42.23% of the total area had a very poor yield of −0.20–0.20, while 39.18% of the total area had a poor yield. But the remaining 28.23% was relatively good yield land. As of 2020, 20.58% of the total area had very low vegetation cover or −0.20–0.20, while 46.77% of the total area had poor yields. But, the remaining 32.66% was relatively good yield land. In arid and semi-arid regions, vegetation cover growth is likely to be directly related to temperature and precipitation distribution. For example, from 2000 to 2009, there was an increase in air temperature and the amount of precipitation significantly decreased during this period. This overlaps with the period of low vegetation cover in the 2000 NDVI map. In addition, in 2020, when the increase in air temperature is relatively stable and the total annual precipitation increases, the NDVI of the basin will have a relatively high value. Therefore, it can be seen from the above comparison that air temperature and precipitation are directly related to NDVI growth.

3.2.2. EVI Analysis

The EVI values of the study area were divided into five levels: no vegetation, low vegetation, medium vegetation, medium and higher vegetation, and higher vegetation. A significant feature of EVI was the degradation of vegetation cover in areas near large lakes and rivers. The amount of vegetation cover also showed similar trends, temporal and spatial variations, to NDVI (Figure 6).

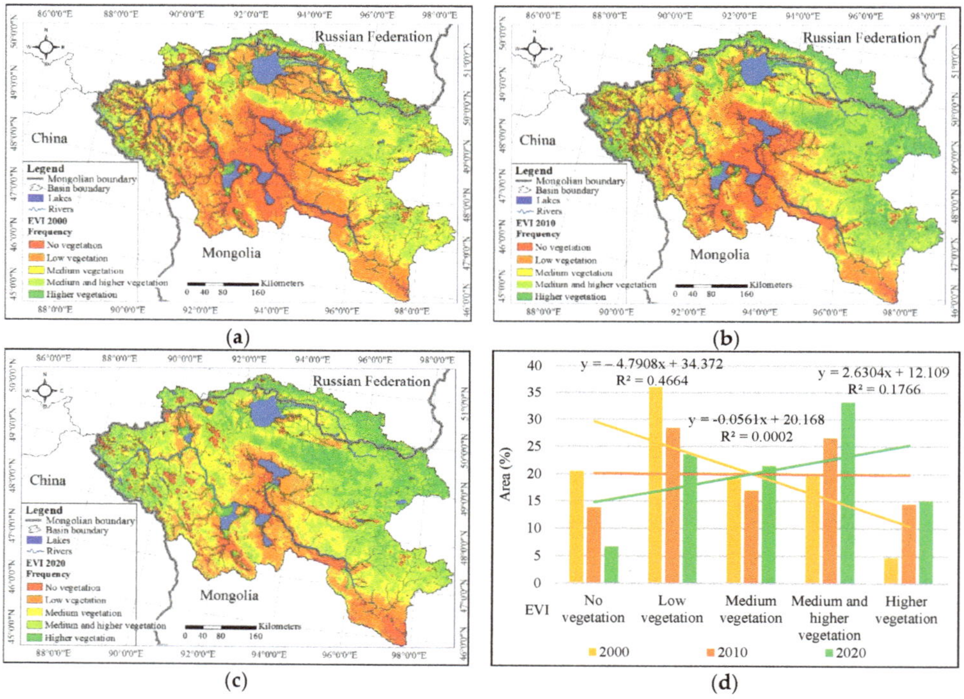

Figure 6. EVI changes in Great Lakes Depression region of Mongolia. (**a**) EVI in 2000; (**b**) EVI in 2010; (**c**) EVI in 2020; and (**d**) percentage of change in EVI area.

As of 2000, 20.45% of non-vegetated land in the basin was the largest, with 36.07% of the land with low vegetation. On the other hand, 43.48% of areas had medium and large vegetation, respectively. Plant growth in 2010 was average compared to the 2000 and 2020 figures. As of 2020, the place without vegetation was 6.68%, while the area with low vegetation was 23.63%. The site with medium and large vegetation increased to 69.69%. The five categories of changes in the study area showed that the area with vegetation was less in 2000 ($y = -4.7908x + 34.372$, $R^2 = 0.4664$), and the region with vegetation increased in 2020 ($y = 2.6304x + 12.109$, $R^2 = 0.1766$). These changes were associated with changes in temperature and precipitation inputs, similar to changes in NDVI. In addition, vegetation cover degradation was significant in areas near surface water in the basin. This may be related to surface water use and possibly livestock grazing.

3.3. Land Cover Analysis

The land cover change was determined using satellite image data during August 1990, 2000, 2010, and 2020 when vegetation growth was stable. Not all land cover categories were separated and classified, but only three typologies were considered: vegetation, water, and empty land without vegetation. When calculating the change in vegetation cover, forest, cropland, shrubland, grassland, and spare vegetation types were included in the vegetated area. But, in the water section, water bodies, wetlands, permanent ice, and snow were included. In the bare areas section, impervious surfaces, bare areas, and consolidated bare area levels were calculated (Figure 7).

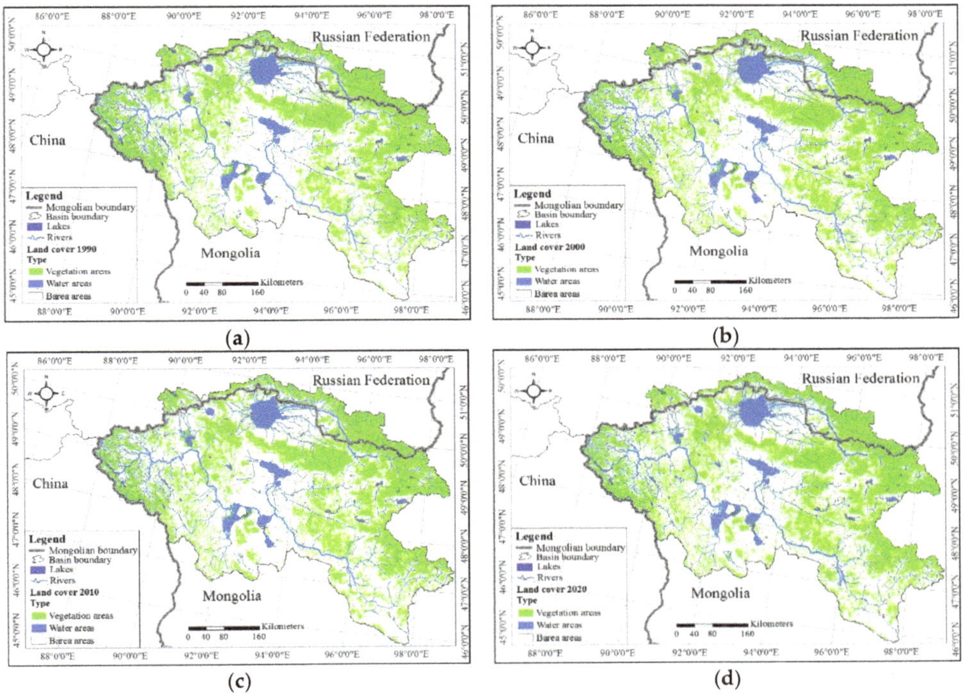

Figure 7. Map of land cover changes in the Great Lakes Depression region: (**a**) 1990 land cover map; (**b**) 2000 land cover map; (**c**) 2010 land cover map; and (**d**) 2020 land cover map.

Vegetation change in land cover showed relatively little spatial variation. Vegetated surface area decreased in 2000 and increased by 2.4% between 2010 and 2020 due to changes in air temperature and precipitation, while water surface area decreased by 1.2% from 1990 to 2020. Bare area also decreased by 3.1% from 1990 to 2020. Vegetation and wetlands, as well as bare areas, changed slightly in relatively small areas between 1990 and 2020 (Table 2).

The land cover change increased from 65% to 66.9% from 1990 to 2020 by almost 2% in areas with vegetation area. The amount of water increased by 1.2%, from 3.7% to 4.9%. Bare area decreased by 3.1%, from 31.3% to 28.2%. This change was similar to the changes in NDVI and EVI.

Table 2. Area and percentage of land cover change.

Land Cover Type	1990		2000		2010		2020	
	Area (km^2)	(%)	Area (km^2)	(%)	Area (km^2)	(%)	Area (km^2)	(%)
Vegetation area	174,452.7	65.0	173,673.8	64.7	175,344.8	65.4	179,485.9	66.9
Water area	9829.5	3.7	10,978.4	4.1	11,276.0	4.2	13,250.0	4.9
Bare area	84,027.3	31.3	83,657.3	31.2	81,688.7	30.4	75,573.7	28.2
Total area	268,309.5	100	268,309.5	100	268,309.5	100	268,309.6	100

3.4. Statistical Analysis

Considering the time and space relationship between vegetation dynamics and hydro-climatic, the relationship between the most sensitive parameters studied was confirmed using statistical analysis. For this purpose, the growth area of vegetation cover was divided into three main areas in the study area: low vegetation area, medium vegetation area, and high vegetation area, and the relationship between air temperature, precipitation, and river flow was calculated in each section (Figure 8).

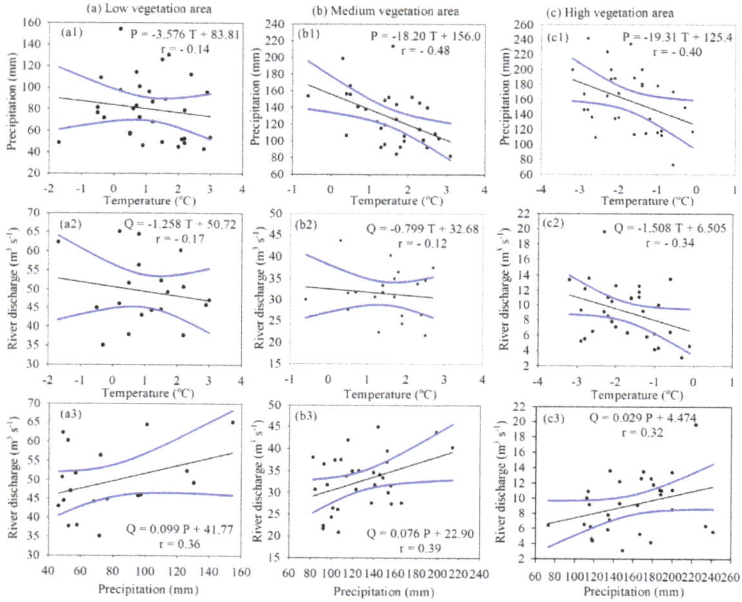

Figure 8. Relationship between hydro-climatic (temperature, precipitation, and river discharge) factors in the Great Lakes Depression region: (**a**) low vegetation area; (**b**) medium vegetation area; and (**c**) high vegetation area.

When considering the relationship between air temperature and precipitation by dividing the study area into low vegetation area (a1–a3), medium vegetation area (b1–b3), and high vegetation area (c1–c3), when the air temperature increases, the pattern of gradual decrease in precipitation was observed at all points. Considering the relationship between precipitation and vegetation cover, vegetation cover increases as precipitation increases. For example, the correlation in (a1) became r = −0.14, while the correlation in (b1) increased to r = −0.48. Vegetation changes are most closely related to climate factors. Considering how the river flow depends on the difference in air temperature, a relatively weak relationship is revealed. The highest is r = −34 in areas with high vegetation area. This indicates that the air temperature in areas with high vegetation area may have an indirect effect on the flow of river water. Considering how the river discharge water depends on the increase in precipitation in the low vegetation area (a1–a3), medium vegetation area (b1–b3), and high vegetation area (c1–c3) in the research area, it is positively moderate at all points where related. The river discharge in the study area is affected by many factors, such as increasing air temperature, the melting of high mountain permafrost, and changing precipitation, and it can be seen that the influence of precipitation is significant. Therefore, in arid and semi-arid regions, the amount of precipitation positively and strongly influences the river discharge. The increase in air temperature will increase evaporation and change precipitation, and the sum of air temperature and precipitation will significantly impact water and vegetation growth that year.

4. Discussion

In the Great Lakes Depression region in western Mongolia, we can detect the changes in vegetation cover related to climate change. Studying these changes may provide opportunities for the development of appropriate land management techniques and predict how vegetation may change in a warmer future. Climate change and its associated hydrologic changes are taking place very strongly in the Great Lakes Depression region of Mongolia [32]. In this region, climate indicators such as temperature increase, precipitation decrease, water surface area decrease, and permafrost glacier area decrease have all been observed in recent years [33]. In particular, due to increased air temperature, snow and glaciers have steadily decreased in recent years [34], leading to less surface water.

During the last 30 years, the air temperature in the basin has warmed by 1 °C from −0.5 °C to +0.5 °C. From 1995 to 2005, the air temperature gradually increased, and after 2008, the warming intensity stabilized slightly. This increase in temperature increases the melting of high mountain permafrost and glaciers. The meltwater flows through surface water, like rivers, and accumulates in low-lying areas. The sudden increase in air temperature in the spring and summer seasons increases the melting of permafrost and glaciers on high mountain tops, causing a sharp rise in river water levels and the expansion of lake water areas in the early season. Conversely, considering the general trend of hydro-climate, the increase in air temperature in this region will decrease the amount of precipitation, and the decrease in precipitation impacts the flow of water in the river. Despite more water earlier in the season, average lake levels steadily declined over time. The change in the water consumption of the river likely also caused the water level of the lake to change. In our study, the decline in precipitation was most strongly linked to lower lake levels.

The expansion and reduction in vegetation cover are intimately connected to alterations in both precipitation and air temperature [35]. Changes in air temperature, the most prominent climate parameter, will indirectly have the most significant impact on vegetation growth. These changes are especially evident in sensitive areas of arid and semi-arid regions [36]. For example, in 2000, when the air temperature increased, the amount of precipitation decreased, and in the years 2010 and 2020, when the precipitation increased, the amount of vegetation cover increased in the basin, especially in the years when the NDVI and EVI values were high. An increase in the sensitivity of plants to water availability could lead to a decrease in tolerance to water changes and an increase in the conflict between plants and human societies related to water [37].

The processing of satellite images is useful in many ways when considering the factors influencing the detection of changes in vegetation cover [38,39]. Consideration of changes in land cover changes in vegetated surfaces, water surfaces, and bare land without vegetation is important for estimating the impact of vegetation cover changes [40]. The biggest change in the area of the basin was the bare area between 1990 and 2020, which was 8453.6 km^2 or 3.1% of the total area of the basin. This area is quite extensive, especially for arid and semi-arid regions. While the area of wetlands in that area is increasing by a small percentage, the bare area, land without vegetation, is increasing, which may indicate a strong natural transition due to natural factors and other factors, especially climate.

5. Conclusions

In this study, hydro-climatic trend analysis, satellite image analysis, land cover determination, and statistical analysis methods were used to calculate the spatio-temporal relationship between vegetation dynamics and hydro-climate changes in the Great Lakes Depression region of Mongolia from 1990 to 2020.

The average annual air temperature increased by +1 °C from −0.5 °C to +0.5 °C during the research period. This increase substantially affected plant growth processes but differed depending on elevation and proximity to water. Mean annual air temperature tended to increase at all stations ($Z = +1.16$). The mean yearly precipitation within the research amounted to 134.6 mm. The amount of precipitation rapidly decreased from 1994 to 2010, and from 2000 to 2014, it decreased with statistical significance. During the study period, the river discharge changed significantly due to the interrelationship between the increased air temperature and the decrease in precipitation. During the study period, average river discharge significantly decreased ($Z = -3.51$). Also, the water level of the studied lake significantly decreased ($Z = -2.03$).

Vegetation cover changes due to hydro-climatic changes were different in high mountains, near the water surface of rivers and lakes, and in arid and semi-arid regions. Vegetation yield is determined by changes in air temperature and precipitation of the same year, and the amount of precipitation in the summer months from 2000 to 2010 and 2020 was high, so the increase in the height of vegetation cover during that period was relatively good. That is why the amount of vegetation cover in NDVI, EVI, and land cover types tends to increase during this period. This change in vegetation cover is one of the paradoxes of climate change in the region that we must continue to study.

The growth of vegetation cover is highly dependent on hydro-climatic changes in the study area. In terms of distance, mainly water and vulnerable parts of vegetation cover in arid and semi-arid areas have changed. The most sensitive areas to climate change were those with moderate-to-high vegetation cover. This suggests that climate change significantly affects changes in vegetation cover in studies conducted in arid and semi-arid regions of Central Asia.

This research can serve as the primary data for researching the relationship between climate change, vegetation cover, land cover change, and ecosystem change in the Great Lakes Depression region of Mongolia, which is representative of the semi-arid region. In the future, studying human factors affecting vegetation cover is essential, as is studying human responses to vegetation change in the region.

Author Contributions: Conceptualization, B.D. and D.Y.; methodology, B.D.; software, N.B. and S.D.; formal analysis, B.D.; investigation, V.A.Z.; resources, S.D.; data curation, N.B.; writing—original draft preparation, B.D.; writing—review and editing, B.D.; visualization, H.Z.; supervision, V.A.Z.; project administration, B.D.; funding acquisition, D.Y. All authors have read and agreed to the published version of the manuscript.

Funding: This research was funded by the Mongolian Science and Technology Foundation (Grant: CHN-2022/274) and the Advanced Grant from the National University of Mongolia (Grant: P2022-4376). Also, it was funded by the National Key Research and Development Project of China (Grant: 2022YFE0119400). And this research was funded by the Chinese Academy of Sciences PIFI Fellowship for Visiting Scientists (Grant: 2024VCB0010).

Data Availability Statement: Data sharing not applicable.

Acknowledgments: Special thanks to reviewers and editors who reviewed and provided valuable feedback on the quality of the article's final version.

Conflicts of Interest: The authors declare no conflict of interest.

References

1. Jiang, W.; Niu, Z.; Wang, L.; Yao, R.; Gui, X.; Xiang, F.; Ji, Y. Impacts of Drought and Climatic Factors on Vegetation Dynamics in the Yellow River Basin and Yangtze River Basin, China. *Remote Sens.* **2022**, *14*, 930. [CrossRef]
2. Barhoum, A.; Jeevanandam, J.; Rastogi, A.; Samyn, P.; Boluk, Y.; Dufresne, A.; Danquah, M.K.; Bechelany, M. Plant celluloses, hemicelluloses, lignins, and volatile oils for the synthesis of nanoparticles and nanostructured materials. *Nanoscale* **2020**, *12*, 22845–22890. [CrossRef] [PubMed]
3. Tramblay, Y.; Koutroulis, A.; Samaniego, L.; Vicente-Serrano, S.M.; Volaire, F.; Boone, A.; Le Page, M.; Llasat, M.C.; Albergel, C.; Burak, S.; et al. Challenges for drought assessment in the Mediterranean region under future climate scenarios. *Earth Sci. Rev.* **2020**, *210*, 103348. [CrossRef]
4. Jiang, L.; Jiapaer, G.; Bao, A.; Kurban, A.; Guo, H.; Zheng, G.; De Maeyer, P. Monitoring the long-term desertification process and assessing the relative roles of its drivers in Central Asia. *Ecol. Indic.* **2019**, *104*, 195–208. [CrossRef]
5. Kappas, M.; Degener, J.; Klinge, M.; Vitkovskaya, I.; Batyrbayeva, M. A Conceptual Framework for Ecosystem Stewardship Based on Landscape Dynamics: Case Studies from Kazakhstan and Mongolia. In *Landscape Dynamics of Drylands across Greater Central Asia: People, Societies and Ecosystems*; Gutman, G., Chen, J., Henebry, G.M., Kappas, M., Eds.; Springer International Publishing: Cham, Switzerland, 2020; pp. 143–189. [CrossRef]
6. Ahlborn, J.; von Wehrden, H.; Lang, B.; Römermann, C.; Oyunbileg, M.; Oyuntsetseg, B.; Wesche, K. Climate—Grazing interactions in Mongolian rangelands: Effects of grazing change along a large-scale environmental gradient. *J. Arid Environ.* **2020**, *173*, 104043. [CrossRef]
7. Hua, L.; Zhao, T.; Zhong, L. Future changes in drought over Central Asia under CMIP6 forcing scenarios. *J. Hydrol. Regi. Stud.* **2022**, *43*, 101191. [CrossRef]
8. Foroumandi, E.; Nourani, V.; Dąbrowska, D.; Kantoush, S.A. Linking Spatial—Temporal Changes of Vegetation Cover with Hydroclimatological Variables in Terrestrial Environments with a Focus on the Lake Urmia Basin. *Land* **2022**, *11*, 115. [CrossRef]
9. Yuchen, L.; Zongxing, L.; Xiaoping, Z.; Juan, G.; Jian, X. Vegetation variations and its driving factors in the transition zone between Tibetan Plateau and arid region. *Ecol. Indic.* **2022**, *141*, 109101. [CrossRef]
10. Zhang, T.; Xu, X.; Jiang, H.; Qiao, S.; Guan, M.; Huang, Y.; Gong, R. Widespread decline in winds promoted the growth of vegetation. *Sci. Total Environ.* **2022**, *825*, 153682. [CrossRef]
11. Noori, R.; Maghrebi, M.; Mirchi, A.; Tang, Q.; Bhattarai, R.; Sadegh, M.; Noury, M.; Torabi Haghighi, A.; Kløve, B.; Madani, K. Anthropogenic depletion of Iran's aquifers. *Proc. Natl. Acad. Sci. USA* **2021**, *118*, e2024221118. [CrossRef]
12. Maghrebi, M.; Noori, R.; Mehr, A.D.; Lak, R.; Darougheh, F.; Razmgir, R.; Farnoush, H.; Taherpour, H.; Moghaddam, S.M.R.A.; Araghi, A.; et al. Spatiotemporal changes in Iranian rivers' discharge. *Elem. Scie. Anthr.* **2023**, *11*, 00002. [CrossRef]
13. Lei, C.; Wagner, P.D.; Fohrer, N. Effects of land cover, topography, and soil on stream water quality at multiple spatial and seasonal scales in a German lowland catchment. *Ecol. Indic.* **2021**, *120*, 106940. [CrossRef]
14. Tola, S.Y.; Shetty, A. Land cover change and its implication to hydrological regimes and soil erosion in Awash River basin, Ethiopia: A systematic review. *Environ. Monit. Assess.* **2021**, *193*, 836. [CrossRef] [PubMed]
15. Chen, H.; Liu, H.; Chen, X.; Qiao, Y. Analysis on impacts of hydro-climatic changes and human activities on available water changes in Central Asia. *Sci. Total Environ.* **2020**, *737*, 139779. [CrossRef] [PubMed]
16. Javadinejad, S.; Eslamian, S.; Ostad-Ali-Askari, K. The analysis of the most important climatic parameters affecting performance of crop variability in a changing climate. *Int. J. Hydro. Scie. Techno.* **2021**, *11*, 1–25. [CrossRef]
17. Guevara-Ochoa, C.; Medina-Sierra, A.; Vives, L. Spatio-temporal effect of climate change on water balance and interactions between groundwater and surface water in plains. *Sci. Total Environ.* **2020**, *722*, 137886. [CrossRef] [PubMed]
18. Zhao, Y.; Miao, Y.; Fang, Y.; Li, Y.; Lei, Y.; Chen, X.; Dong, W.; An, C. Investigation of factors affecting surface pollen assemblages in the Balikun Basin, central Asia: Implications for palaeoenvironmental reconstructions. *Ecol. Indic.* **2021**, *123*, 107332. [CrossRef]
19. Baasanmunkh, S.; Oyuntsetseg, B.; Urgamal, M.; Norris, J.; Shiga, T.; Choi, H.J. Notes on the taxonomy of Nymphaeaceae and Menyanthaceae in Mongolia. *J. Asia-Pac. Biodivers.* **2022**, *15*, 129–137. [CrossRef]
20. Baasanmunkh, S.; Batlai, O.; Tsegmed, Z.; Khurelpurev, O.; Magsar, U.; Batdelger, G.; Chuluunbat, J.; Nyamjantsan, N.; Petr, K.; Jae, C.H. Distribution of vascular plants in Mongolia—I Part. *Mong. J. Biol. Sci.* **2022**, *20*, 3–28. [CrossRef]
21. Zhang, X.; Liu, L.; Chen, X.; Gao, Y.; Xie, S.; Mi, J. GLC_FCS30: Global land-cover product with fine classification system at 30 m using time-series Landsat imagery. *Earth Syst. Sci. Data* **2021**, *13*, 2753–2776. [CrossRef]
22. Dorjsuren, B.; Batsaikhan, N.; Yan, D.; Yadamjav, O.; Chonokhuu, S.; Enkhbold, A.; Qin, T.; Weng, B.; Bi, W.; Demberel, O.; et al. Study on Relationship of Land Cover Changes and Ecohydrological Processes of the Tuul River Basin. *Sustainability* **2021**, *13*, 1153. [CrossRef]
23. Dorjsuren, B.; Yan, D.; Wang, H.; Chonokhuu, S.; Enkhbold, A.; Yiran, X.; Girma, A.; Gedefaw, M.; Abiyu, A. Observed Trends of Climate and River Discharge in Mongolia's Selenga Sub-Basin of the Lake Baikal Basin. *Water* **2018**, *10*, 1436. [CrossRef]
24. Li, X.; Fang, G.; Wen, X.; Xu, M.; Zhang, Y. Characteristics analysis of drought at multiple spatiotemporal scale and assessment of CMIP6 performance over the Huaihe River Basin. *J. Hydrol. Reg. Stud.* **2022**, *41*, 101103. [CrossRef]
25. Dorjsuren, B.; Yan, D.; Wang, H.; Chonokhuu, S.; Enkhbold, A.; Davaasuren, D.; Girma, A.; Abiyu, A.; Jing, L.; Gedefaw, M. Observed trends of climate and land cover changes in Lake Baikal basin. *Environ. Earth Sci.* **2018**, *77*, 725. [CrossRef]
26. Sen, P.K. Estimates of the Regression Coefficient Based on Kendall's Tau. *J. Am. Stat. Assoc.* **1968**, *63*, 1379–1389. [CrossRef]

27. Rouse, J.W.; Haas, R.H.; Schell, J.A.; Deering, D.W. Monitoring vegetation systems in the Great Plains with ERTS. *NASA Spec. Publ.* **1974**, *351*, 309–313.
28. Barboza, T.O.C.; Ardigueri, M.; Souza, G.F.C.; Ferraz, M.A.J.; Gaudencio, J.R.F.; Santos, A.F.d. Performance of Vegetation Indices to Estimate Green Biomass Accumulation in Common Bean. *Agric. Eng.* **2023**, *5*, 840–854. [CrossRef]
29. Ray, R.; Das, A.; Hasan, M.S.U.; Aldrees, A.; Islam, S.; Khan, M.A.; Lama, G.F.C. Quantitative Analysis of Land Use and Land Cover Dynamics using Geoinformatics Techniques: A Case Study on Kolkata Metropolitan Development Authority (KMDA) in West Bengal, India. *Remote Sens.* **2023**, *15*, 959. [CrossRef]
30. Ren, Z.; Tian, Z.; Wei, H.; Liu, Y.; Yu, Y. Spatiotemporal evolution and driving mechanisms of vegetation in the Yellow River Basin, China during 2000–2020. *Ecol. Indic.* **2022**, *138*, 108832. [CrossRef]
31. Xue, L.; Kappas, M.; Wyss, D.; Wang, C.; Putzenlechner, B.; Thi, N.P.; Chen, J. Assessment of Climate Change and Human Activities on Vegetation Development in Northeast China. *Sensors* **2022**, *22*, 2509. [CrossRef]
32. Demberel, O.; Munkhbat, B.; Dorjsuren, B.; Callaghan, T.V.; Tsogoo, B.; Zemtsov, V.A.; Shaarav, O.; Gongor, E.; Jargalsaikhan, Z.; Ganhuyag, N.; et al. Relationship between Dynamics of Modern Glaciers of the Mt. Munkhkhairkhan (Mongolian Altai) and Climate. *Water* **2023**, *15*, 1921. [CrossRef]
33. Klinge, M.; Schlütz, F.; Zander, A.; Hülle, D.; Batkhishig, O.; Lehmkuhl, F. Late Pleistocene lake level, glaciation and climate change in the Mongolian Altai deduced from sedimentological and palynological archives. *Quat. Res.* **2020**, *99*, 168–189. [CrossRef]
34. Pan, C.G.; Kamp, U.; Munkhjargal, M.; Halvorson, S.J.; Dashtseren, A.; Walther, M. An Estimated Contribution of Glacier Runoff to Mongolia's Upper Khovd River Basin in the Altai Mountains. *Mt. Res. Dev.* **2019**, *39*, R12–R20. [CrossRef]
35. Qu, S.; Wang, L.; Lin, A.; Yu, D.; Yuan, M.; Li, C.a. Distinguishing the impacts of climate change and anthropogenic factors on vegetation dynamics in the Yangtze River Basin, China. *Ecol. Indic.* **2020**, *108*, 105724. [CrossRef]
36. Li, G.; Yu, L.; Liu, T.; Jiao, Y.; Yu, J. Modeling Potential Impacts on Regional Climate Due to Land Surface Changes across Mongolia Plateau. *Remote Sens.* **2022**, *14*, 2947. [CrossRef]
37. Jiang, T.; Wang, X.; Afzal, M.M.; Sun, L.; Luo, Y. Vegetation Productivity and Precipitation Use Efficiency across the Yellow River Basin: Spatial Patterns and Controls. *Remote Sens.* **2022**, *14*, 5074. [CrossRef]
38. Talukdar, S.; Singha, P.; Mahato, S.; Shahfahad; Pal, S.; Liou, Y.-A.; Rahman, A. Land-Use Land-Cover Classification by Machine Learning Classifiers for Satellite Observations—A Review. *Remote Sens.* **2020**, *12*, 1135. [CrossRef]
39. Alvarez-Vanhard, E.; Corpetti, T.; Houet, T. UAV & satellite synergies for optical remote sensing applications: A literature review. *Sci. Remote Sens.* **2021**, *3*, 100019.
40. Xie, Z.; Phinn, S.R.; Game, E.T.; Pannell, D.J.; Hobbs, R.J.; Briggs, P.R.; McDonald-Madden, E. Using Landsat observations (1988–2017) and Google Earth Engine to detect vegetation cover changes in rangelands—A first step towards identifying degraded lands for conservation. *Remote Sens. Environ.* **2019**, *232*, 111317. [CrossRef]

water

MDPI

Article

Water Protection Zones—Impacts on Weed Vegetation of Arable Soil

Jan Winkler [1], Tomáš Řičica [1], Věra Hubačíková [2], Eugeniusz Koda [3], Magdalena Daria Vaverková [2,3], Ladislav Havel [1] and Mariusz Żółtowski [3,*]

1 Department of Plant Biology, Faculty of AgriSciences, Mendel University in Brno, Zemědělská 1, 613 00 Brno, Czech Republic; jan.winkler@mendelu.cz (J.W.); tomas.ricica@mendelu.cz (T.Ř.); ladislav.havel@mendelu.cz (L.H.)

2 Department of Applied and Landscape Ecology, Faculty of AgriSciences, Mendel University in Brno, Zemědělská 1, 613 00 Brno, Czech Republic; vera.hubacikova@mendelu.cz (V.H.); magdalena.vaverkova@mendelu.cz or magdalena_vaverkova@sggw.edu.pl (M.D.V.)

3 Department of Revitalization and Architecture, Institute of Civil Engineering, Warsaw University of Life Sciences, Nowoursynowska 159, 02 776 Warsaw, Poland; eugeniusz_koda@sggw.edu.pl

* Correspondence: mariusz_zoltowski@sggw.edu.pl

Abstract: The aim of this study is to evaluate the occurrence of weeds under conditions of limited herbicide use due to the protection zone of water resources. A total of 23 weed species were found in maize stands, 19 species were found in wheat stands, and 16 species were found in rapeseed stands. The redundancy analysis (RDA) results show significant differences in weed occurrence and composition due to herbicide regulation in each crop. Changes in weed composition induced by herbicide application limitations lead to a preference for more specialized weed species (specialists) at the expense of widespread species (generalists). Limiting the use of pesticides in sensitive and vulnerable areas, such as water sources, bodies, and watercourses, is justified from the perspective of protecting the aquatic environment and biodiversity. However, such measures can cause weed growth that is difficult to control, and therefore, it is important to search for new methods for weed control in field crops. Determining a balance between safeguarding water resources and addressing agricultural challenges remains crucial for sustainable land and water management.

Keywords: water protection; herbicides; biodiversity; agriculture; weed flora

1. Introduction

Human activity is often the primary source of pollution, bringing potentially hazardous substances into atmospheric, terrestrial, and aquatic ecosystems [1]. Agriculture is one of the largest sources of emissions in surface water and aquatic ecosystems [2,3]. Hoever, aquatic ecosystems are often contaminated with herbicides [4]. This contamination depends on many parameters, such as watershed structure and land use (anthropic activity, surface, and slope), soil structure and composition, environmental factors, hydrology (rainfall, flow, and wind), aquatic environmental conditions (pH, organic matter, and suspended particles), and the chemical properties of herbicides [5,6]. Many European studies have reported the detection of active herbicides such as diuron, metolachlor, isoproturon, terbutryn, and atrazine in aquatic environments [7,8]. Owing to their phytotoxic effects, their behavior in freshwater ecosystems poses a risk to non-target organisms, such as benthic diatoms (*Diatomeae*) [9]. Herbicides generally inhibit various vital functions in photosynthetic organisms [10]. Weeds have been suppressed since the beginning of agriculture because they are the main biotic cause of yield losses in field crops [11]. A revolution in weed control started in the early 1950s, owing to the use of herbicides, which laid the foundation for industrial agriculture [12,13]. In recent years, increased use of herbicides has been reported in various countries [14–16]. During the 12 years of the monitoring of pilot farms in Switzerland, an increase from 0.034 kg/ha to 0.141 kg/ha in glyphosate use in arable land was recorded [17].

The use of pesticides is one of the most debated aspects of agricultural intensification, given their potential direct and indirect consequences at the individual, population, and ecosystem levels [18,19]. The direct toxic effects of pesticides can appear quickly after their application [20], whereas the indirect effects of pesticides can occur one or several years later [21,22]. The toxicity of pesticides can vary substantially among affected organisms [23,24], and the ecological impacts of pesticides can persist several years after their application [25]. Herbicide use can also reduce shelter for bees, insects, reptiles, etc., in agricultural landscapes [26–28].

The leakage of herbicides, or the concept of herbicide loss, is based on spatial heterogeneity, where hydrological factors of the environment dominate, for example, the nature of the first rain event after herbicide application. Several studies have demonstrated that factors such as topography and soil hydrological properties can clarify the spatial patterns of herbicide loss [29]. Some herbicides can be transported from agricultural land to the surface and groundwater, where they can affect the ecosystem [30–32].

Pesticides, together with nitrates, currently form the main pollution sources of drinking water in Europe [33]. To limit or at least mitigate the negative effects of agricultural emissions on water, in recent decades, the European Union (EU) has developed an extensive and interconnected regulatory political framework that influences water management, agricultural practices, and environmental protection [34]. For example, the Nitrates Directive (91/676/EEC) [35] builds on the objectives of the Drinking Water Directive (98/83/EC) [36] and the Water Framework Directive (2000/60/EC) [37] to form an overarching framework for EU directives on the specific functions of water, the use of chemicals, and their effect on environmental protection and the state of European waters. Apart from international treaties and EU law, the basic legal regulation in the Czech Republic (CR) governing water protection is Act No. 254/2001 Sb. [38] (water law). Although the directives are linked, their implementation has side effects that prevent the effective protection of drinking water sources. For example, fertilizer use rules are not always beneficial to groundwater and drinking water quality [34]. The adopted EU framework directives reflect a growing awareness of the complexity of the water and river basin issue [39–41]. The new provisions related to the Water Directive require interdisciplinary cooperation within individual EU member states and international cooperation [34]. The pesticide regulatory framework in Europe creates strong incentives for growers to reduce herbicide application. In Denmark, since the late 1980s, the demand for herbicides has been increasing; therefore, several pesticide action plans have been initiated [42].

In CR water protection, the use of water and the right to access it are regulated by the Water Act [38]. The Water Act is specified or elaborated upon by subsequent regulations (government regulations and decrees) issued by the Ministry of the Environment together with the Ministry of Agriculture. Subsequent regulations have defined zones of hygienic protection for surface water sources, including rules and restrictions for the use of pesticides.

It should be emphasized that the pollution of aquatic ecosystems by herbicides and pesticides has far-reaching consequences for biodiversity, ecosystem services, and human health [43–45]. Pesticides can also negatively affect the environment in ponds, rivers, and other freshwater and marine habitats [46,47].

Restrictions on the application of herbicides in areas of surface water protection have led to a limited choice of herbicides. Different weed regulations create diverse selection pressures on weed vegetation. The aims of this study are (i) to assess the effect of herbicide limitations on the species composition of weeds in selected crops, (ii) to determine the weed species that meet the conditions of limited weed regulation, and (iii) to determine the proportions of neophytes and species with invasive status. Water surface protection based on herbicide use reduction can lead to challenges in the regulation of certain weed species. Increased weed infestation by specific types of weeds without effective regulation can exacerbate conflict between farmers and water management authorities.

2. Materials and Methods

2.1. Study Area

Selected plots are in the cadastral territory of Zarazice (South Moravian Region, Czech Republic). The monitored agricultural holding operates over an area of 200 ha and includes flint maize (*Zea mays*), winter wheat (*Triticum aestivum*), winter rapeseed (*Brassica napus*), spring barley (*Hordeum vulgare*), and alfalfa (*Medicago sativa*).

The study plots are situated in two geomorphologically different parts. Plots in the southern part of the study area lie near the Morava River. The soil types are mainly fluvial soils formed from water sediments of the river, and the altitude is 170 m above sea level. This area falls within the groundwater protection zone, where the use of herbicides listed in the regulations is restricted. The plots in the eastern part of the study area are located at an altitude of 220 m above sea level. The most common soil types are cambizemes and regozemes. Both parts belong to the area, which is characterized by an average annual temperature exceeding 10 °C with an annual rainfall of up to 500 mm [48–50]. The Morava River flows through the center of the cadastral territory of Zarazice. Furthermore, there is an artificial water canal—the Baťův kanál on the right bank of the Morava—which is used for recreation purposes. The western border of the cadastral territory is formed by the relief arm of the Morava River, which was built in the 1930s to divert flood flows. The entire area between the relief arm and the Baťa canal is delimited as a protection zone of the water source–Bzenec complex. In this complex, underground water is collected from wells and serves as a mass source of drinking water. The investigated plots in the southern part fall into the second degree Protection Zone, in which the requirements and prohibitions based on the Decision on the Determination of Water Resource Protection Zones from 1990 are respected.

No herbicide limitations (full regulation): In the plots, it is permissible to use all registered herbicides that are approved for the given crop, and their application is governed by general regulations for the use of plant protection products. Herbicides applied to maize crops include BALATON (Terbuthylazine) at a rate of 3.0 L/ha and STORY (Florasulam and Mesotrione) at a rate of 0.3 L/ha. Herbicides applied to winter wheat crops consist of GLEAN 75 PX (Chlorsulfuron) at a rate of 0.02 kg/ha and HURICANE (Aminopyralid, Florasulam, and Pyroxsulam) at a rate of 0.2 kg/ha. Herbicides applied to winter rapeseed crops include BUTISAN STAR (Chlormequat and Metazachlor) at a rate of 2.0 L/ha, GARLAND FORTE (Propaquizafop) at a rate of 0.8 L/ha, and GARLAND FORTE (Propaquizafop) at a rate of 0.5 L/ha.

Herbicide limitations (limitation): Within the designated protective zone plots, chemical weed control is restricted. Only herbicides with granted exceptions are allowed to be used on these plots. Herbicides applied to maize crops include EQUIP ULTRA (foramsulfuron) at a rate of 2.0 L/ha and STORY (Florasulam and Mesotrione) at a rate of 0.3 L/ha. Herbicides applied to winter wheat crops consist of HURICANE (Aminopyralid, Florasulam, and Pyroxsulam) at a rate of 0.2 kg/ha. Herbicides applied to winter rapeseed crops include GLEAN 75 PX (Chlorsulfuron) at a rate of 0.02 kg/ha and HURICANE (Aminopyralid, Florasulam, and Pyroxsulam) at a rate of 0.2 kg/ha.

2.2. Method of Vegetation Assessment

Weed assessments were performed in six plots. Three of these are located in the surface water protection zone. The other three plots were located in a zone in which herbicide use was not restricted. Flint maize, winter wheat, and winter rapeseed were grown in the plots. The plots' characteristics are listed in Table 1. Weeds were assessed using a numerical method. Eight sampling areas with a size of 1 m^2 were demarcated on each monitored plot. Sampling areas were evenly distributed in the plots. After demarcating the sampling area, the weed species were identified and counted. The taxonomic nomenclature of the plants follows Kaplan et al. [51]. The observations were conducted between March and June 2018.

Table 1. Characteristics of selected plots.

Water Protection Zone	Crop	Surface Area (ha)	GPS
No herbicide limitations (full regulation)	Maize	7.12	48.9282369 N, 17.3842381 E
	Winter wheat	11.11	48.9334244 N, 17.3749683 E
	Rapeseed	9.08	48.9294775 N, 17.3726725 E
Herbicide limitations (limitation)	Maize	7.07	48.9418253 N, 17.3554419 E
	Winter wheat	15.95	48.9441083 N, 17.3509789 E
	Rapeseed	8.05	48.9502931 N, 17.3555419 E

The identified weed species were categorized into groups based on their biological characteristics. The spring weed group includes annual species that germinate at temperatures of 0–8 °C and do not survive the winter. The summer weed group encompasses annual species that germinate at temperatures of 8 °C and higher and do not survive the winter period. The winter weed group includes annual species that survive winter. The perennial weed group comprises species with multiple growing seasons over several years. The weeding crops group encompasses crops that behave like weeds and grow within other crops as a consequence of previous cropping.

The results were processed using multivariate analyses of ecological data. The first analysis performed was the segmental Detrended Correspondence Analysis (DCA), which calculated the length of the gradient (Lengths of Gradient) and was followed by RDA. Statistical significance was tested using the Monte Carlo test with 999 calculated permutations. All multivariate analyses were performed using CANOCO 5.0 [52]. The necessary calculations were performed using Canoco 5.0 computer program.

3. Results

A total of 23 weed taxa were identified in corn stands, 19 were identified in winter wheat stands, and 16 were identified in winter rapeseed stands within the assessment. The average number of weeds is shown in Figure 1. The representations of the individual weed taxa are shown in Figure 2 (for maize stands), Figure 3 (for wheat stands), and Figure 4 (for rapeseed stands).

The results of the RDA, which evaluated the number of individuals of weed taxa in wheat stands, were significant at the level of significance of $\alpha = 0.063$ for all canonical axes. However, the differences in weed abundance were statistically inconclusive. A graphical illustration of the RDA results is shown in Figure 6.

The results of the RDA, which evaluated the number of individuals of weed taxa in maize, were significant at the significance level of $\alpha = 0.023$ for all canonical axes. The differences in weed abundance were statistically significant. A graphical illustration of the results of the RDA is shown in Figure 5.

The results of the RDA, which assessed the number of weed taxa in rapeseed stands, were significant at the significance level of $\alpha = 0.023$ for all canonical axes. The differences in weed abundance were statistically significant. A graphical illustration of the RDA is shown in Figure 7.

Based on the RDA, the identified plant taxa were divided into three groups. The distribution of taxa into groups according to the RDA is shown in Table 2.

The results presented in Figure 1 demonstrate significant differences among the weed groups within the monitored crops. However, the representation of weed groups under the conditions of varied herbicide regulation limitations was not as pronounced.

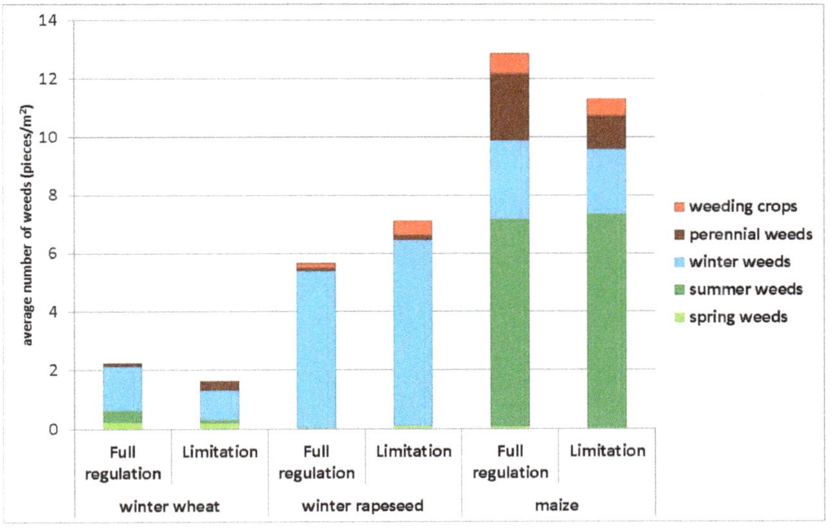

Figure 1. Average number of weeds in monitored crops.

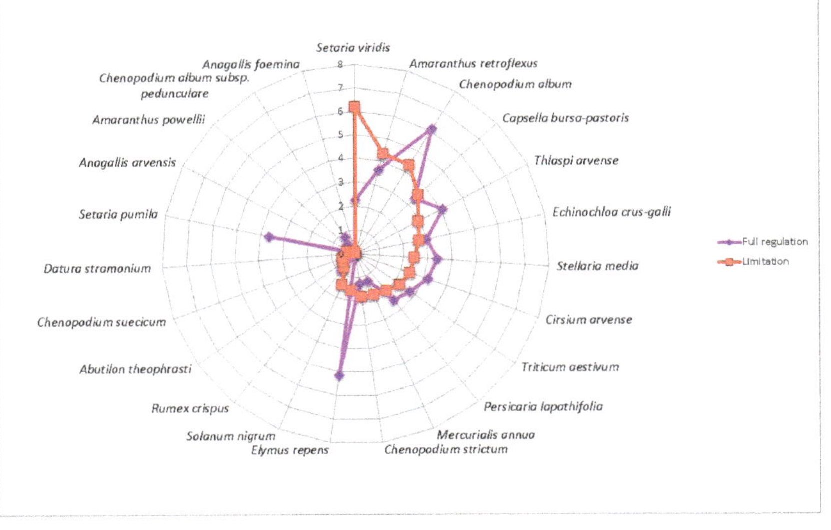

Figure 2. Representation of weed taxa found in maize stands (pieces/m^2).

From the results presented in Figures 2 and 5, it is evident that the limitation of herbicide usage alters the weed species composition in maize. In maize crops, the use of only specific herbicides (limitation) creates more favorable conditions for weed species, such as *Setaria viridis*, *Mercurialis annua*, *Chenopodium strictum*, *Solanum nigrum*, *Abutilon theophrasti*, *Datura stramonium*, and *Chenopodium suecicum*. Under full regulation conditions, weed species such as *Chenopodium album*, *Setaria pumila*, and *Elymus repens* find it easier to establish themselves.

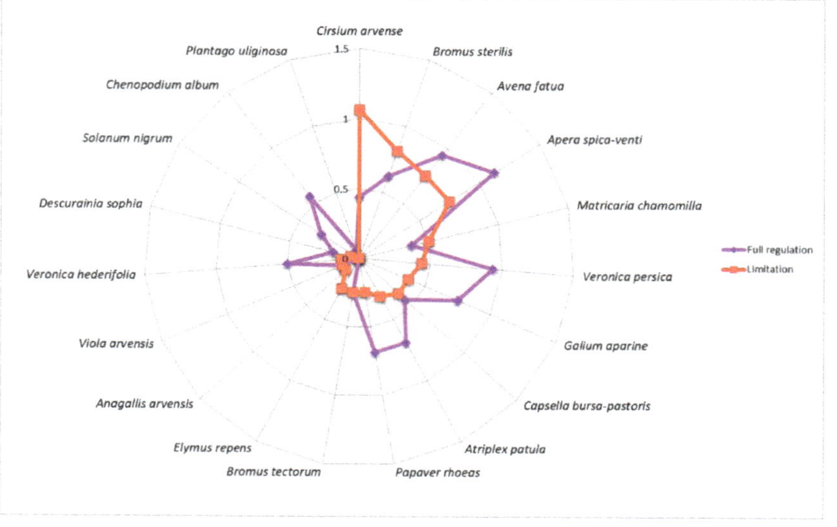

Figure 3. Representation of weed taxa found in wheat stands (pieces/m^2).

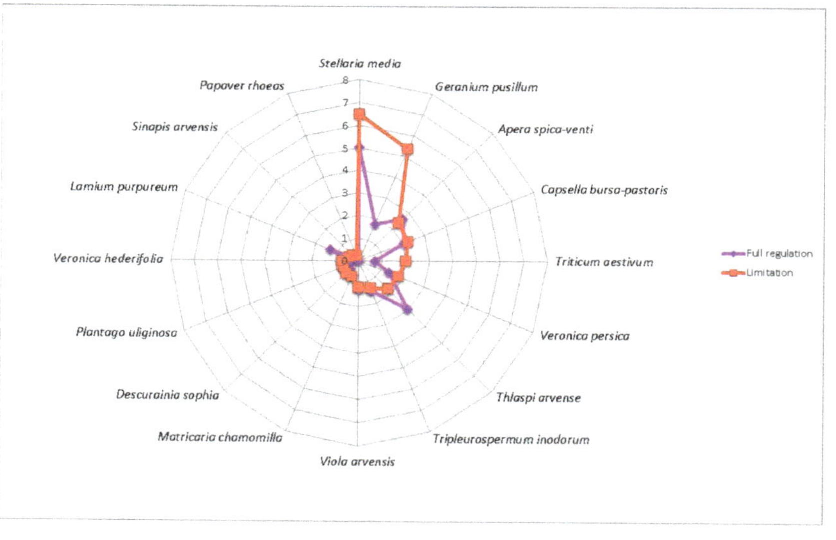

Figure 4. Representation of weed taxa found in rapeseed stands (pieces/m^2).

From the results depicted in Figures 3 and 6, it is evident that the limitation of herbicide usage also alters the weed species composition in winter wheat. The restricted use of herbicides (limitation) creates more favorable conditions for weed species such as *Cirsium arvense* and *Elymus repens*. Under full regulation conditions, weed species such as *Apera spica-venti, Atriplex patula, Avena fatua, Descurainia Sophia, Galium aparine, Papaver rhoeas,* and *Veronica persica* find it easier to establish themselves.

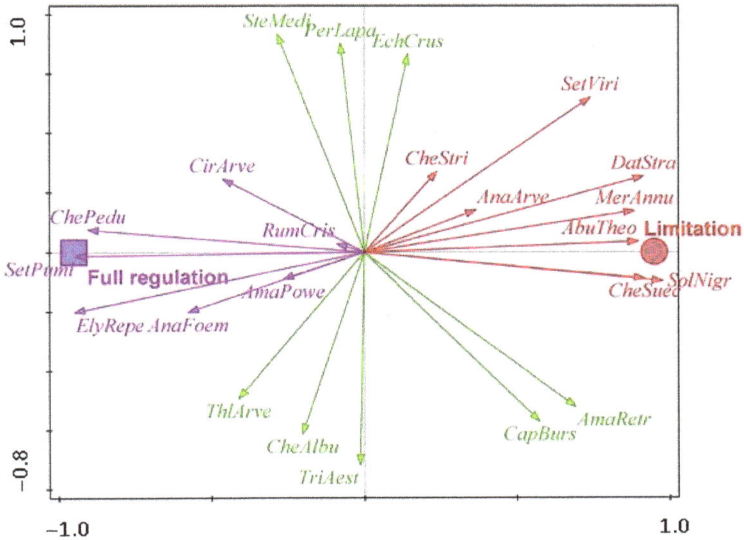

Figure 5. Response of weed occurring in maize stands to limited herbicide regulation (RDA result; F-ratio = 8.1; p-value = 0.023); purple color indicates species preferring full regulation, green color indicates species without preference, and red color indicates species preferring limitation.

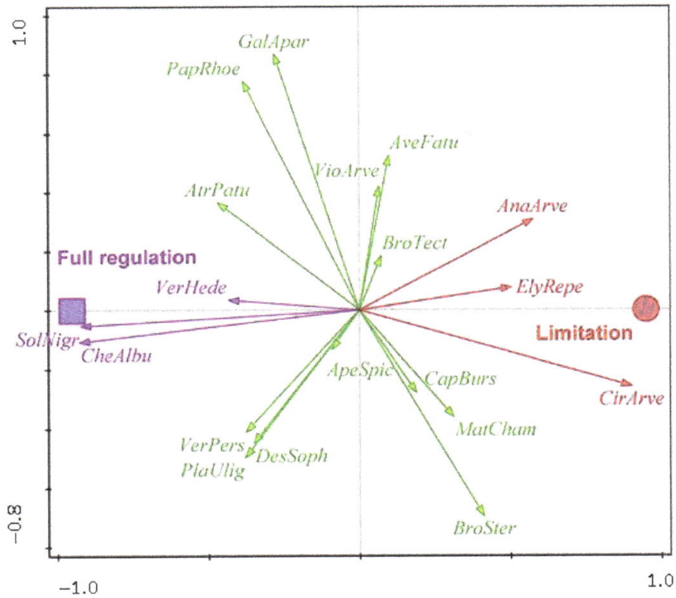

Figure 6. Response of weeds occurring in maize stands to limited herbicide regulation (result of RDA; F-ratio = 2.4.; p-value = 0.063); purple color indicates species preferring full regulation, green color indicates species without preference, and red color indicates species preferring limitation.

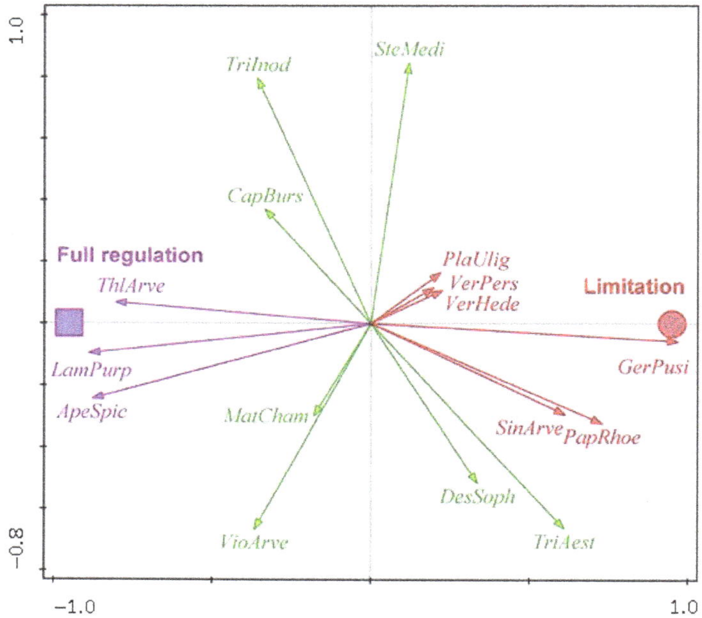

Figure 7. Response of weeds occurring in rapeseed stands to limited herbicide regulation (RDA result; F-ratio = 8.0.; *p*-value = 0.023); purple color indicates species preferring full regulation, green color indicates species without preference, and red color indicates species preferring limitation.

From the results presented in Figures 4 and 7, it is evident that the limitation of herbicide usage also alters the weed species composition in winter rapeseed. The restricted use of herbicides (limitation) creates more favorable conditions for weed species such as *Geranium pusillum*, *Papaver rhoeas*, *Plantago uliginosa*, *Sinapis arvensis*, *Veronica hederifolia*, and *V. persica*. Under full regulation conditions, weed species such as *Apera spica-venti*, *Lamium purpureum*, and *Thlaspi arvense* are established more easily.

Table 2. Groups of weeds according to their reaction to limitation of herbicide use (RDA).

Occurence in Crop Stand	Groups of Weeds		
	Species Preferring Herbicide Limitation	**Species without Any Preference, Affected by Other Factors**	**Species Preferring Full Herbicide Regulation**
Flint maize	*AbuTheo—Abutilon theophrasti*, *AnaArve—Anagallis arvensis*, *DatStra—Datura stramonium*, *CheStri—Chenopodium strictum*, *CheSuec—Chenopodium suecicum*, *MerAnnu—Mercurialis annua*, *SetViri—Setaria viridis*, *SolNigr—Solanum nigrum*	*AmaRetr—Amaranthus retroflexus*, *CapBurs—Capsella bursa-pastoris*, *EchCrus—Echinochloa crus-galli*, *CheAlbu—Chenopodium album*, *PerLapa—Persicaria lapathifolia*, *SteMedi—Stellaria media*, *ThlArve—Thlaspi arvense*, *TriAest—Triticum aestivum*	*AmaPowe—Amaranthus powellii*, *AnaFoem—Anagallis foemina*, *CirArve—Cirsium arvense*, *ElyRepe—Elymus repens*, *ChePedu—Chenopodium album* subsp. *Pedunculare*, *RumCris—Rumex crispus*, *SetPumi—Setaria pumila*

Table 2. *Cont.*

Occurence in Crop Stand	Groups of Weeds		
	Species Preferring Herbicide Limitation	**Species without Any Preference, Affected by Other Factors**	**Species Preferring Full Herbicide Regulation**
Winter wheat	*AnaArve—Anagallis arvensis, CirArve—Cirsium arvense, ElyRepe—Elymus repens*	*CheAlbu—Chenopodium album, SolNigr—Solanum nigrum, VerHede—Veronica hederifolia*	*ApeSpic—Apera spica-venti, AtrPatu—Atriplex patula, AveFatu—Avena fatua, BroSter—Bromus sterilis, BroTect—Bromus tectorum, CapBurs—Capsella bursa-pastoris, DesSoph—Descurainia Sophia, GalApar—Galium aparine, MatCham—Matricaria chamomilla, PapRhoe—Papaver rhoeas, PlaUlig—Plantago uliginosa, VerPers—Veronica persica, VioArve—Viola arvensis*
Winter rapeseed	*GerPusi—Geranium pusillum, PapRhoe—Papaver rhoeas, PlaUlig—Plantago uliginosa, SinArve—Sinapis arvensis, VerHede—Veronica hederifolia, VerPers—Veronica persica*	*CapBurs—Capsella bursa-pastoris, DesSoph—Descurainia Sophia, MatCham—Matricaria chamomilla, SteMedi—Stellaria media, TriAest—Triticum aestivum, TriInod—Tripleurospermum inodorum, VioArve—Viola arvensis*	*ApeSpic—Apera spica-venti, LamPurp—Lamium purpureum, ThlArve—Thlaspi arvense*

4. Discussion

The weeding of selected crops under real conditions varies based on chemical regulation limitations, which are given by the rules of the Water Protection Zones. The reactions of weeds are manifested primarily in species composition rather than in the intensity of weeding.

Higher weeding and a higher proportion of the weeds *Setaria viridis, Mercurialis annua, Chenopodium strictum*, and *Solanum nigrum* were observed in the maize stands where the use of herbicides was restricted. Species that are considered neophytes or species with an invasive status (*Abutilon theophrasti, Datura stramonium*, and *Chenopodium suecicum*) were also found in these stands to a greater extent.

The differences in the representation of weeds in the wheat stands were statistically significant according to the RDA. This might have been caused by the low weeding rate, which was implied by the high efficiency of the herbicides allowed in the water protection zones. Nevertheless, perennial weeds (*Cirsium arvense* and *Elymus repens*), which are difficult to control and very harmful in wheat stands, were more common in plots with herbicide limitations. The occurrence of perennial weed species is mainly associated with land. Although they spread slowly, there is a risk of spreading from a long-term perspective.

The winter rapeseed stands differed in terms of weed species composition. *Geranium pusillum, Papaver rhoeas, Plantago uliginosa, Sinapis arvensis, Veronica hederifolia*, and *V. persica* occurred more frequently in plots with herbicide limitation. These weeds are sensitive to common herbicides but are less sensitive to herbicides permitted in areas with herbicide limitations.

Changes in the species composition of weeds evoked by the limitation of herbicide applications lead to the promotion of more specialized weeds (specialists) and to the exclusion of generally widespread species (generalists). According to Clavel et al. [53], pesticides, especially herbicides, can contribute to the process of "biotic homogenization," which is the representation of weeds that are resistant to herbicide increase. The heterogeneity of habitats in the landscape is increasing by introducing limitations on the application of herbicides. The reduction in pesticide doses is a fundamental contributor to taxonomic and/or functional diversity in the broader agricultural landscape, according to Chiron et al. [25].

Herbicides co-create a selection pressure on field vegetation [54,55]. Different regulation creates a specific pressure on weed vegetation, to which the plants also respond in the longer term by changing their life strategy [56]. Weeds that can survive this pressure occupy this niche and remain part of the agricultural landscape. Field vegetation under herbicide regulation is important for the abundance of birds [27,57] and other animals that are capable of settling in intensively managed fields [58].

Limiting the use of pesticides in sensitive areas (near water bodies) appears to be justified from the perspective of water protection and biodiversity [59]. The ecosystem response of agricultural land to environmental changes remains understudied [60,61]. Biodiversity and soil fertility are crucial for humanity [62]; therefore, they must be given proper attention. The massive use of pesticides around the world leads to the stabilization and increase in crop yields, but also raises strong concerns about the impacts on biodiversity [63–65]. A recent review summarized the results of 394 studies and found that a large number of studies reported significant negative effects of pesticides on soil invertebrates [66,67], which constitutes a significant part of global biodiversity [60,68,69]. They are also an essential part of terrestrial food webs and play an important role in ecosystem services for many terrestrial animals, including vertebrates [70–74]. A loss of biodiversity destabilizes farmland ecosystems, which potentially threatens the sustainable intensification of agriculture [75].

A floristic analysis of the site is a well-known way of assessing the quality of the environment, because the entire flora of the site is a bioindicator [76]. According to Van Kleunen et al. [77], the presence of non-native taxa is an indicator of poor environmental quality. It can be assumed that a simple limitation of herbicide use is not a guarantee of improving the environment or water quality. Several studies have addressed the impacts of government measures and implemented management (for example, ES 2018 and EC 2019) on water quality [78], but little empirical research has been conducted to contribute to improving the quality of underground and surface water [79].

The limitations of chemical weed control resulting from regulations on water protection creates a need for non-chemical weed control in field crop stands. According to Melander et al. [80], research on direct non-chemical methods of weed control uses policy initiatives that aim to reduce the reliance on pesticides and promote organic crop production. The need for non-chemical methods of weed regulation can also be attributed to the lack of new active substances of chemical herbicides, which results from stricter requirements for the registration of pesticides and environmental regulations. This situation caused a drastic decrease in available pesticides, with the loss being the greatest in Europe, with 945 active substances in 1999 compared to 336 in 2009, which resulted in a 64% reduction [81]. Our results show that on plots with herbicide limitations in real conditions, there is an increase in difficult-to-control weeds (*Cirsium arvense*, *Elymus repens*, *Geranium pusillum*, and *Setaria viridis*) or new species of weeds (*Abutilon theophrasti*, *Mercurialis annua*, and *Solanum nigrum*). Their occurrence requires changes in weed regulation by farmers, which are often reluctantly accepted.

Regulations aimed at water protection are of paramount importance to safeguard the environment and human health. The impact on other ecosystem components must be monitored and evaluated. Limiting the use of herbicides reduces surface water pollution but simultaneously introduces challenges for farmers. Protective and regulatory measures must be understood on a broader scale of the entire landscape. These measures should be assessed not only from the perspective of water protection, but also by considering other landscape components influenced by these regulations. In formulating protective and regulatory measures, multidisciplinary discourse is necessary along with efforts to find collaborative solutions. To ensure the viability and effectiveness of these measures, it is crucial to avoid a scenario in which safeguarding one landscape element (water) significantly disadvantages the other (agriculture).

Therefore, it is essential to look for other techniques and methods to protect surface water. In the entire landscape, buffer zones are very important, and according to Mykrä et al. [82], they are very effective in protecting aquatic environments because of the diversity of plant communities in floodplain forests. Buffer zones or coastal zones are commonly considered to be effective filters of nutrients (N, P), as well as pesticides [83–86]. However, the relative effectiveness of these buffer zones depends on topography, vegetation and soil type, climate, and the extent of nutrient loading, and probably chiefly on their width [87–90].

Riparian vegetation affects stream water quality and biodiversity in several ways [87,91,92]. The vegetated riparian zone is important for the degradation of organic agrochemicals, including pesticides [93]. An appropriate design of the buffer zone based on site characteristics and landscape restrictions offers experts a tool to take steps to limit the negative effects of agriculture [94,95].

The need to feed billions of people necessitate reliable and predictable crop yields, with a parallel requirement to preserve conditions for biodiversity. This can be achieved using a model of agricultural production that includes water bodies and respects the multi-function use of productive landscapes [75].

5. Conclusions

Restrictions on weed management within water protection zones, as applied in the Czech Republic, led to alterations in the weed spectrum of the chosen crops. The reaction of field weeds can be seen primarily in the weed species composition. Changes in the vegetation composition were more evident in the stands of maize and winter rapeseed. In winter wheat stands, the differences in the weed composition were not statistically significant. Places with limited herbicide application are mainly suitable for weeds that are less sensitive to permitted herbicides, perennial species, and neophytes. The proportion of neophytes and invasive species was higher, especially in maize stands in plots with herbicide limitation. The responses of weeds to the reduction in pesticide use can help in planning appropriate weed control measures for individual crops.

Water protection rules can cause problems for the regulation of certain weed species. The limited assortment of herbicides leads to the repeated application of herbicides with the same active substance, which can result in the emergence of resistance in weeds. In areas designated for water protection, a complete ban on the use of pesticides appears to be more convenient, and in agricultural production areas, there is a transition to the organic farming regime.

The planning and implementation of appropriate measures for weed control in individual crops should consider specific conditions of a given location and the consequences of the chosen practices. A comprehensive approach to this issue should be based on scientific knowledge, monitoring of the development of weed resistence, and the ongoing optimization of strategies to control them. By adhering to this approach, sustainable agriculture that is in harmony with environmental protection and long-term sustainability can be achieved.

Alternative biological and physical weed control methods could represent a promising direction for further research. A thorough analysis of the economic and environmental aspects of these methods would help in finding a balanced and sustainable solution that would minimize negative impacts on the environment and agricultural production. Educating farmers about optimal weed control practices is also key to achieving sustainable agriculture and the long-term sustainable use of land and water resources.

Author Contributions: Conceptualization, J.W., E.K. and L.H.; methodology, J.W. and T.Ř.; validation, J.W., V.H. and M.Ž.; formal analysis, J.W.; investigation, J.W. and T.Ř.; resources, J.W.; data curation, J.W. and V.H.; writing—original draft preparation, J.W., V.H. and M.D.V.; writing—review and editing, J.W., M.D.V. and M.Ž.; visualization, J.W. and M.Ž.; supervision, J.W. and M.D.V.; project administration, J.W.; funding acquisition, M.Ž. All authors have read and agreed to the published version of the manuscript.

Funding: This work was created as a result of the project TACR TH04030244, Increasing biodiversity and promoting ecosystem services in the agricultural landscape utilizing alternative meadows and pasture management.

Data Availability Statement: Not applicable.

Conflicts of Interest: The authors declare no conflict of interest.

References

1. Passariello, B.; Giuliano, V.; Quaresima, S.; Barbaro, M.; Caroli, S.; Forte, G.; Carelli, G.; Iavicoli, I. Evaluation of the Environmental Contamination at an Abandoned Mining Site. *Microchem. J.* **2002**, *73*, 245–250. [CrossRef]
2. Carpenter, S.R.; Caraco, N.F.; Correll, D.L.; Howarth, R.W.; Sharpley, A.N.; Smith, V.H. Nonpoint pollution of surface waters with phosphorus and nitrogen. *Ecol. Appl.* **1998**, *8*, 559–568. [CrossRef]
3. Hanifzadeh, M.; Nabati, Z.; Longka, P.; Malakul, P.; Apul, D.; Kim, D.S. Life cycle assessment of superheated steam drying technology as a novel cow manure management method. *J. Environ. Manag.* **2017**, *199*, 83–90. [CrossRef]
4. Gilliom, R.J. Pesticides in U.S. streams and groundwater. *Environ. Sci. Technol.* **2007**, *41*, 3408–3414. [CrossRef]
5. Blanchoud, H.; Moreau-Guigon, E.; Farrugia, F.; Chevreuil, M.; Mouchel, J.M. Contribution by urban and agricultural pesticide uses to water contamination at the scale of the Marne watershed. *Sci. Total Environ.* **2007**, *375*, 168–179. [CrossRef]
6. Ulrich, U.; Dietrich, A.; Fohrer, N. Herbicide transport via surface runoff during intermittent artificial rainfall: A laboratory plot scale study. *Catena* **2013**, *101*, 38–49. [CrossRef]
7. Meyer, B.; Pailler, J.Y.; Guignard, C.; Hoffmann, L.; Krein, A. Concentrations of dissolved herbicides and pharmaceuticals in a small river in Luxembourg. *Environ. Monit. Assess.* **2011**, *180*, 127–146. [CrossRef] [PubMed]
8. Köck-Schulmeyer, M.; Ginebreda, A.; González, S.; Cortina, J.L.; de Alda, M.L.; Barceló, D. Analysis of the occurrence and risk assessment of polar pesticides in the Llobregat River Basin (NE Spain). *Chemosphere* **2012**, *86*, 8–16. [CrossRef] [PubMed]
9. Akerblom, N. *Agricultural Pesticide Toxicity to Aquatic Organisms—A Literature Review*, 1st ed.; Department of Environmental Assessment, Swedish University of Agricultural Sciences: Uppsala, Sweden, 2004; p. 31. Available online: http://webstar.vatten.slu.se/IMA/Publikationer/internserie/2004-16.pdf (accessed on 10 February 2023).
10. Wakabayashi, K.; Böger, P. Target sites for herbicides: Entering the 21st century. *Pest Manag. Sci. Former. Pestic. Sci.* **2002**, *58*, 1149–1154. [CrossRef]
11. Délye, C.; Jasieniuk, M.; Le Corre, V. Deciphering the evolution of herbicide resistance in weeds. *Trends Genet.* **2013**, *29*, 649–658. [CrossRef]
12. Liebman, M.; Mohler, C.; Staver, C. *Ecological Management of Agricultural Weeds*; Cambridge University Press: Cambridge, UK, 2001. [CrossRef]
13. Merotto, A., Jr.; Gazziero, D.L.; Oliveira, M.C.; Scursoni, J.; Garcia, M.A.; Figueroa, R.; Turra, G. Herbicide use history and perspective in South America. *Adv. Weed Sci.* **2022**, *40*, 1–18. [CrossRef] [PubMed]
14. Chauvel, B.; Tschudy, C.; Munier-Jolain, N. Gestion intégrée de la flore adventice dans les systèmes de culture sans labour. *Cah. Agric.* **2011**, *20*, 194–203. [CrossRef]
15. Thompson, M.; Chauhan, B.S. History and perspective of herbicide use in Australia and New Zealand. *Adv. Weed Sci.* **2022**, *40*, 1–12. [CrossRef]
16. Li, Q.; Wang, J.; Wu, J.; Zhai, Q. The dual impacts of specialized agricultural services on pesticide application intensity: Evidence from China. *Pest Management Science* **2023**, *79*, 76–87. [CrossRef] [PubMed]
17. Dugon, J.; Favre, G.; Zimmermann, A.; Charles, R. Pratiques phytosanitaires dans un réseau d'exploitations de grandes cultures de 1992 a 2004. *Rech. Agron. Suisse* **2010**, *1*, 416–423. (In French). Available online: https://www.agrarforschungschweiz.ch/wp-content/uploads/pdf_archive/2010_1112_f_1615.pdf (accessed on 10 February 2023).
18. Arancibia, F.; Motta, R.C.; Clausing, P. The neglected burden of agricultural intensification: A contribution to the debate on land-use change. *J. Land Use Sci.* **2020**, *15*, 235–251. [CrossRef]
19. Blettler, D.; Manresa, J.A.B.; Fagúndez, G.A. A review of the effects of agricultural intensification and the use of pesticides on honey bees and their products and possible palliatives. *Span. J. Agric. Res.* **2022**, *20*, e03R02. [CrossRef]
20. Mitra, A.; Chatterjee, C.; Mandal, F.B. Synthetic chemical pesticides and theireffects on birds. *Res. J. Environ. Toxicol.* **2011**, *5*, 81–96. [CrossRef]
21. Hole, D.G.; Perkins, A.J.; Wilson, J.D.; Alexander, I.H.; Grice, P.V.; Evans, A.D. Does organic farming benefit biodiversity? *Biol. Conserv.* **2005**, *122*, 113–130. [CrossRef]
22. Bruggisser, O.T.; Schmidt-Entling, M.H.; Bacher, S. Effects of vineyardmanagement on biodiversity at three trophic levels. *Biol. Conserv.* **2010**, *143*, 1521–1528. [CrossRef]
23. Serrão, J.E.; Plata-Rueda, A.; Martínez, L.C.; Zanuncio, J.C. Side-effects of pesticides on non-target insects in agriculture: A mini-review. *Sci. Nat.* **2022**, *109*, 17. [CrossRef] [PubMed]
24. Tudi, M.; Li, H.; Li, H.; Wang, L.; Lyu, J.; Yang, L.; Tong, S.; Yu, Q.J.; Ruan, H.D.; Atabila, A.; et al. Exposure routes and health risks associated with pesticide application. *Toxics* **2022**, *10*, 335. [CrossRef] [PubMed]
25. Chiron, F.; Chargé, R.; Julliard, R.; Jiguet, F.; Muratet, A. Pesticide doses, landscape structure and their relative effects on farmland birds. *Agric. Ecosyst. Environ.* **2014**, *185*, 153–160. [CrossRef]

26. Kleijn, D.; Kohler, F.; Báldi, A.; Batáry, P.; Concepción, E.D.; Clough, Y.; Díaz, M.; Gabriel, D.; Holzschuh, A.; Knop, E.; et al. On the relationship between farmland biodiversity and land-use intensityin Europe. *Proc. R. Soc.* **2009**, *276*, 903–909. [CrossRef]
27. Guerrero, I.; Morales, M.B.; Oñate, J.J.; Geiger, F.; Berendse, F.; de Snoo, G.; Eggers, S.; Pärt, T.; Bengtsson, J.; Clement, L.W.; et al. Response of ground-nesting farmland birds to agricultural intensificationacross Europe: Landscape and field level management factors. *Biol. Conserv.* **2012**, *152*, 74–80. [CrossRef]
28. de Montaigu, C.T.; Goulson, D. Habitat quality, urbanisation & pesticides influence bird abundance and richness in gardens. *Sci. Total Environ.* **2023**, *870*, 161916. [CrossRef]
29. Leu, C.; Singer, H.; Stamm, C.; Müller, S.R.; Schwarzenbach, R.P. Simultaneous assessment of sources, processes, and factors influencing herbicide losses to surface waters in a small agricultural catchment. *Environ. Sci. Technol.* **2004**, *38*, 3827–3834. [CrossRef]
30. Liess, M.; Von der Ohe, P.C. Analyzing effects of pesticides on invertebrate communities in streams. *Environ. Toxicol. Chem.* **2005**, *24*, 954–965. [CrossRef]
31. Dugan, S.T.; Muhammetoglu, A.; Uslu, A. A combined approach for the estimation of groundwater leaching potential and environmental impacts of pesticides for agricultural lands. *Sci. Total Environ.* **2023**, *901*, 165892. [CrossRef]
32. Nahar, K.; Baillie, J.; Zulkarnain, N.A. Herbicide Fate and Transport in the Great Barrier Reef: A Review of Critical Parameters. *Water* **2023**, *15*, 237. [CrossRef]
33. EEA. *European Waters—Assessment of Status and Pressures 2018*; European Environmental Agency: Luxembourg, 2018; p. 90. [CrossRef]
34. Wuijts, S.; Claessens, J.; Farrow, L.; Doody, D.G.; Klages, S.; Christophoridis, C.; Cvejić, R.; Glavan, M.; Nesheim, I.; Platjouw, F.; et al. Protection of drinking water resources from agricultural pressures: Effectiveness of EU regulations in the context of local realities. *J. Environ. Manag.* **2021**, *287*, 112270. [CrossRef] [PubMed]
35. EEC. Council Directive 91/676/EEC of 12 December 1991 concerning the protection of waters against pollution caused by nitrates from agricultural sources. 1991. Available online: https://eur-lex.europa.eu/eli/dir/1991/676/oj (accessed on 10 February 2023).
36. EC. Council Directive 98/83/EC on the quality of water intended for human consumption. 1998. Available online: https://eur-lex.europa.eu/eli/dir/1998/83/oj (accessed on 10 February 2023).
37. EC. Directive 2000/60/EC of the European Parliament and of the Council of 23 October 2000 establishing a framework for Community action in the field of water policy. 2000. Available online: https://eur-lex.europa.eu/legal-content/EN/TXT/?uri=celex%3A32000L0060 (accessed on 10 February 2023).
38. Act No. 254/2001 Coll. Water Act [Zákon č. 254/2001 Sb., vodní zákon]. Available online: https://eagri.cz/public/web/mze/legislativa/pravni-predpisy-mze/tematicky-prehled/Legislativa-MZe_uplna-zneni_zakon-2001-254-viceoblasti.html (accessed on 8 June 2023). (In Czech).
39. OECD. *OECD INVENTORY Water Governance Indicators and Measurement Frameworks*; OECD: Paris, France, 2015; p. 44. Available online: https://www.oecd.org/cfe/regionaldevelopment/Inventory_Indicators.pdf (accessed on 10 February 2023).
40. OECD. *OECD Principles on Water Governance (Daegu Declaration)*; OECD: Paris, France, 2015.
41. Wuijts, S.; Driessen, P.P.J.; Van Rijswick, H.F.M.W. Towards More Effective Water Quality Governance: A Review of Social-Economic, Legal and Ecological Perspectives and Their Interactions. *Sustainability* **2018**, *10*, 914. [CrossRef]
42. Jørgensen, W.L.N.; Kudsk, P.; Ørum, J.E. Links between pesticide use pattern and crop production in Denmark with special reference to winter heat. *Crop Prot.* **2019**, *119*, 147–157. [CrossRef]
43. Devi, P.I.; Manjula, M.; Bhavani, R.V. Agrochemicals, environment, and human health. *Annu. Rev. Environ. Resour.* **2022**, *47*, 399–421. [CrossRef]
44. Adedibu, P.A. Ecological problems of agriculture: Impacts and sustainable solutions. *ScienceOpen* **2023**. *preprints*. [CrossRef]
45. Zahoor, I.; Mushtaq, A. Water Pollution from Agricultural Activities: A Critical Global Review. *Int. J. Chem. Biochem. Sci.* **2023**, *23*, 164–176.
46. Islam, M.A.; Amin, S.N.; Rahman, M.A.; Juraimi, A.S.; Uddin, M.K.; Brown, C.L.; Arshad, A. Chronic effects of organic pesticides on the aquatic environment and human health: A review. *Environ. Nanotechnol. Monit. Manag.* **2022**, *18*, 100740. [CrossRef]
47. Le Gal, A.S.; Priol, P.; Georges, J.Y.; Verneau, O. Population structure and dynamics of the Mediterranean Pond Turtle Mauremys leprosa (Schweigger, 1812) in contrasted polluted aquatic environments. *Environ. Pollut.* **2023**, *330*, 121746. [CrossRef]
48. Culek, M. (Ed.) *Biogeographical Division of the Czech Republic (Biogeografické Členění České Republiky)*, 1st ed.; Enigma: Prague, Czech Republic, 1996; p. 347. (In Czech)
49. CGS. *Geological Map of the Czech Republic, 1:50 000*; Czech Geological Society: Prague, Czech Republic, 2018. Available online: https://mapy.geology.cz/geocr50/ (accessed on 8 November 2022).
50. CGS. *Map of Soil Types of the Czech Republic, 1:50 000*; Czech Geological Society: Prague, Czech Republic, 2017. Available online: https://mapy.geology.cz/pudy/ (accessed on 8 November 2022).
51. Kaplan, Z.; Danihelka, J.; Chrtek, J.; Kirschner, J.; Kubát, K.; Štech, M.; Štěpánek, J. (Eds.) *Key to the Flora of the Czech Republic [Klíč ke Květeně České Republiky]*, 2nd ed.; Academia: Prague, Czech Republic, 2019; p. 1168. (In Czech)
52. Ter Braak, C.J.F.; Šmilauer, P. *Canoco Reference Manual and User's Guide: Software for Ordination*; Version 5.0; Microcomputer Power: Ithaca, NY, USA, 2012.
53. Clavel, J.; Julliard, R.; Devictor, V. Worldwide decline of specialistspecies: Towards a global functional homogenization? *Front. Ecol. Environ.* **2011**, *9*, 222–228. [CrossRef]

54. Kolářová, M.; Piskáčková, T.A.R.; Tyšer, L.; Hoová, T.T. Characterisation of Czech arable weed communities according to management and production area considering the prevalence of herbicide-resistant species. *Weed Res.* **2023**, *63*, 57–67. [CrossRef]
55. Winkler, J.; Dvořák, J.; Hosa, J.; Martínez Barroso, P.; Vaverková, M.D. Impact of Conservation Tillage Technologies on the Biological Relevance of Weeds. *Land* **2023**, *12*, 121. [CrossRef]
56. Winkler, J.; Vaverková, M.D.; Havel, L. Anthropogenic life strategy of plants. *Anthr. Rev.* **2023**, *10*, 455–462. [CrossRef]
57. Geiger, F.; Bengtsson, J.; Berendse, F.; Weisser, W.W.; Emmerson, M.; Morales, M.B.; Ceryngier, P.; Liira, J.; Tscharntke, T.; Winqvist, C.; et al. Persistent negative effects of pesticides onbiodiversity and biological control potential on European farmland. *Basic Appl. Ecol.* **2010**, *11*, 97–105. [CrossRef]
58. Chiron, F.; Filippi-Codaccioni, O.; Jiguet, F.; Devictor, V. Effects of non croppedlandscape diversity on spatial dynamics of farmland birds in intensive farmingsystems. *Biol. Conserv.* **2010**, *43*, 2609–2616. [CrossRef]
59. Frelih-Larsen, A.; Chivers, C.A.; Herb, I.; Mills, J.; Reed, M. The role of public consultations in decision-making on future agricultural pesticide use: Insights from European Union's Farm to Fork Strategy public consultation. *J. Environ. Policy Plan.* **2023**, *25*, 476–492. [CrossRef]
60. FAO; ITPS; GSBI; SCBD; EC. *State of Knowledge of Soil Biodiversity—Status, Challenges and Potentialities*, 1st ed.; FAO: Rome, Italy, 2020; p. 618. [CrossRef]
61. Phillips, H.R.P.; Cameron, E.K.; Ferlian, O.; Türke, M.; Winter, M.; Eisenhauer, N. Red list of a black box. *Nat. Ecol. Evol.* **2017**, *1*, 103. [CrossRef]
62. Wall, D.H.; Nielsen, U.N.; Six, J. Soil biodiversity and human health. *Nature* **2015**, *528*, 69–76. [CrossRef] [PubMed]
63. Bernhardt, E.S.; Rosi, E.J.; Gessner, M.O. Synthetic chemicals as agents of global change. *Front. Ecol. Environ.* **2017**, *15*, 84–90. [CrossRef]
64. Pelosi, C.; Bertrand, C.; Daniele, G.; Coeurdassier, M.; Benoit, P.; Nélieu, S.; Lafay, F.; Bretagnolle, V.; Gaba, S.; Vulliet, E.; et al. Residues of currently used pesticides in soils and earthworms: A silent threat. *Agric. Ecosyst. Environ.* **2021**, *305*, 107167. [CrossRef]
65. Wang, Z.; Walker, G.W.; Muir, D.C.G.; Nagatani-Yoshida, K. Toward a global understanding of chemical pollution: A first comprehensive analysis of national and regional chemical inventories. *Environ. Sci. Technol.* **2020**, *54*, 2575–2584. [CrossRef]
66. Gunstone, T.; Cornelisse, T.; Klein, K.; Dubey, A.; Donley, N. Pesticides and soil invertebrates: A Hazard assessment. *Front. Environ. Sci.* **2021**, *9*, 643847. [CrossRef]
67. Beaumelle, L.; Tison, L.; Eisenhauer, N.; Hines, J.; Malladi, S.; Pelosi, C.; Thouvenot, L.; Phillips, H.R.P. Pesticide effects on soil fauna communities—A meta-analysis. *J. Appl. Ecol.* **2023**, *60*, 1239–1253. [CrossRef]
68. Bardgett, R.D.; van der Putten, W.H. Belowground biodiversity and ecosystem functioning. *Nature* **2014**, *515*, 505–511. [CrossRef] [PubMed]
69. Eisenhauer, N.; Bonn, A.; Guerra, C.A. Recognizing the quiet extinction of invertebrates. *Nat. Commun.* **2019**, *10*, 50. [CrossRef] [PubMed]
70. Barnes, A.E.; Robinson, R.A.; Pearce-Higgins, J.W. Collation of a century of soil invertebrate abundance data suggests long-term declines in earthworms but not tipulids. *PLoS ONE* **2023**, *18*, e0282069. [CrossRef]
71. Scherber, C.; Eisenhauer, N.; Weisser, W.W.; Schmid, B.; Voigt, W.; Fischer, M.; Schulze, E.D.; Roscher, C.; Weigelt, A.; Allan, E.; et al. Bottom-up effects of plant diversity on multitrophic interactions in a biodiversity experiment. *Nature* **2010**, *468*, 553–556. [CrossRef]
72. Schuldt, A.; Assmann, T.; Brezzi, M.; Buscot, F.; Eichenberg, D.; Gutknecht, J.; Härdtle, W.; He, J.S.; Klein, A.M.; Kühn, P.; et al. Biodiversity across trophic levels drives multifunctionality in highly diverse forests. *Nat. Commun.* **2018**, *9*, 2989. [CrossRef] [PubMed]
73. Soliveres, S.; Van Der Plas, F.; Manning, P.; Prati, D.; Gossner, M.M.; Renner, S.C.; Alt, F.; Arndt, H.; Baumgartner, V.; Binkenstein, J.; et al. Biodiversity at multiple trophic levels is needed for ecosystem multifunctionality. *Nature* **2016**, *536*, 456–459. [CrossRef] [PubMed]
74. Wagg, C.; Bender, S.F.; Widmer, F.; Heijden, M.G. Soil biodiversity and soil community composition determine ecosystem multifunctionality. *Proc. Natl. Acad. Sci. USA* **2016**, *111*, 5266–5270. [CrossRef]
75. Schiesari, L.; Saito, V.; Ferreira, J.; Freitas, L.S.; Goebbels, A.J.; Leite, J.P.C.B.; Oliveira, J.C.; Pelinson, R.M.; Querido, B.B.; Carmo, J.; et al. Community reorganization stabilizes freshwater ecosystems in intensively managed agricultural fields. *J. Appl. Ecol.* **2023**, *60*, 1327–1339. [CrossRef]
76. Braga, L.; Furia, E.; Buldrini, F.; Mercuri, A.M. Pollen and Flora as Bioindicators in Assessing the Status of Polluted Sites: The Case Study of the Mantua Lakes (SIN "Laghi di Mantova e Polo Chimico"; N Italy). *Sustainability* **2023**, *15*, 9414. [CrossRef]
77. Van Kleunen, M.; Dawson, W.; Maurel, N. Characteristics of Successful Alien Plants. *Mol. Ecol.* **2015**, *24*, 1954–1968. [CrossRef] [PubMed]
78. Newig, J.; Fritsch, O. Environmental governance: Participatory, multi-level—And effective? *Environ. Policy Gov.* **2009**, *19*, 18. [CrossRef]
79. Wuijts, S.; Driessen, P.P.J.; Van Rijswick, H.F.M.W. Governance Conditions for Improving Quality Drinking Water Resources: The Need for Enhancing Connectivity. *Water Resour. Manag.* **2018**, *32*, 1245–1260. [CrossRef]
80. Melander, B.; Rasmussen, I.; Bàrberi, P. Integrating physical and cultural methods of weed control—Examples from European research. *Weed Sci.* **2005**, *53*, 369–381. [CrossRef]
81. Beckie, H.J.; Tardif, F.J. Herbicide cross resistance in weeds. *Crop Prot.* **2012**, *35*, 15–28. [CrossRef]

82. Mykrä, H.; Annala, M.; Hilli, A.; Hotanen, J.P.; Hokajärvi, R.; Jokikokko, P.; Karttunen, K.; Kesälä, M.; Kuoppala, M.; Leinonen, A.; et al. GIS-based planning of buffer zones for protection of boreal streams and their riparian forests. *For. Ecol. Manag.* **2023**, *528*, 120639. [CrossRef]
83. Arora, K.; Mickelson, S.K.; Baker, J.L.; Tierney, D.P.; Peters, C.J. Herbicide retention by vegetative buffer strips from runoff under natural rainfall. *Trans. ASAE* **1996**, *39*, 2155–2162. [CrossRef]
84. Decamps, H.; Pinay, G.; Naiman, R.J.; Petts, G.E.; McClain, M.E.; Hillbricht-Ilkowska, A.H.T.A.; Hanley, T.A.; Holmes, R.M.; Quinn, J.; Gibert, J.; et al. Riparian zones: Where biogeochemistry meets biodiversity in management practice. *Pol. J. Ecol.* **2004**, *52*, 3–18. Available online: https://miiz.waw.pl/pliki/article/ar52_1_01.pdf (accessed on 10 February 2023).
85. Mankin, K.; Daniel, R.; Ngandu, M.; Barden, C.J.; Hutchinson, S.L.; Geyer, W.A. Grass-shrub riparian buffer removal of sediment, phosphorus, and nitrogen from simulated runoff. *JAWRA* **2007**, *43*, 1108–1116. [CrossRef]
86. Sieczka, A.; Bujakowski, F.; Falkowski, T.; Koda, E. Morphogenesis of a Floodplain as a Criterion for Assessing the Susceptibility to Water Pollution in an Agriculturally Rich Valley of a Lowland River. *Water* **2018**, *10*, 399. [CrossRef]
87. Mander, Ü.; Kuusemets, V.; Hayakawa, Y. Purification processes, ecological functions, planning and design of riparian buffer zones in agricultural watersheds. *Ecol. Eng.* **2005**, *24*, 21–432. [CrossRef]
88. Dosskey, M.G.; Helmers, M.J.; Eisenhauer, D.E. An approach for using soil surveys to guide the placement of water quality buffers. *J. Soil Water Conserv.* **2006**, *61*, 344–354. Available online: https://www.srs.fs.usda.gov/pubs/ja/ja_dosskey003.pdf (accessed on 10 February 2023).
89. Lam, Q.D.; Schmalz, B.; Fohrer, N. The impact of agricultural Best Management Practices on water quality in a North German lowland catchment. *Environ. Monit Assess.* **2011**, *183*, 351–379. [CrossRef]
90. Winkler, J.; Jeznach, J.; Koda, E.; Sas, W.; Mazur, Ł.; Vaverková, M.D. Promoting Biodiversity: Vegetation in a Model Small Park Located in the Research and Educational Centre. *J. Ecol. Eng.* **2022**, *23*, 146–157. [CrossRef]
91. Renouf, K.; Harding, J.S. Characterizing riparian buffer zones of an agriculturally modified landscape. *New Zealand J. Mar. Freshw. Res.* **2015**, *49*, 323–332. [CrossRef]
92. Liu, X.; Zhang, X.; Zhang, M. Major factors influencing the efficacy of vegetated buffers on sediment trapping: A review and analysis. *J. Environ. Qual.* **2008**, *37*, 1667–1674. [CrossRef] [PubMed]
93. Unger, I.M.; Goyne, K.W.; Kremer, R.J.; Kennedy, A.C. Microbial community diversity in agroforestry and grass vegetative filter strips. *Agrofor. Syst.* **2013**, *87*, 395–402. [CrossRef]
94. Frey, M.P.; Schneider, M.K.; Dietzel, A.; Reichert, P.; Stamm, C. Predicting critical source areas for diffuse herbicide losses to surface waters: Role of connectivity and boundary conditions. *J. Hydrol.* **2009**, *365*, 23–36. [CrossRef]
95. Lind, L.; Hasselquist, E.M.; Laudon, H. Towards ecologically functional riparian zones: A meta-analysis to develop guidelines for protecting ecosystem functions and biodiversity in agricultural landscapes. *J. Environ. Manag.* **2019**, *249*, 109391. [CrossRef]

water

MDPI

Article

Impact of the Construction of Water Conservation Projects on Runoff from the Weigan River

Jingwen Su [1,2], Aihua Long [1,2,3,*], Fulong Chen [1,*], Cai Ren [1,2], Pei Zhang [2,3], Ji Zhang [2,4], Xinchen Gu [2,4] and Xiaoya Deng [2,3]

1 College of Water Conservancy & Architectural Engineering, Shihezi University, Shihezi 832000, China; jeaven2022@163.com (J.S.); cyrus1837@163.com (C.R.)
2 China Institute of Water Resources and Hydropower Research, Beijing 100038, China; zhangpei-cool@163.com (P.Z.); zhangji940319@tju.edu.cn (J.Z.); gxc@tju.edu.cn (X.G.); dengxy@iwhr.com (X.D.)
3 State Key Laboratory of Simulation and Regulation of Water Cycle in River Basin, China Institute of Water Resources and Hydropower Research, Beijing 100038, China
4 School of Civil Engineering, Tianjin University, Tianjin 300072, China
* Correspondence: ahlong@iwhr.com (A.L.); cfl103@shzu.edu.cn (F.C.)

Abstract: In order to use water resources more efficiently, the construction of water conservation projects in dryland watersheds has changed the natural water cycle processes. This study used the SWAT (Soil and Water Assessment Tool) model coupled with the glacier module to simulate the hydrological processes in the upper reaches of the Weigan River estuary from 1965 to 1991, to restore and quantitatively evaluate the conditions of the estuarine runoff in the no-reservoir scenario, and to analyse the impact of the construction of water conservation projects on the estuarine runoff based on this model. The results show that the SWAT model has good applicability in the study area, with 41.45% and 58.55% of the increase in runoff due to increased precipitation and temperature, respectively, over the 52 years study period. The degree of influence of the construction of water conservation projects on runoff from the mountain in different seasons was spring > autumn > winter > summer, with 83.28% of the spring runoff being influenced by artificial regulation. The construction of water conservation projects has alleviated water shortage problems to a certain extent, and is an effective measure for achieving the efficient allocation of water resources in arid areas.

Keywords: human activities; water conservation construction; SWAT model; runoff configuration; glacial runoff simulation; Weigan River Basin

1. Introduction

The study of the impact of human activities on the natural water cycle is a hot topic in current research [1–3]. Natural and anthropogenic factors are two major drivers of the water cycle in a basin. The natural factors are mainly reflected in changes in vertical water circulation due to climate change and changes in the spatial and temporal distribution of water resources in the horizontal direction. Anthropogenic factors mainly include changes in the basin substrate conditions due to human activities and the exploitation of water resources by humans [4]. In recent years, with the continuous social and economic progress and development, the demand for water resources for population growth, industrial, agricultural, and urban development has increased significantly [5,6]. The increasing human activities and exploitation of water resources have led to various forms of anthropogenic disturbances in the rivers, resulting in changes in the driving conditions and influencing factors of natural hydrological processes in the basin, which have a significant impact on water cycle processes [7].

This research area has received increasing attention from scholars around the world in recent years [8]. For example, Chawla et al. [9] used the VIC model to simulate the hydrological processes in the upper Ganges River Basin in India to analyse the effects of land use and climate change on runoff. Moldir Rakhimova et al. [10] assessed the impact of climate change and human activities on runoff from the Buktirma River Basin in Kazakhstan using various methods such as the climate elasticity method and circulation models. Lei Hou et al. [11] quantitatively assessed the impact of climate change and human activities on runoff changes in the upper reaches of the Yongding River Basin based on the Budyko hypothesis of the climate elasticity approach. Jinping Liu [12], Hongguang Chen [13], Jianyu Liu [14], and other scholars used hydrological models and other methods to analyse the contribution of climate change and human activities to runoff changes in different time periods; the results showed that the contribution of climate change and human activities to runoff varied greatly in different river sections and time periods, and the dominant factors affecting runoff changes were not the same. Lei Wang et al. [15] established a SWAT hydrological model of the Qingshui River Basin in Zhangjiakou and quantitatively analysed the impact of land use scenario changes on runoff in the study area. Cai Ren [16] and Zubaida Muyibul [17] used the SWAT model to simulate the runoff processes in the Yarkant River basin and Urumqi River basin, and quantitatively analysed the degree of influence of climate and subsurface changes on runoff; their studies concluded that the degree of influence of climate change on runoff was greater than that of subsurface changes (i.e., human activities). The research method to investigate the influence of basin water cycle drivers on runoff is more mature; the research conclusion can guide the actual production.

Most of the current studies on water cycle influences are limited to climate change and land use changes on the water cycle [18]. However, research on the impact of human exploitation of water resources in watersheds is less well documented, especially in arid zones. The construction of hydraulic projects, as the main component of human activities in water resource exploitation, refers to the human modification of the substratum structure in a strict sense [19], and also changes the distribution process of water resources in space and time, which, in turn, affects the water cycle processes in basins. As research progresses, scholars have found that the impact of water conservation construction on runoff has changed the original water cycle process in basins; however, how to quantify this impact needs further in-depth study.

In this study, we selected the upper reaches of the Weigan River outlet as the study area. A SWAT distributed hydrological model with a coupled glacier module was constructed to simulate the monthly runoff processes in the study area from 1965 to 1991 (27 years). The study area was investigated for the overall runoff evolution in the last 52 years, and the components of the runoff from the Weigan River were traced, while runoff from the study area from 1992 to 2016 (25 years) without the reservoir was predicted, and the impact of the construction of the Kizil Reservoir on the runoff from the Weigan River was subsequently analysed. This study can provide more comprehensive decision support for the sustainable development, use, and management of regional water resources.

2. Data Sources and Methods

2.1. Overview of the Study Area

The upstream area of the Weigan River outlet (Figure 1) is located in the Baicheng Basin in the Aksu region of Xinjiang, connected to the middle of the southern foothills of the Tianshan Mountains in the north and adjacent to the Queletage Mountains in the south, with a geographical location between 80°15′~83°02′ E and 41°31′~42°39′ N, covering an area of 16,792.56 km^2. The overall topography of the study area slopes from northwest to southeast, with elevations ranging from 1100 to 6778 m. The Weigan River is a typical dry inland glacial snowmelt recharge river. According to the Glacier Catalogue of China, 853 glaciers with a total area of 1783.86 km^2 are distributed in the upper headwater area of the Weigan River, and the river runoff exhibits clear seasonal changes due to the influence of glacial meltwater [20]. The five tributaries (Muzati River, Kapuslang River, Tylervichuk River, Karasu River, and Heizi River) in the area are distributed in the shape of a comb, and each tributary flows from north to south and from west to east, along the southern edge of the Baicheng Basin to the southeast corner of the basin near the Kizil Thousand Buddha Cave [21]. Here, it is called the Weigan River after gathering out of the mountains and passing southward through the southern edge of the basin in the Queletage Mountains [22], and finally flowing to the northern edge of the Tarim Basin through the Weigan River Canyon. Before the construction of the Kizil Reservoir, the hydrological station of the Kizil Reservoir was located 3 km above the dam site of the Kizil Reservoir. In 1985, construction officially started on the Kizil Reservoir; the hydrological station of Heizi Reservoir was moved down to 1 km below the cross-section of the dam site and named Heizi Reservoir (II) Station. In August 1991, the main project of the reservoir was completed and water storage operations began. The Kizil Reservoir de-risking and strengthening project started in 2009, and the highest reservoir storage level has gradually transitioned to the designed storage level of 1149.57 m after de-risking and strengthening [23], which has greatly reduced the flood control pressure in the downstream cities and counties of Kuche, Xinhe, and Shaya [24].

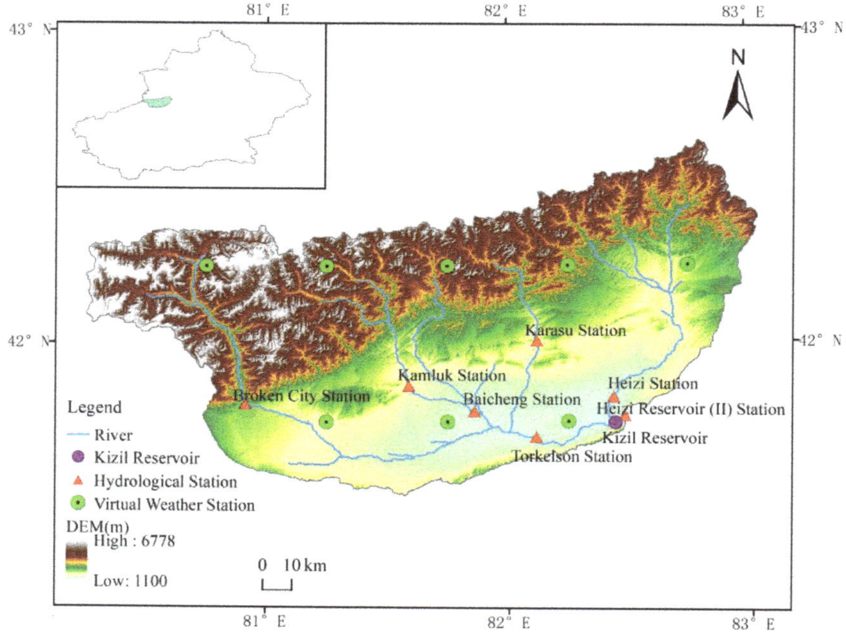

Figure 1. Overview of the area upstream of the Weigan River outlet.

2.2. Data Sources

The basic data used in this study included digital elevation model (DEM) data, glacier cataloguing data, soil data, land use data, and meteorological data. Among them, the DEM data were obtained from the Geospatial Data Cloud, and original SRTMDEM elevation data with a resolution of 90 m were used. The glacier cataloguing data were obtained from the National Cryosphere Desert Data Center, including the first glacier cataloguing data produced from aerial topographic maps from 1987 to 2004 [25] and the second glacier cataloguing data extracted from Landsat TM/ETM+ and ASTER remote sensing images from 2006 to 2013 [26]. In this study, the initial glacier inventory data were applied to the model calibration period (1965–1978) and validation period (1979–1991), and the secondary glacier inventory data were applied to the model prediction period (1992–2016). Soil data were obtained from the 1:1 million soil data provided by the Second National Land Survey of Institute of Soil Science, Chinese Academy of Sciences. The land use data were obtained from the national land use datasets of 1980 and 2000 with a resolution of 30 m provided by the Geospatial Data Cloud. The glacial snow land use types in the 1980 and 2000 land use datasets and the glacial soil types in the soil data were replaced with the glacial distribution in the first and second glacial cataloguing data, respectively. The hydrometeorological data were obtained from the China Meteorological Data Service Centre using the $0.5° \times 0.5°$ grid point dataset (V2.0) of daily values of surface air temperature and precipitation in China. In this study, the time series of daily maximum temperature and daily minimum temperature with daily precipitation data from 1961 to 2016 were selected to drive the SWAT model simulation to restore the hydrological processes in the study area.

2.3. Research Methodology

2.3.1. Glacier Module Algorithm

The glacier module consists of three main modules: a glacier ablation algorithm, glacier area change, and the glacier accumulation rate [27]. Among them, the glacier ablation algorithm is mainly based on the modified temperature index method to simulate the glacier melting process [28,29], which corrects spatial heterogeneity with the influence of solar radiation factor and topography factor, and then uses the linear relationship between the ablation factor and temperature to finally find the amount of glacier melting. The calculation formulae are as follows:

$$M = \begin{cases} (F_M + R_{ice} I_{pot}) \times (T - T_{mlt,ice}), & T > T_{mlt,ice} \\ 0, & T \leq T_{mlt,ice} \end{cases} \quad (1)$$

where M is the daily-scale glacier ablation (mm), F_M is the glacier temperature ablation factor, R_{ice} is the glacier radiation ablation factor, I_{pot} is the potential direct solar radiation ($W \cdot m^{-2}$), T is the daily-scale mean temperature (°C), and $T_{mlt,ice}$ is the temperature threshold reached at glacier ablation (°C).

$$V = cS^{\gamma} \quad (2)$$

where V is the glacier volume (m^3), S is the glacier surface area (m^2), c is a constant, and γ is a dimensionless scale factor.

$$F = W_s \times \beta_o \left\{ 1 + \sin\left[\frac{2\pi}{365}(t - 81) \right] \right\} \quad (3)$$

where F is the glacial material accumulation, W_s is the snow water equivalent, β_o is the base accumulation factor, and t is the ordinal number of a given day.

2.3.2. Hydrological Process Simulation and Evaluation of Results

(1) SWAT-based hydrological process simulation

The SWAT model is a distributed watershed hydrological model that uses the daily scale as the unit running step, and is based on the GIS platform to simulate and construct the hydrological cycle process under the changes in different influencing factors [30]. In the 1990s, the United States Department of Agriculture (USDA) Institute of Agriculture developed the SWAT model. Its predecessor is the SWRRB model, which is based on the integration of the features of CREAMS, EPIC, and GLEAMS [31], and has been continuously modified and developed to form the most representative distributed hydrological model [32,33] which can be used as a tool for large-scale watershed runoff simulation [34]. This study took the upper area of the Weigan River outlet as an example, and constructed a SWAT model to restore the hydrological processes in the study area before the reservoir had been built. The modelling process is mainly divided into the following steps.

First, the DEM data were imported, and the Heizi Reservoir (II) station was set as the basin outlet. By drawing up different discretization schemes with different catchment area thresholds, it was finally determined that the best runoff simulation was achieved when the catchment threshold was 300 km^2. The input DEM data were then subjected to operations such as water system extraction, watershed boundary depiction, sub-basin delineation, and parameter calculation to discretize the study watershed into 33 sub-basins. Second, the database of soil types, the land use database, and slope types of the Weigan River Basin established in advance were entered in turn. Notably, the period of 1965–1978 was selected as the model calibration period and the period of 1979–1991 was the validation period in this study. Among them, the first glacial inventory dataset and 1980 land use data were used in the calibration period intra model, and the second glacial inventory dataset and 2000 land use data were used in the validation period. Meanwhile, the glacial snow land use types in the land use data and the glacial soil types in the soil data were replaced with the glacial distribution in the first and second glacier cataloguing data, respectively. To reflect the differences in the hydrologic responses of different soil types, land uses, and slope combinations in the model, the SWAT model with 419 hydrologic response units (Hrus) was divided. Finally, the meteorological data required for the model operation and the prepared weather generator were imputed, and the model warm-up period was set to 1961–1964 to minimize the influence of the initial model conditions on the simulation results. Through the parameter sensitivity analysis, the more sensitive parameters (Table 1) were automatically or manually adjusted if necessary, until the model evaluation results met the simulation criteria to determine the model parameter, and complete model calibration and validation work.

Table 1. Results of parameter determination.

Parameter Module	Parameters	Definition	Scope	Optimum Value
Runoff	CN2	Initial SCS runoff curve number for moisture condition II	−0.2~0.2	0.0650
	SOL_AWC	Available water capacity of the first soil layer (mm·mm^{-1})	0~1	0.8250
	ESCO	Soil evaporation compensation factor	0~1	0.4450
	CH_K2	Effective hydraulic conductivity in the main channel alluvium (mm·h^{-1})	−0.01~500	13.8742
	ALPHA_BF	Baseflow alpha factor (days)	0~1	0.0167
	REVAPMN	Threshold depth of water in the shallow aquifer for "revap" to occur (mm H_2O)	0~1000	79.5000
	GW_DELAY	Groundwater delay (days)	0~500	417.5000
	GWQMIN	Threshold depth of water in the shallow aquifer for return flow to occur (mm H_2O)	0~5000	103.0000
	GW_REVAP	Groundwater "revap" coefficient	0.02~0.2	0.0659
Snow	SFTMP	Snowfall temperature (°C)	−5~5	4.1150
	SMTMP	Snow melt base temperature (°C)	−5~5	4.5700
	SMFMX	Maximum melt rate for snow during the year (mm H_2O·°C^{-1}·day^{-1})	0~10	7.2500
	SMFMN	Minimum melt rate for snow during the year (mm H_2O·°C^{-1}·day^{-1})	0~10	1.4050
	TIMP	Snow pack temperature lag factor	0~1	0.0225
	TLAPS	Temperature lapse rate (°C·km^{-1})	−50~50	−8.7500
Glacier	B_{melt6}	Maximum melt rate for the glacier during the year (mm H_2O·°C^{-1}·day^{-1})	1.4~16	1.9000
	B_{melt12}	Minimum melt rate for the glacier during the year (mm H_2O·°C^{-1}·day^{-1})	1.4~16	2.8000
	gmlt_tmp	Glacier ablation threshold temperature (°C)	−5~5	1.0000

(2) Evaluation of simulation results

To verify the reliability of the model, the Nash–Sutcliffe efficiency coefficient (*NSE*), the ratio of the root mean square error to the standard deviation of the measured values (*RSR*), the percent bias (*PBIAS*), and the coefficient of determination (R^2) were selected to evaluate the model fitting effect in this study; the formulae are shown below:

$$NSE = 1 - \frac{\sum_{i=1}^{n}(Q_{oi} - Q_{si})^2}{\sum_{i=1}^{n}(Q_{oi} - \overline{Q}_o)^2} \tag{4}$$

$$RSR = \frac{RMSE}{STDEV_o} = \frac{\sqrt{\sum_{i=1}^{n}(Q_{oi} - Q_{si})^2}}{\sqrt{\sum_{i=1}^{n}(Q_{oi} - \overline{Q}_o)^2}} \tag{5}$$

$$PBIAS = \sum_{i=1}^{n}\frac{Q_{si} - Q_{oi}}{Q_{oi}} \times 100 \tag{6}$$

$$R^2 = \frac{\left(\sum_{i=1}^{n}(Q_{oi} - \overline{Q}_o)(Q_{si} - \overline{Q}_s)\right)^2}{\sum_{i=1}^{n}(Q_{oi} - \overline{Q}_o)^2 \sum_{i=1}^{n}(Q_{si} - \overline{Q}_s)^2} \quad (7)$$

where Q_{si} denotes the model simulated runoff, m^3, and Q_{oi} denotes the actual observed runoff, m^3.

The *NSE* indicates the degree of fit between the simulated and measured values, with a range from $-\infty$ to 1. The closer to 1, the better the simulation. *RSR* standardizes the standard deviation of the measured values, and the closer to 0, the better the simulation. *PBIAS* reflects the cumulative deviation between simulated and measured values. A *PBIAS* value greater than 0 indicates that the model underestimates the deviation of measured values, a value less than 0 indicates that the model overestimates the deviation of measured values, and a *PBIAS* value equal to 0 is the optimal value, indicating that the model simulation is accurate. R^2 indicates the degree of linear correlation between simulated and measured values, and the closer it is to 1, the better the simulation effect is, but the response to the overall deviation of high or low simulated values is not very obvious. The model simulation results are generally accepted when $NSE > 0.5$, $RSR \leq 0.7$, and $PBIAS < \pm 25\%$, and the model simulation results are excellent when $NSE > 0.75$, $RSR \leq 0.5$, and $PBIAS < \pm 10\%$.

2.3.3. The Process of Predicting Natural Mountain Runoff and its Response to the Construction of Water Conservation Projects

Since the Kizil Reservoir started to lower its gates for storage in August 1991, it has caused the flow measured at the Heizi Reservoir (II) station, which is not far downstream, to change from natural flow to artificial regulated flow. Therefore, 1992 was taken as the dividing point between natural flow and artificial regulated flow in this study. In this study, we selected 1992–2016 as the model prediction period, based on the constructed and validated reliable SWAT model with the required input data, to predict the natural outflow runoff in the reduced study area without human activities, and to explore the response of the Weigan River outflow runoff to human activities from seasonal-scale and monthly scale comparisons. In recent years, the area of arable land within the study region (Baicheng County) has grown rapidly (Figure 2), with a 29.0% increase during the 30 years period from 1990 to 2020 (1348 km^2 in 2020); arable water use in Baicheng County increased from 6.89×10^8 m^3 to 8.86×10^8 m^3 from 2000 to 2020. However, the actual measured runoff into the Kizil Reservoir did not decrease (Table 2): the multi-year (1961–2000) average runoff (27.37×10^8 m^3) of the five tributaries in the area was basically consistent with the multi-year average runoff into the Kizil Reservoir of 27.10×10^8 m^3 in this study, indicating the existence of a more complex and special water cycle relationship in Baicheng County. Therefore, this study only considered the impact of human activities on runoff from the Weigan River outflow from the perspective of the construction of water conservation projects. The overall research process approach is shown in Figure 3.

Table 2. Major river conditions in the study area.

River	Hydrological Station	Measured Annual Runoff (10^8 m^3)
Muzati River	Broken City Station	14.66
Kapuslang River	Kamluk Station	6.65
Tylervichuk River	Baicheng Station	0.80
Karasu River	Karasu Station	2.19
Heizi River	Heizi Station	3.08
Total		27.37

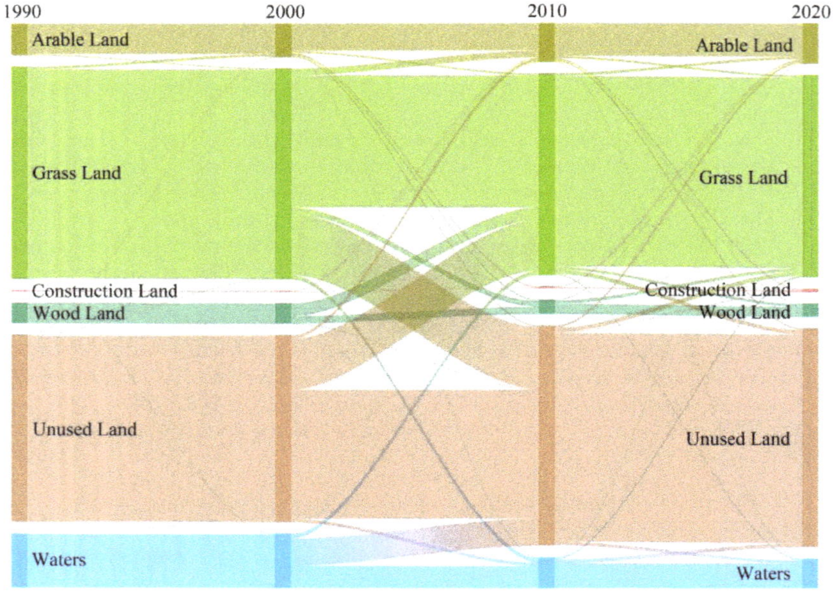

Figure 2. Land use status and interannual transfer changes in the upper reaches of the mountain pass of Weigan River from 1990 to 2020 based on remote sensing.

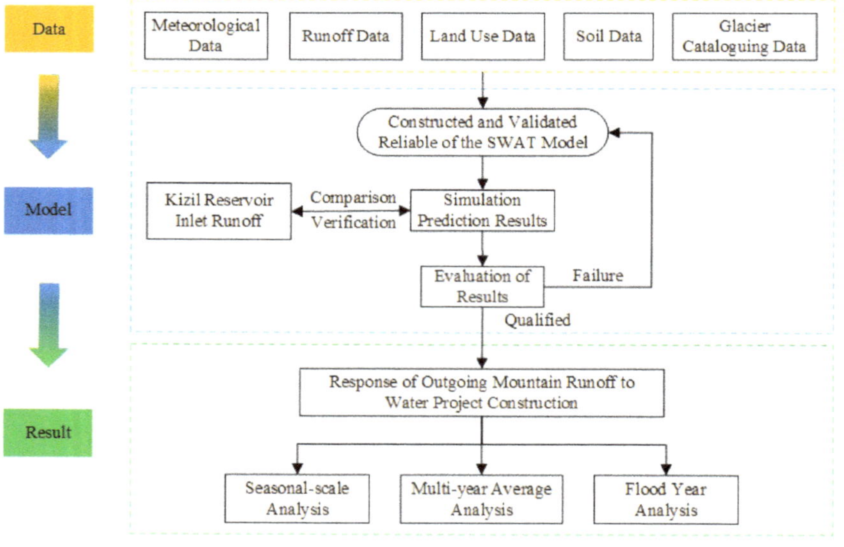

Figure 3. Flow chart of the study on the prediction of natural mountain runoff and its response to the construction of water conservation projects.

3. Results and Analysis

3.1. Simulation Prediction Results and Evaluation Analysis

First, monthly runoff data from 1965 to 1991 at the Heizi Reservoir hydrological station were selected to calibrate and validate the SWAT model for the upstream area of the Weigan

River outlet. The simulation results of the model calibration period and validation period are shown in Figure 4a; the simulation results with the glacier module added were better than those without the glacier module, so the glacier module was included in the simulation when reverting the prediction of the runoff from the Weigan River outlet without human activities. The simulated values with the glacier module are in good agreement with the measured values at the Heizi Reservoir hydrological station during the flat and dry periods of 1965–1991. Except for 1971, 1979, 1983, and 1986, when the simulated values of the glacier module were lower than the measured values in the same period, and 1973, 1974, 1977, 1980, and 1981, when the simulated values of the glacier module were higher than the measured values in the same period, the overall fit was good in the other years. The simulation results of the calibration period and validation period were evaluated, and the simulation results were evaluated with reference to the model evaluation criteria (Table 3). The evaluation of the simulation results of the Heizi reservoir hydrological station in the calibration period and the validation period were excellent, and the simulation effect was good, which showed that the model could accurately reflect the runoff process of the Weigan River from the mountain.

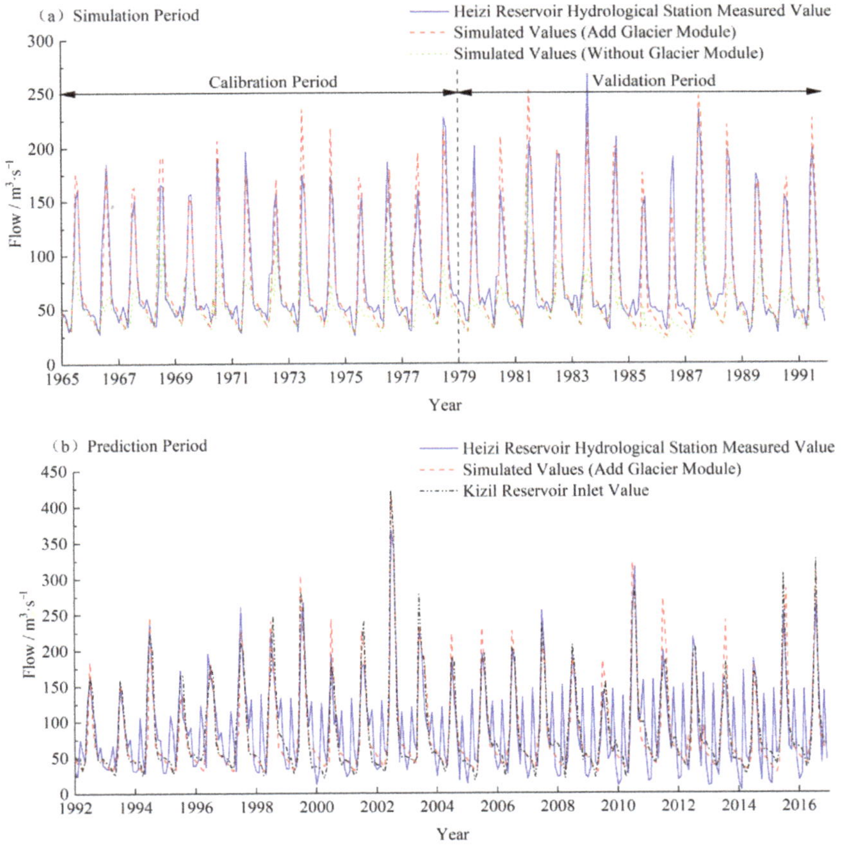

Figure 4. Simulation (**a**) and prediction (**b**) results of monthly average flow of SWAT model, 1965–2016.

Table 3. Evaluation of monthly scale runoff simulation results in the upstream area of the Weigan River outlet.

Station Name	Time Period	Scenario	NSE	RSR	PBIAS (%)	R^2	Simulation Results
Heizi Reservoir Hydrological Station	Calibration Period (1 January 1965–31 December 1978)	No Glacier Module	0.08	0.96	29.54	0.35	Unqualified
		Glacier Module	0.84	0.40	−0.16	0.86	Excellent
	Validation Period (1 January 1979–31 December 1991)	No Glacier Module	0.09	0.96	34.02	0.39	Unqualified
		Glacier Module	0.82	0.42	2.90	0.85	Excellent
	Prediction Period (1 January 1992–31 December 2016)	Glacier Module	0.88	0.34	1.11	0.89	Excellent

Based on the SWAT model that was constructed and verified to be reliable, combined with the input data from 1992 to 2016, the natural mountain runoff in the study area from 1992 to 2016 under no human activity conditions was obtained. The measured flow at Heizi Reservoir (II) station has been influenced by the artificially regulated storage and release of water from the reservoir since 1992; therefore, the predicted runoff from the mountain without human activities from 1992 to 2016 was compared with the runoff into the Kizil Reservoir during the same period (Figure 4b). The results show that the relative errors between the two are small, except for the floods in 2000, 2011, and 2013, when the model simulation results were larger than the incoming flow of Kizil reservoir in the same period. At the same time, in 2002, 2010, and 2016, when floods in the study area broke out and the incoming water from the upper reaches increased compared with the other years, the flow predicted by the model simulation for the same period was also closer to the actual situation. The simulation results of the model in the prediction period are presented in Table 3, in which R^2 reached 0.89 and NSE was 0.88, the model simulation prediction results are excellent, and the simulation effect is good. The model had considerable reliability in predicting the hydrological process in the mountain area before the reservoir was built in the restored study area, and could more realistically reflect the runoff from the mountain when there was no human activity in the study area.

3.2. Portrayal and Analysis of the Weigan River Outflow Runoff Group Structure

From the actual measured runoff at the Heizi reservoir hydrological station (Figure 5a), it can be seen that the multi-year average runoff from the Weigan River from 1965 to 2016 was 26.63×10^8 m^3, of which the multi-year average glacial runoff was 8.17×10^8 m^3, accounting for 32.40% of the total runoff from the mountain. The actual measured month-by-month runoff from the Heizi reservoir hydrological station from 1965 to 2016 showed an overall increasing trend, increasing from 21.41×10^8 m^3 in 1965 to 32.47×10^8 m^3 in 2016, at a rate of approximately 2.1×10^8 m^3/10a, an increase of 51.66% compared to 1965. This indicated that the water inflow from the outlet of the Weigan River showed a significant increasing trend in the last 52 years, which has, to a certain extent, relieved the water stress in the middle and lower reaches of the Weigan River and the irrigation area. Separating the simulation results of the outflow from the mountains showed that glacier runoff increased from 6.15×10^8 m^3 in 1965 to 8.77×10^8 m^3 in 2016, an increase of 2.62×10^8 m^3 or 42.60% over 1965, with an increase rate of about 0.5×10^8 m^3/10a, and the contribution of glacier runoff fluctuated from 27.09% in 1965 to 28.49% in 2016, which was as high as 44.84% in 2008.

From the annual runoff simulation results (Figure 5a), it can be seen that the overall trends in the simulated and measured values were generally the same during the simulation period, except for 1985 and 1986, when the simulated values were lower than the measured values at the Heizi reservoir hydrological station. In the prediction, although the simulated values without human activities were consistent with the overall runoff trend in the measured values, the model simulated runoff from 1994 to 2001 was lower than the measured runoff by 4.91×10^8 m^3 per year. If the simulated runoff without human activities was compared with the incoming runoff from the Kizil Reservoir, the average annual deviation between the two was only 1.21×10^8 m^3, and the relative error did not exceed 5%. This indicates that the runoff from the mountain without human activities predicted in this study had a high degree of confidence. Meanwhile, comparing the runoff from the Weigan River before and after the presence of human activities, the annual average increase in incoming water from 1992 to 2016 compared with 1965 to 1991 was 2.99 × 10^8 m^3, and the quantitative separation showed that the increase in runoff caused by precipitation was 1.24 × 10^8 m^3, accounting for 41.45% of the increase in incoming water; the increase in runoff due to glacial melt caused by temperature rise was 1.75 × 10^8 m^3, accounting for 58.55%. This shows that under global climate change conditions, the increase in runoff due to temperature rise was predominant in the study area.

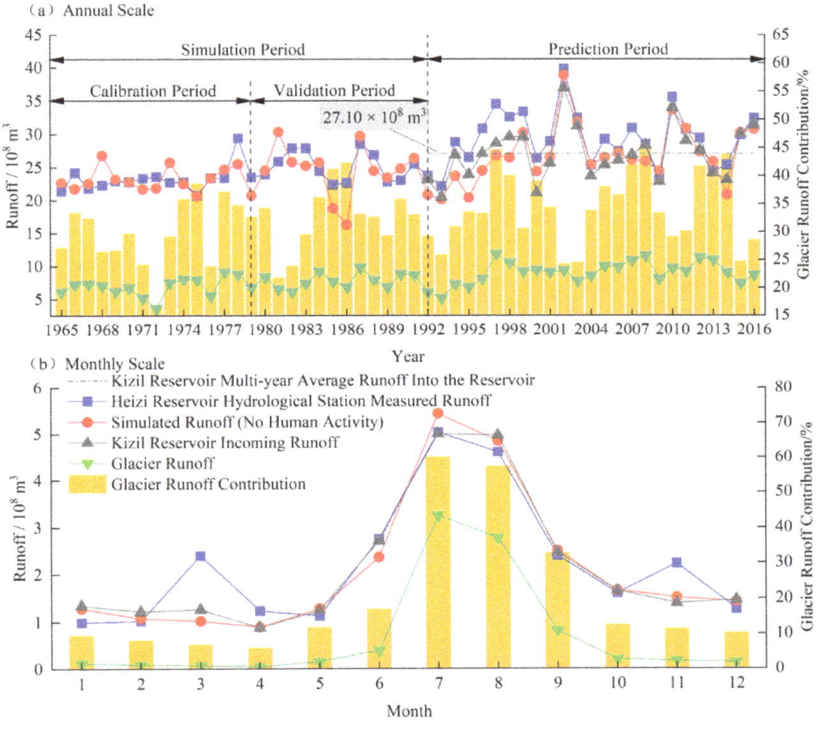

Figure 5. Annual scale (**a**) and monthly scale (**b**) simulation predictions of mountain runoff and glacier runoff contributions.

From the monthly scale analysis (Figure 5b), the hydrological processes in the study area mainly occurred in June, July, August, and September. According to the actual measured values at the hydrological station of the Heizi Reservoir, the flow production in June to September accounted for 55.39% of the annual flow production. The glacier flow production mainly occurred in July, August, and September, accounting for 82.98% of the annual glacier runoff. Comparing the simulated results of runoff from the mountain without human activity with the inlet runoff of the Kizil Reservoir, the simulated values were closer to the natural runoff values in the remaining months except for February, March, and June, when the simulated runoff was slightly lower than the inlet runoff of the Kizil Reservoir, and July and November, when the simulated runoff was slightly higher than the inlet runoff of Kizil Reservoir; the overall trends in the two curves were basically the same.

3.3. Impacts of the Construction of Water Projects on Runoff from the Mountains

The measured data from the Heizi reservoir hydrological station showed an increasing trend in runoff from the upper reaches of the Weigan River from 1965 to 2016 in spring, summer, and autumn, and a decreasing trend in winter. The Weigan River outflow runoff was mainly concentrated in summer, and according to the simulation results of the natural outflow runoff, incoming water in the summer accounted for 51.06% of the annual accumulation of water. The response of the runoff from the Weigan River to the artificial regulation of reservoir storage was analysed by comparing the measured and model simulated predictions from 1992 to 2016 at the Heizi Reservoir (II) station (Figure 6). The cut-off year between natural runoff and runoff influenced by artificial regulation was 1992, and the simulated natural runoff in summer and spring was greater than the measured runoff influenced by artificial regulation after the reservoir began storing water. The analysis shows that, under the influence of artificial regulation, the multi-year average runoff values in summer and winter were 0.15×10^8 m^3 and 0.41×10^8 m^3 greater than the simulated natural runoff, accounting for 3.33% and 7.08% of the natural runoff in summer and winter, respectively. However, comparing the simulated natural runoff in spring and autumn with the measured runoff affected by the artificial adjustment, we found that the simulated natural runoff in spring and autumn was lower than the measured runoff affected by the artificial adjustment, and the natural outflow in spring and autumn was increased by an additional 0.92×10^8 m^3 and 0.49×10^8 m^3, respectively, after the artificial adjustment, which accounted for 83.28% and 24.36% of the natural outflow in spring and autumn, respectively. This shows that the degree of impact of the construction of the water conservation project on natural runoff in spring and autumn was greater than its impact on natural runoff in summer and winter.

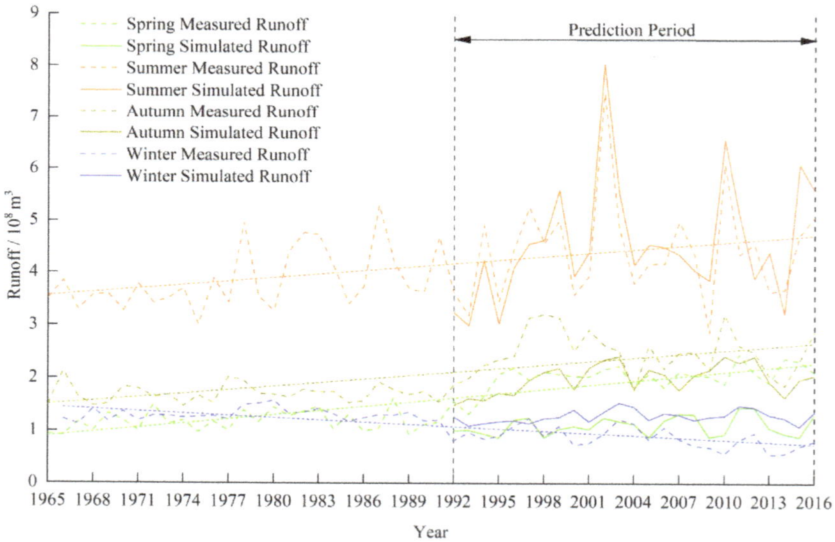

Figure 6. Seasonal-scale comparison of measured and simulated runoff at the outlet of the Weigan River.

To further analyse the impact of the construction of water projects on the intra-annual distribution of runoff, the simulated runoff was compared to the measured runoff at Heizi Reservoir (II) station (Figure 7a). It can be seen that the simulated values in March and November were significantly lower than the measured runoff values, while the simulated runoff in the rest of the months exhibited less deviation from the measured runoff, and the overall trend was basically the same. The reason for this is that in August 1991, the Kizil Reservoir started to store water, and the artificial storage of the reservoir, such as storage at the end of the flood and replenishment during the dry period, made the actual measured runoff in March and November at the Heizi Reservoir (II) station higher than the incoming runoff from the Kizil Reservoir. The flood season of water coming from the outlet of the Weigan River also increased from July and August to March, July, August, and November, solving the problem of insufficient spring irrigation in March and pressurized saltwater resources for winter irrigation in November for the irrigation areas downstream of the Weigan River. At the same time, an average of 2.74 × 10^8 m^3 more water was impounded annually, and 5.28 × 10^8 m^3 more water was regulated across seasons through storage than before the reservoir was built. Since the reservoir was built and started storing water, the Kizil Reservoir has held the equivalent of a 10-year flood 13 times, the equivalent of a 40-year flood once ("1 August 2016" flood peak flow 3420 $m^3 \cdot s^{-1}$ [35]), and the equivalent of a 100-year flood twice ("23 July 2002" flood peak flow 3677 $m^3 \cdot s^{-1}$ [21] and "29 July 2010" flood peak flow 3360 $m^3 \cdot s^{-1}$ [36]). After the construction of the reservoir, the actual measured runoff from the Heizi Reservoir (II) station, compared with the simulated runoff values without human activities (Figure 7b–d), the temporal distribution of water resources increased by an additional 0.92 × 10^8 m^3, 2.04 × 10^8 m^3, and 1.62 × 10^8 m^3 of water in 2002, 2010, and 2016, respectively, which strongly alleviated the problem of the uneven temporal distribution of water resources.

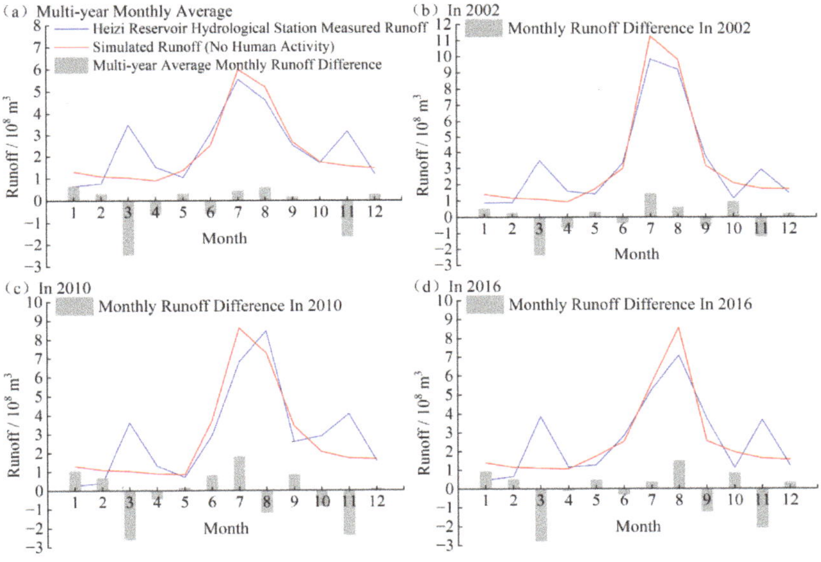

Figure 7. Comparison of monthly measured and simulated runoff for multi-year average (**a**), 2002 (**b**), 2010 (**c**), and 2016 (**d**).

4. Discussion

4.1. Model Applicability and Simulation Results

In this study, the SWAT model with a coupled glacier module was used to simulate the mountain runoff in the Weigan River Basin from 1965 to 1978 (calibration period) and from 1979 to 1991 (validation period). NSE coefficients of 0.84 and 0.82 were achieved in the simulation evaluation results; the simulated values fit the measured runoff curves relatively well, and the relative error was small. Therefore, the model can not only simulate the restoration of the outflow runoff process in the study area without reservoir construction more accurately, but also predict the restoration of the natural outflow runoff in the study area without human activities from 1992 to 2016 based on the model. This study compared the model prediction results with the NSE coefficient of 0.88 for the prediction period using the runoff from the Kizil Reservoir in the same period instead of the runoff measured at the Heizi Reservoir (II) station without the influence of artificial regulation. This shows that the SWAT model has good applicability in the area upstream of the Weigan River outlet and can be further used for the study of related problems.

The multi-year average runoff of the Weigan River outflow from 1965 to 2016 was 26.63×10^8 m^3, with an overall upward trend. Its increase rate was approximately 2.1×10^8 m^3/10a, which is closer to the increase in runoff from the Weigan River Basin from 1960 to 2013 (1.8×10^8 m^3/10a) calculated by Peng Qin [37]. During the 52 years, the glacial runoff of the upper Weigan River Basin showed a fluctuating yet stable upward trend, with an average multi-year glacier runoff of 8.17×10^8 m^3, accounting for 32.40% of the total runoff from the mountain, similar to the results of Minxia Ni (30.8%) [38]. The result of 51.06% of the annual incoming water in summer in this study is more consistent with the range of 50~70% evident in the statistics of Jianjun Duan [39] and others, and the calculation results of Peng Qin [37] and others (56%). The simulated reduced runoff in this paper better reflects the actual runoff process in the study area, and also showed that the model has some reliability.

4.2. Analysis of the Response of Outgoing Mountain Runoff to the Construction of Water Conservation Projects

In a related study of the Kizil Reservoir, Mingwang Zhang [40] analysed the effects of precipitation and human activities on factors such as runoff from the perspective of water and sedimentary sequences, using methods such as double accumulation curves. However, this study was an analysis of the relationship between the response of runoff from the Kizil Reservoir to the construction of the water project from the perspective of artificial storage in the Kizil Reservoir. The construction of the Kizil Reservoir has expanded the flood season of water coming from upstream of the study area from July and August to March, July, August, and November. During the runoff discharge process, 2.74×10^8 m^3 more water was stored each year, and 5.28×10^8 m^3 more water was released in the time distribution of water resources than before the reservoir was built. Notably, there are many irrigation areas around the Kizil Reservoir, and the irrigation return water produced by watering the irrigation areas will be delayed, which makes the measured runoff of the Heizi Reservoir (II) station larger than the simulated natural runoff. At the same time, the degree of impact of the construction of water conservation projects on runoff from the mountain varied from season to season (in descending order, spring > autumn > winter > summer), where the proportion of natural runoff affected by artificial regulation in spring was as high as 83.28%. The construction of water conservation projects has, to a certain extent, regulated the distribution of runoff from the mountain over time, alleviating the problem of insufficient water resources for downstream irrigation areas and ecological inter-seasonal diversion, spring irrigation salt washing, and autumn irrigation overwintering.

4.3. Shortcomings of this Study

The response of outgoing mountain runoff to human activities in this study was explored and analysed based on the SWAT distributed hydrological model. However, the existing hydrological model is only a general description of the water cycle process [41], which cannot fully and objectively restore the real hydrological situation. This conclusion was also reached by Zhenliang Yin et al. [42] in their discussion of the progress of hydrological simulation studies in the mountainous region of the main stream of the Heihe River in the Qilian Mountains. At the same time, because the outgoing runoff at the Heizi Reservoir (II) station before the reservoir was constructed could not actually be measured, the incoming runoff from the Kizil Reservoir during the same period was used for the evaluation of the simulation results of the outgoing runoff under no human activity conditions during the prediction period. Although the distance between the Heizi Reservoir (II) station and Kizil Reservoir is not far, it cannot be denied that the errors between the two caused by seepage losses in the runoff process, as well as reservoir seepage and surface evaporation, still exist. This aspect should be studied in more depth in future simulations to provide a reference for watershed water resource management.

5. Conclusions

In order to make more efficient use of water resources, water conservation projects are constructed in arid watersheds. This study considered the impact of human activities on the runoff from the Weigan River from the perspective of water conservation construction, and thus explored the extent to which water conservation projects affect natural water cycle processes. In this study, the SWAT model was used to simulate and predict runoff for the 1965–2016 scenario without a reservoir in the upstream area of the Weigan River outlet, and then analysed the component composition of runoff from the study area, and compared and explored the degree of influence of the presence or absence of human activities on the runoff from the study area to obtain the following conclusions:

(1) The SWAT model with the coupled glacier module was used to simulate the outflow runoff from 1965 to 1991 in the Weigan River Basin, and to predict natural runoff in the study area without human activities from 1992 to 2016 based on the constructed SWAT model predictions; its NSE coefficients all reached above 0.8. This shows that the SWAT model with a coupled glacier module has good applicability in the upper Weigan River Basin and can be further used for studies of glacial-runoff-related problems in arid zones.

(2) During the 52 years from 1965 to 2016, the overall runoff volume from the Weigan River showed an increasing trend, among which glacial runoff exhibited a stable increasing trend, and runoff due to precipitation also exhibited a fluctuating increasing trend. The average value of the increase in incoming water from 1992 to 2016 compared with 1965 to 1991 was 2.99×10^8 m^3, with 41.45% and 58.55% of the increase in incoming water was caused by the increases in precipitation and temperature, respectively.

(3) The runoff from the Weigan River is mainly concentrated in summer, with 51.06% of the annual water intake in summer. The construction of hydraulic projects has translated to human activities having a very different impact on the runoff from the mountain in different seasons (in descending order spring > autumn > winter > summer), with 83.28% of the natural runoff in spring being influenced by artificial regulation.

Author Contributions: Conceptualization, A.L., C.R. and P.Z.; methodology, J.S. and C.R.; software, J.S. and C.R.; validation, A.L. and X.G.; formal analysis, J.S., C.R. and J.Z.; investigation, J.S. and A.L.; resources, A.L.; data curation, J.S. and A.L.; writing—original draft preparation, J.S.; writing—review and editing, J.S., C.R., A.L. and X.G.; visualization, J.S., J.Z. and X.G.; supervision, A.L., F.C. and P.Z.; project administration, A.L.; funding acquisition, A.L. and X.D. All authors have read and agreed to the published version of the manuscript.

Funding: This work was financially supported by the Third Xinjiang Scientific Expedition (Grant No. 2022xjkk0103; Grant No. 2021xjkk0406) and the National Natural Science Foundation of China (Grant No. 52179028).

Data Availability Statement: Not applicable.

Conflicts of Interest: The authors declare no conflict of interest.

References

1. Barnett, T.P.; Pierce, D.W.; Hidalgo, H.G.; Bonfils, C.; Santer, B.D.; Das, T.; Bala, G.; Wood, A.W.; Nozawa, T.; Mirin, A.A.; et al. Human-Induced Changes in the Hydrology of the Western United States. *Science* **2008**, *319*, 1080–1083. [CrossRef] [PubMed]
2. Song, X.; Zhang, J.; Zhang, C.; Liu, C. Review for impacts of climate change and human activities on water cycle. *J. Hydraul. Eng.* **2013**, *44*, 779–790. [CrossRef]
3. Xie, Z.; Chen, S.; Qin, P.; Jia, B.; Xie, J. Research on Climate Feedback of Human Water Use and Its Impact on Terrestrial Water Cycles—Advances and Challenges. *Adv. Earth Sci.* **2019**, *34*, 801–813.
4. Qiu, Y. Study on Comprehensive Assessment and Evolution Law of Water Resources. Ph.D. Thesis, China Institute of Water Resources & Hydropower Research, Beijing, China, 2006.

5. Wada, Y.; Flörke, M.; Hanasaki, N.; Eisner, S.; Fischer, G.; Tramberend, S.; Satoh, Y.; van Vliet, M.T.H.; Yillia, P.; Ringler, C.; et al. Modeling global water use for the 21st century: The Water Futures and Solutions (WFaS) initiative and its approaches. *Geosci. Model Dev.* **2016**, *9*, 175–222. [CrossRef]
6. Veldkamp, T.I.E.; Wada, Y.; Aerts, J.C.J.H.; Döll, P.; Gosling, S.N.; Liu, J.; Masaki, Y.; Oki, T.; Ostberg, S.; Pokhrel, Y.; et al. Water scarcity hotspots travel downstream due to human interventions in the 20th and 21st century. *Nat. Commun.* **2017**, *8*, 15697. [CrossRef]
7. Montanari, A.; Young, G.; Savenije, H.H.G.; Hughes, D.; Wagener, T.; Ren, L.L.; Koutsoyiannis, D.; Cudennec, C.; Toth, E.; Grimaldi, S.; et al. "Panta Rhei—Everything Flows": Change in hydrology and society—The IAHS Scientific Decade 2013–2022. *Hydrol. Sci. J.* **2013**, *58*, 1256–1275. [CrossRef]
8. Liu, J.; Zhang, Q.; Chen, X.; Gu, X. Quantitative evaluations of human- and climate-induced impacts on hydrological processes of China. *Acta Geogr. Sin.* **2016**, *71*, 1875–1885.
9. Chawla, I.; Mujumdar, P. Isolating the impacts of land use and climate change on streamflow. *Hydrol. Earth Syst. Sci.* **2015**, *19*, 3633–3651. [CrossRef]
10. Rakhimova, M.; Liu, T.; Bissenbayeva, S.; Mukanov, Y.; Gafforov, K.S.; Bekpergenova, Z.; Gulakhmadov, A. Assessment of the Impacts of Climate Change and Human Activities on Runoff Using Climate Elasticity Method and General Circulation Model (GCM) in the Buqtyrma River Basin, Kazakhstan. *Sustainability* **2020**, *12*, 4968. [CrossRef]
11. Hou, L.; Peng, W.; Qu, X.; Chen, Q.; Fu, Y.; Dong, F.; Zhang, H. Runoff Changes Based On Dual Factors In the Upstream Area of Yongding River Basin. *Ecol. Environ. Conserv.* **2019**, *28*, 143–152. [CrossRef]
12. Liu, J.; Ren, Y.; Zhang, W.; Tao, H.; Yi, L. Study on the influence of climate and underlying surface change on runoff in the Yarlung Zangbo River basin. *J. Glaciol. Geocryol.* **2022**, *44*, 275–287.
13. Chen, H.; Meng, F.; Sa, C.; Luo, M.; Wang, M.; Liu, G. Analysis of the characteristics of runoff evolution and its driving factors in a typical inland river basin in arid regions. *Arid. Zone Res.* **2023**, *40*, 39–50. [CrossRef]
14. Liu, J.; Zhang, Q.; Deng, X.; Ci, H.; Chen, X. Quantitative analysis the influences of climate change and human activities on hydrological processes in Poyang Basin. *J. Lake Sci.* **2016**, *28*, 432–443.
15. Wang, L.; Liu, T.; Xie, J. Study on the Effect of Different Land Use Scenarios on Runoff in Qingshuihe Basin of Zhangjiakou Based on SWAT Model. *Res. Soil Water Conserv.* **2019**, *26*, 245–251. [CrossRef]
16. Ren, C.; Long, A.; Yu, J.; Yin, Z.; Zhang, J. Effects of climate and underlying surface changes on runoff of Yarkant River Source. *Arid. Land Geogr.* **2021**, *44*, 1373–1383.
17. Muybra, Z.; Shi, Q.; Mohetar, P.; Zhang, R. Land use and climate change effects on runoff in the upper Urumqi River watershed: A SWAT model based analysis. *Acta Ecol. Sin.* **2018**, *38*, 5149–5157.
18. Xu, Z.; Jiang, Y. Studies on runoff evolution mechanism under changing environment: A state-of-the-art review. *Hydro-Sci. Eng.* **2022**, *191*, 9–18. [CrossRef]
19. Yang, X.; Wu, W.; Zheng, C.; Wang, Q. Attribution Identification of Runoff Change in Yihe River Basin Based on Budyko Hypothesis. *Res. Soil Water Conserv.* **2023**, *30*, 100–106. [CrossRef]
20. Cao, X. Analysis of the "8.1" Catastrophic Flood in the Weigan River Basin of Xinjiang. *Ground Water* **2017**, *39*, 101–102+108.
21. Qian, X. Comparison between characteristics of 7.19 flood and 7.23 flood in Weigan river basin. *Dam Saf.* **2013**, *76*, 15–18.
22. Li, H. Prospects and Thoughts on Science and Technology Demand of Water Project in Aksu Area. *Hongshui River* **2022**, *41*, 16–19.
23. Wang, F. Discussion on the Current Situation of Sediment Deposition in Kizil Reservoir and the Operation Mode of Reservoir Joint Operation. *Water Conserv. Sci. Technol. Econ.* **2015**, *21*, 82–84.
24. Wu, J. Comparative analysis of regulation and storage capacity of Kizil reservoir before and after reinforcement. *Water Resour. Plan. Des.* **2015**, *142*, 66–67.
25. Wu, L. *The First Glacier Inventory Dataset of China*; National Cryosphere Desert Data Center: Lanzhou, China, 2020. [CrossRef]
26. Liu, S.; Guo, W.; Xu, J. *The Second Glacier Inventory Dataset of China(V1.0)*; National Cryosphere Desert Data Center: Lanzhou, China, 2019. [CrossRef]
27. Liu, J.; Long, A.; Deng, X.; Yin, Z.; Deng, M.; An, Q.; Gu, X.; Li, S.; Liu, G. The Impact of Climate Change on Hydrological Processes of the Glacierized Watershed and Projections. *Remote Sens.* **2022**, *14*, 1314. [CrossRef]
28. Yin, Z.; Liu, S.; Zou, S.; Li, J.; Yang, L.; Deo, R. The Spatial and Temporal Contribution of Glacier Runoff to Watershed Discharge in the Yarkant River Basin, Northwest China. *Water* **2017**, *9*, 159. [CrossRef]
29. Hock, R. Temperature index melt modelling in mountain areas. *J. Hydrol.* **2003**, *282*, 104–115. [CrossRef]
30. Lou, C. Analysis of Water Resource Availability in Yanghe River Basin Based on SWAT Model. Master's Thesis, Hubei University of Technology, Wuhan, China, 2020.
31. White, M.J.; Gambone, M.; Haney, E.; Arnold, J.; Gao, J. Development of a Station Based Climate Database for SWAT and APEX Assessments in the US. *Water* **2017**, *9*, 437. [CrossRef]
32. Arnold, J.G.; Allen, P.M. Estimating Hydrologic Budgets for Three Illinois Watersheds. *J. Hydrol.* **1996**, *176*, 57–77. [CrossRef]
33. Arnold, J.G.; Srinivasan, R.; Muttiah, R.S.; Williams, J.R. Large Area Hydrologic Modeling and Assessment Part I: Model Development. *J. Am. Water Resour. Assoc.* **1998**, *34*, 73–89. [CrossRef]
34. Liu, K.; Li, M.; Lv, Z.; Yin, Z.; Liu, Z.; Liang, L. Application of SWAT Model in Large-scale Watershed. *Water Resour. Power* **2023**, *41*, 35–38. [CrossRef]

35. Zhu, Y. Analysis on 2016 '8·1'flood situation and reservoir scheduling condition in Xinjiang Kizil Reservoir. *Water Conserv. Constr. Manag.* **2018**, *38*, 81–85. [CrossRef]
36. Li, X. Analysis of the "2010 · 7.29" catastrophic flood in the Weigan River Basin. *Inn. Mong. Water Resour.* **2013**, *146*, 180–182.
37. Qin, P.; Zhao, C.; Sheng, Y.; Dong, Y. Runoff Change Characteristics of Weiganhe River in Recent 54 Years and Their Influencing Factors Analyzing. *J. China Hydrol.* **2016**, *36*, 85–91.
38. Ni, M.; Duan, Z.; Xia, J. Melting of Mountain Glacier and Its Risk to Future Water Resources in Southern Xinjiang, China. *Mt. Res.* **2022**, *40*, 329–342. [CrossRef]
39. Duan, J.; Cao, X.; Shen, Y.; Gao, Q.; Wang, S. Surface Water Resources and Its Trends in Weigan River Basin on the South Slope of Tianshan, China during 1956–2007. *J. Glaciol. Geocryol.* **2010**, *32*, 1211–1219.
40. Zhang, M.; Shi, K. Comprehensive analysis of water-sediment variation characteristics at the confluence of the upper reaches of the Weigan River and Heizi River with multiple methods and multiple influencing factors. *Water Supply* **2021**, *22*, 1275–1292. [CrossRef]
41. Wang, H.; Li, Y.; Ren, L.; Wang, J.; Yan, D.; Lu, F. Uncertainty of hydrologic model and general framework of ensemble simulation. *Water Resour. Hydropower Eng.* **2015**, *46*, 21–26. [CrossRef]
42. Yin, Z.; Xiao, H.; Zou, S.; Lu, Z.; Wang, W. Progress of the Research on Hydrological Simulation in the Mainstream of the Heihe River, Qilian Mountains. *J. Glaciol. Geocryol.* **2013**, *35*, 438–446.

MDPI AG
Grosspeteranlage 5
4052 Basel
Switzerland
Tel.: +41 61 683 77 34

Water Editorial Office
E-mail: water@mdpi.com
www.mdpi.com/journal/water

www.ingramcontent.com/pod-product-compliance
Lightning Source LLC
Chambersburg PA
CBHW040748250926
51850CB00007BB/1439